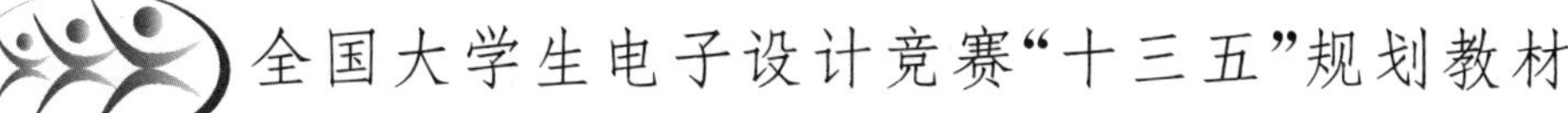

全国大学生电子设计竞赛
ARM 嵌入式系统应用设计与实践
（第 2 版）

黄智伟　李月华　主编

北京航空航天大学出版社

内容简介

针对全国大学生电子设计竞赛的特点和要求编写的《全国大学生电子设计竞赛 ARM 嵌入式系统应用设计与实践(第 2 版)》共分 9 章，内容包括：LPC214x ARM 微控制器最小系统的设计与制作，可选择的 ARM 微处理器，STM32F 系列 32 位微控制器最小系统的设计与制作，LED、LCD 和触摸屏显示电路的设计与制作，ADC 和 DAC 电路的设计与制作，直流电机、步进电机和舵机的驱动电路，光电、超声波、图像识别、色彩识别、电子罗盘、倾角传感器、角度传感器的应用，E^2PROM 电路及应用无线数据传输与 CAN 总线应用的设计制作与编程，系统应用设计，MDK 集成开发环境以及 ISP 下载方法。

本书可作为高等院校电子信息、通信工程、自动化、电气控制等专业学生参加全国大学生电子设计竞赛的培训教材，也可作为各类电子制作、课程设计、毕业设计的教学参考书，还可作为电子工程技术人员进行电子电路设计与制作的参考书。

图书在版编目(CIP)数据

全国大学生电子设计竞赛 ARM 嵌入式系统应用设计与实践 / 黄智伟，李月华主编. -- 2 版. -- 北京 ：北京航空航天大学出版社，2016.9

ISBN 978-7-5124-2244-5

Ⅰ. ①全… Ⅱ. ①黄… ②李… Ⅲ. ①微处理器—系统设计—高等学校—教材 Ⅳ. ①TP332

中国版本图书馆 CIP 数据核字(2016)第 215752 号

全国大学生电子设计竞赛
ARM 嵌入式系统应用设计与实践(第 2 版)
黄智伟　李月华　主编
责任编辑　杨　昕
*
北京航空航天大学出版社出版发行
北京市海淀区学院路 37 号(邮编 100191)　http://www.buaapress.com.cn
发行部电话：(010)82317024　传真：(010)82328026
读者信箱：emsbook@buaacm.com.cn　邮购电话：(010)82316936
北京泽宇印刷有限公司印装　各地书店经销
*
开本：710×1 000　1/16　印张：22.5　字数：480 千字
2016 年 9 月第 2 版　2016 年 9 月第 1 次印刷　印数：3 000 册
ISBN 978-7-5124-2244-5　定价：49.00 元

若本书有倒页、脱页、缺页等印装质量问题，请与本社发行部联系调换。联系电话：(010)82317024

序

全国大学生电子设计竞赛是教育部倡导的四大学科竞赛之一，是面向大学生的群众性科技活动，目的在于促进信息与电子类学科课程体系和课程内容的改革；促进高等院校实施素质教育以及培养大学生的创新能力、协作精神和理论联系实际的学风；促进大学生工程实践素质的培养，提高针对实际问题进行电子设计与制作的能力。

1. 规划教材由来

全国大学生电子设计竞赛既不是单纯的理论设计竞赛，也不仅仅是实验竞赛，而是在一个半封闭的、相对集中的环境和限定的时间内，由一个参赛队共同设计、制作完成一个有特定工程背景的作品。作品成功与否是竞赛能否取得好成绩的关键。

为满足高等院校电子信息工程、通信工程、自动化、电气控制类等专业学生参加全国大学生电子设计竞赛的需要，我们修订并编写了这套规划教材：《全国大学生电子设计竞赛系统设计(第 3 版)》、《全国大学生电子设计竞赛电路设计(第 3 版)》、《全国大学生电子设计竞赛技能训练(第 3 版)》、《全国大学生电子设计竞赛制作实训(第 3 版)》、《全国大学生电子设计竞赛常用电路模块制作(第 2 版)》、《全国大学生电子设计竞赛 ARM 嵌入式系统应用设计与实践(第 2 版)》、《全国大学生电子设计竞赛基于 TI 器件的模拟电路设计》。该套规划教材从 2006 年出版以来，已多次印刷，一直是全国各高等院校大学生电子设计竞赛训练的首选教材之一。随着全国大学生电子设计竞赛的深入发展，特别是 2007 年以来，电子设计竞赛题目要求的深度、难度都有很大的提高。2009 年竞赛的规则与要求也出现了一些变化，如对"最小系统"的定义、"性价比"与"系统功耗"指标要求等。为适应新形势下全国大学生电子设计竞赛的要求与特点，我们对该套规划教材的内容进行了修订与补充。

2. 规划教材内容

《全国大学生电子设计竞赛系统设计(第 3 版)》在详细分析了历届全国大学生电子设计竞赛题目类型与特点的基础上，通过 48 个设计实例，系统介绍了电源类、信号源类、无线电类、放大器类、仪器仪表类、数据采集与处理类以及控制类 7 大类赛题的变化与特点、主要知识点、培训建议、设计要求、系统方案、电路设计、主要芯片、程序设计等内容。通过对这些设计实例进行系统方案分析、单元电路设计、集成电路芯片选择，可使学生全面、系统地掌握电子设计竞赛作品系统设计的基本方法，培养学生系统分析、开发创新的能力。

《全国大学生电子设计竞赛电路设计(第 3 版)》在详细分析了历届全国大学生电子设计竞赛题目的设计要求及所涉及电路的基础上，精心挑选了传感器应用电路、信号调理电路、放大器电路、信号变换电路、射频电路、电机控制电路、测量与显示电路、电源电路、ADC 驱动和 DAC 输出电路 9 类共 180 多个电路设计实例，系统介绍了每个电路设计实例所采用的集成电路芯片的主要技术性能与特点、芯片封装与引脚功能、内部结构、工作原理和应用电路等内容。通过对这些电路设计实例的学习，学生可以全面、系统地掌握电路设计的基本方法，培养电路分析、设计和制作的能力。由于各公司生产的集成电路芯片类型繁多，限于篇幅，本书仅精选了其中很少的部分以"抛砖引玉"。读者可根据电路设计实例举一反三，并利用参考文献中给出的大量的公司网址，查询更多的电路设计应用资料。

《全国大学生电子设计竞赛技能训练(第 3 版)》从 7 个方面系统介绍了元器件的种类、特性、选用原则和需注意的问题；印制电路板设计的基本原则、工具及其制作；元器件、导线、电缆、线扎和绝缘套管的安装工艺和焊接工艺；电阻、电容、电感、晶体管等基本元器件的检测；电压、分贝、信号参数、时间和频率、电路性能参数的测量，噪声和接地对测量的影响；电子产品调试和故障检测的一般方法，模拟电路、数字电路和整机的调试与故障检测；设计总结报告的评分标准，写作的基本格式、要求与示例，以及写作时应注意的一些问题等内容；赛前培训、赛前题目分析、赛前准备工作和赛后综合测评实施方法、综合测评题及综合测评题分析等。通过上述内容的学习，学生可以全面、系统地掌握在电子竞赛作品制作过程中必需的一些基本技能。

《全国大学生电子设计竞赛制作实训(第 3 版)》指导学生完成 SPCE061A 16 位单片机、AT89S52 单片机、ADμC845 单片数据采集、PIC16F882/883/884/886/887 单片机等最小系统的制作；运算放大器运算电路、有源滤波器电路、单通道音频功率放大器、双通道音频功率放大器、语音录放器、语音解说文字显示系统等模拟电路的制作；FPGA 最小系统、彩灯控制器等数字电路的制作；射频小信号放大器、射频功率放大器、VCO(压控振荡器)、PLL－VCO 环路、调频发射器、调频接收机等高频电路的制作；DDS AD9852 信号发生器、MAX038 函数信号发生器等信号发生器的制作；DC－DC 升压变换器、开关电源、交流固态继电器等电源电路的制作；GU10 LED 灯驱动电路、A19 LED 灯驱动电路、AC 输入 0.5 W 非隔离恒流 LED 驱动电路等 LED 驱动电路的制作。介绍了电路组成、元器件清单、安装步骤、调试方法、性能测试方法等内容，可使学生提高实际制作能力。

《全国大学生电子设计竞赛常用电路模块制作(第 2 版)》以全国大学生电子设计竞赛中所需要的常用电路模块为基础，介绍了 AT89S52、ATmega128、ATmega8、C8051F330/1 单片机，LM3S615 ARM Cortex－M3 微控制器，LPC2103 ARM7 微控制器 PACK 板的设计与制作；键盘及 LED 数码管显示器模块、RS－485 总线通信模块、CAN 总线通信模块、ADC 模块和 DAC 模块等外围电路模块的设计与制作；放大器模块、信号调理模块、宽带可控增益直流放大器模块、音频放大器模块、D 类放大器模块、菱形功率放大器模块、宽带功率放大器模块、滤波器模块的设计与制作；反射式光电传感器模块、超声波发射与接收模块、温湿度传感器模块、阻抗测量模块、音频信号检测模块的设计与制作；直流电机驱动模块、步进电机驱动模块、函数信号发生器模块、DDS 信号发生器模块、压频转换模块的设计与制作；线性稳压电源模块、DC/DC 电路模块、Boost 升压模块、DC－AC－DC 升压电源模块的设计与制作；介绍了电路模块在随动控制系统、基于红外线的目标跟踪与无线测温系统、声音导引系统、单相正弦波逆变电源、无线环境监测模拟装置中的应用；介绍了地线的定义、接地的分类、接地的方式，接地系统的设计原则、导体的阻抗、地线公共阻抗产生的耦合干扰，模拟前端小信号检测和放大电路的电源电路结构、ADC 和 DAC 的电源电路结构、开关稳压器电路、线性稳压器电路，模/数混合电路的接地和电源 PCB 设计，PDN 的拓扑结构、目标阻抗、基于目标阻抗的 PDN 设计、去耦电容器的组合和容量计算等内容。本书以实用电路模块为模板，叙述简洁清晰，工程性强，可使学生提高常用电路模块的制作能力。所有电路模块都提供电路图、PCB 图和元器件布局图。

《全国大学生电子设计竞赛 ARM 嵌入式系统应用设计与实践（第 2 版）》以 ARM 嵌入式系统在全国大学生电子设计竞赛应用中所需要的知识点为基础，介绍了 LPC214x ARM 微控制器最小系统的设计与制作，可选择的 ARM 微处理器，以及 STM32F 系列 32 位微控制器最小系统的设计与制作；键盘及 LED 数码管显示器电路、汉字图形液晶显示器模块、触摸屏模块、LPC214x 的 ADC 和 DAC、定时器/计数器和脉宽调制器（PWM）、直流电机、步进电机和舵机驱动电路、光电传感器、超声波传感器、图像识别传感器、色彩传感器、电子罗盘、倾角传感器、角度传感器、E^2PROM 24LC256 和 SK－SDMP3 模块、nRF905 无线收发器电路模块、CAN 总线模块电路与 LPC214x ARM 微控制器的连接、应用与编程；基于 ARM 微控制器的随动控制系统、音频信号分析仪、信号发生器和声音导引系统的设计要求、总体方案设计、系统各模块方案论证与选择、理论分析及计算、系统主要单元电路设计和系统软件设计；MDK 集成开发环境、工程的建立、程序的编译、HEX 文件的生成以及 ISP 下载。该书突出了 ARM 嵌入式系统应用的基本方法，以实例为模板，可使学生提高 ARM 嵌入式系统在电子设计竞赛中的应用能力。本书所有实例程序都通过验证，相关程序清单可以在北京航空航天大学出版社网站“下载中心”下载。

《全国大学生电子设计竞赛基于 TI 器件的模拟电路设计》介绍的模拟电路是电子系统的重要组成部分，也是电子设计竞赛各赛题中的一个重要组成部分。模拟电路在设计制作中会受到各种条件的制约（如输入信号微弱、对温度敏感、易受噪声干扰等）。面对海量的技术资料、生产厂商提供的成百上千种模拟电路芯片，以及数据表中几十个参数，如何选择合适的模拟电路芯片，完成自己所需要的模拟电路设计，实际上是一件很不容易的事情。模拟电路设计已经成为电子系统设计过程中的瓶颈。本书从工程设计和竞赛要求出发，以 TI 公司的模拟电路芯片为基础，通过对模拟电路芯片的基本结构、技术特性、应用电路的介绍，以及大量的、可选择的模拟电路芯片、应用电路及PCB设计实例，图文并茂地说明了模拟电路设计和制作中的一些方法、技巧及应该注意的问题，具有很好的工程性和实用性。

3. 规划教材特点

本规划教材的特点：以全国大学生电子设计竞赛所需要的知识点和技能为基础，内容丰富实用，叙述简洁清晰，工程性强，突出了设计制作竞赛作品的方法与技巧。"系统设计"、"电路设计"、"技能训练"、"制作实训"、"常用电路模块制作"、"ARM嵌入式系统应用设计与实践"和"基于TI器件的模拟电路设计"这7个主题互为补充，构成一个完整的训练体系。

《全国大学生电子设计竞赛系统设计(第3版)》通过对历年的竞赛设计实例进行系统方案分析、单元电路设计和集成电路芯片选择，全面、系统地介绍电子设计竞赛作品的基本设计方法，目的是使学生建立一个"系统概念"，在电子设计竞赛中能够尽快提出系统设计方案。

《全国大学生电子设计竞赛电路设计(第3版)》通过对9类共180多个电路设计实例所采用的集成电路芯片的主要技术性能与特点、芯片封装与引脚功能、内部结构、工作原理和应用电路等内容的介绍，使学生全面、系统地掌握电路设计的基本方法，以便在电子设计竞赛中尽快"找到"和"设计"出适用的电路。

《全国大学生电子设计竞赛技能训练(第3版)》通过对元器件的选用、印制电路板的设计与制作、元器件和导线的安装和焊接、元器件的检测、电路性能参数的测量、模拟/数字电路和整机的调试与故障检测、设计总结报告的写作等内容的介绍，培训学生全面、系统地掌握在电子竞赛作品制作过程中必需的一些基本技能。

《全国大学生电子设计竞赛制作实训(第3版)》与《全国大学生电子设计竞赛技能训练(第3版)》相结合，通过对单片机最小系统、FPGA最小系统、模拟电路、数字电路、高频电路、电源电路等30多个制作实例的讲解，可使学生掌握主要元器件特性、电路结构、印制电路板、制作步骤、调试方法、性能测试方法等内容，培养学生制作、装配、调试与检测等实际动手能力，使其能够顺利地完成电子设计竞赛作品的制作。

《全国大学生电子设计竞赛常用电路模块制作(第2版)》指导学生完成电子设计竞赛中常用的微控制器电路模块、微控制器外围电路模块、放大器电路模块、传感器电路模块、电机控制电路模块、信号发生器电路模块和电源电路模块的制作，所制作的模块可以直接在竞赛中使用。

《全国大学生电子设计竞赛ARM嵌入式系统应用设计与实践(第2版)》以ARM嵌入式系统在全国大学生电子设计竞赛应用中所需要的知识点为基础；以LPC214x ARM微控制器最小系统为核心；以LED、LCD和触摸屏显示电路，ADC和DAC电路，直流电机、步进电机和舵机的驱动电路，光电、超声波、图像识别、色彩识别、电子罗盘、倾角传感器、角度传感器，E^2PROM，SD 卡，无线收发器模块，CAN 总线模块的设计制作与编程实例为模板，使学生能够简单、快捷地掌握 ARM 系统，并且能够在电子设计竞赛中熟练应用。

《全国大学生电子设计竞赛基于 TI 器件的模拟电路设计》从工程设计出发，结合电子设计竞赛赛题的要求，以 TI 公司的模拟电路芯片为基础，图文并茂地介绍了运算放大器、仪表放大器、全差动放大器、互阻抗放大器、跨导放大器、对数放大器、隔离放大器、比较器、模拟乘法器、滤波器、电压基准、模拟开关及多路复用器等模拟电路芯片的选型、电路设计、PCB 设计以及制作中的一些方法和技巧，以及应该注意的一些问题。

4. 读者对象

本规划教材可作为电子设计竞赛参赛学生的训练教材，也可作为高等院校电子信息工程、通信工程、自动化、电气控制等专业学生参加各类电子制作、课程设计和毕业设计的教学参考书，还可作为电子工程技术人员和电子爱好者进行电子电路和电子产品设计与制作的参考书。

作者在本规划教材的编写过程中，参考了国内外的大量资料，得到了许多专家和学者的大力支持。其中，北京理工大学、北京航空航天大学、国防科技大学、中南大学、湖南大学、南华大学等院校的电子竞赛指导老师和队员提出了一些宝贵意见和建议，并为本规划教材的编写做了大量的工作，在此一并表示衷心的感谢。

由于作者水平有限，本规划教材中的错误和不足之处，敬请各位读者批评指正。

黄智伟

2016 年 6 月 18 日于南华大学

第 2 版前言

随着全国大学生电子设计竞赛的深入发展，特别是从 2009 年以来，电子设计竞赛从题目要求的深度、广度都有了很大的提高。2009 年竞赛规则与要求出现了一些变化，如对微控制器选型的限制、“最小系统”的定义、“性价比”与“系统功耗”的指标要求等。除单片机、FPGA 外，ARM、DSP 等微控制器及最小系统也开始在电子设计竞赛中得到应用。

针对新形势下全国大学生电子设计竞赛的特点和需要，为高等院校电子信息工程、通信工程、自动化、电气控制等专业学生编写了这本《全国大学生电子设计竞赛 ARM 嵌入式系统应用设计与实践(第 2 版)》，作为培训教材。本书的特点是以 ARM 嵌入式系统在全国大学生电子设计竞赛应用所需要的知识点为基础，以实例为模板，叙述简洁清晰，工程性强，突出了 ARM 嵌入式系统应用的基本方法，培养了学生的竞赛设计与制作、综合分析与开发创新的能力。

全书共分 9 章：第 1 章，LPC214x ARM7 微控制器与最小系统，介绍了 LPC214x 的特性、封装、内部结构与功能，LPC214x ARM7 CPU PACK 板和 LPC214x ARM7 最小系统实验板的设计与制作，可选择的 ARM 微处理器，以及 STM32F 系列 32 位微控制器最小系统的设计与制作。第 2 章，显示器电路，介绍了键盘及 LED 数码管显示器电路，汉字图形液晶显示器模块，触摸屏模块的设计、制作与编程。第 3 章，ADC 和 DAC 电路，介绍了 LPC214x 的 ADC(模/数转换器)和 DAC(数/模转换器)电路设计与制作。第 4 章，电机控制，介绍了 LPC214x 的定时器/计数器和脉宽调制器(PWM)，直流电机、步进电机和舵机驱动电路的设计、制作与编程。第 5 章，传感器电路，介绍了光电传感器、超声波传感器、图像识别传感器、色彩传感器、电子罗盘、倾角传感器、角度传感器的应用与编程。第 6 章，数据存储，介绍了 E^2PROM 24LC256 和 SK－SDMP3 模块的应用与编程。第 7 章，数据传输，介绍了基于 nRF905 的无线收发器电路模块和 CAN 总线模块电路的设计、制作与编程。第 8 章，系统应用，介绍了基于 ARM 微控制器的随动控制系统、音频信号分析仪、正弦波信号发生器和声音导引系统的设计要求、总体方案设计、系统各模块方案论证与选择、理论分析及计算、系统主要单元电路设计和系统软件设计。第 9 章，开发环境及 ISP 下载，介绍了 MDK 集成开发环境和 ISP 下载，包括工程的建立、程序的编译、HEX 文件的生成。

本书可以作为各类电子制作、课程设计、毕业设计的教学参考书，以及电子工程技术人员进行电子电路设计与制作的参考书。

本书所有实例程序都通过验证，相关程序清单可以在北京航空航天大学出版社网站的“下载专区”下载。

本书在编写过程中，参考了大量国内外著作和资料，得到了许多专家和学者的大力支持，听取了多方面的意见和建议。税梦玲、张强、欧科军对本书中的实例进行了设计、编程与验证，李富英高级工程师对本书进行了审阅，南华大学王彦教授、朱卫华副教授、陈文光教授、李圣副教授、湖南师范大学邓月明博士、张翼、李军、戴焕昌、汤玉平、金海锋、李林春、谭仲书、彭湃、尹晶晶、全猛、周到、杨乐、黄俊、伍云政、李维、周望、李文玉、方果、许超龙、姚小明、马明、黄政中、邱海枚、欧俊希、陈杰、彭波、许俊杰等人也为本书的编写做了大量的工作，在此一并表示衷心的感谢。

由于水平有限，不足之处敬请各位读者批评指正。

黄智伟

2016 年 7 月于南华大学

第 1 版前言

随着全国大学生电子设计竞赛的深入和发展，近几年，特别是从 2005 年以后，电子设计竞赛的题目要求从深度和广度上都增加了。2009 年竞赛规则与要求也出现了一些变化，如对微控制器选型的限制、“最小系统”的定义、“性价比”与“系统功耗”的指标要求等。除单片机、现场可编程门阵列（FPGA）外，ARM、数字信号处理器（DSP）等微控制器及最小系统也开始在电子设计竞赛中得到应用。

针对新形势下全国大学生电子设计竞赛的特点和需要，为高等院校电子信息工程、通信工程、自动化、电气控制等专业学生编写了这本《全国大学生电子设计竞赛 ARM 嵌入式系统应用设计与实践》，可作为培训教材。本书的特点是以 ARM 嵌入式系统在全国大学生电子设计竞赛中应用所需要的知识点为基础，以实例为模板，叙述简洁清晰，工程性强，突出了 ARM 嵌入式系统应用的基本方法，培养了学生的竞赛设计与制作、综合分析与开发创新的能力。

全书共分 9 章：第 1 章 LPC214x ARM7 微控制器与最小系统，介绍了 LPC214x 的特性、封装、内部结构与功能，以及 ARM CPU PACK 板和 ARM7 最小系统实验板的设计与制作。第 2 章显示器电路，介绍了键盘及 LED 显示器电路，汉字图形点阵液晶显示模块，触摸屏模块的设计、制作与编程。第 3 章 ADC 和 DAC 电路，介绍了 LPC214x 的 ADC（模/数转换器）和 DAC（数/模转换器）电路设计与制作。第 4 章电机控制，介绍了 LPC214x 的定时器/计数器和脉宽调制器（PWM），直流电机、步进电机和舵机驱动电路的设计、制作与编程。第 5 章传感器电路，介绍了光电传感器、超声波传感器、图像识别传感器、色彩传感器、电子罗盘、倾角传感器和角度传感器的应用与编程。第 6 章数据存储，介绍了 E^2PROM 24LC256 和 SK－SDMP3 模块的应用与编程。第 7 章数据传输，介绍了基于 nRF905 的无线收发器电路模块和 CAN 总线模块电路的设计、制作与编程。第 8 章系统应用，介绍了基于 ARM 微控制器的随动控制系统、音频信号分析仪、正弦波信号发生器和声音导引系统的设计要求、总体方案设计、系统各模块方案论证与选择、理论分析与计算、系统主要单元电路设计和系统软件设计。第 9 章开发环境及 ISP 下载，介绍了 ADS 1.2 和 MDK 集成开发环境，包括工程的建立、程序的编译、HEX 文件的生成，以及 ISP 下载。

本书可以作为各类电子制作、课程设计、毕业设计的教学参考书，以及电子工程技术人员进行电子电路设计与制作的参考书。

本书所有实例程序都通过验证，相关程序清单可以登录北京航空航天大学出版社下载中心 http://www.buaapress.com.cn/buaa/html/download/index.asp 免费下载。

本书在编写过程中,参考了大量国内外著作和资料,得到了许多专家和学者的大力支持,听取了多方面的意见和建议。税梦玲、张强、欧科军对书中的实例进行了设计、编程与验证,李富英高级工程师对本书进行了审阅,南华大学王彦副教授、朱卫华副教授、陈文光副教授、李圣老师、湖南师范大学邓月明老师、张翼、李军、戴焕昌、汤玉平、金海锋、李林春、谭仲书、彭湃、尹晶晶、全猛、周到、杨乐、黄俊、伍云政、李维、周望、李文玉、方果、许超龙、姚小明、马明、黄政中、邱海枚、欧俊希、陈杰、彭波、许俊杰等人也为本书的编写做了大量的工作,在此一并表示衷心的感谢。

由于水平有限,不足之处敬请各位读者批评指正。

黄智伟

2010 年 9 月于南华大学

目　　录

第1章　LPC214x ARM7微控制器与最小系统 …… 1

1.1　LPC214x的特性与封装 …… 1

1.1.1　LPC214x的主要特性 …… 1

1.1.2　LPC214x的封装形式与引脚功能 …… 3

1.2　LPC214x的内部结构与功能 …… 6

1.2.1　LPC214x的内部结构 …… 6

1.2.2　LPC214x的内部结构功能描述 …… 6

1.3　LPC214x ARM7最小系统设计与制作 …… 14

1.3.1　LPC214x ARM7 CPU PACK板电路 …… 14

1.3.2　LPC214x ARM7最小系统实验板电路 …… 18

1.4　可选择的ARM微处理器 …… 25

1.4.1　ARM体系结构简介 …… 25

1.4.2　ARM7微处理器 …… 28

1.4.3　ARM9微处理器 …… 29

1.4.4　ARM11微处理器 …… 30

1.4.5　Cortex－A微处理器 …… 33

1.4.6　Cortex－R微处理器 …… 35

1.4.7　Cortex－M微处理器 …… 37

1.4.8　SecurCore微处理器 …… 39

1.5　STM32F系列32位微控制器最小系统的设计与制作 …… 40

1.5.1　STM32系列32位微控制器简介 …… 40

1.5.2　STM32F103xx系列微控制器的主要特性 …… 42

1.5.3　STM32F103xx系列微控制器的内部结构 …… 44

1.5.4　STM32F系列32位微控制器系统板简介 …… 45

1.5.5　STM32F系统板电原理图和PCB图 …… 50

1.5.6　STM32F系统板的应用设计与实践 …… 53

目　录

第2章　显示器电路 …… 55

2.1　键盘及LED数码管显示器电路的设计与制作 …… 55
2.1.1　ZLG7290B的主要特性 …… 55
2.1.2　ZLG7290B的应用电路 …… 57
2.1.3　ZLG7290B应用中应注意的一些问题 …… 57
2.1.4　ZLG7290B显示键盘应用程序设计 …… 60
2.2　液晶显示器模块的连接与编程 …… 65
2.2.1　FYD12864－0402B汉字图形点阵液晶显示模块简介 …… 65
2.2.2　LPC2148最小系统开发板与FYD12864－0402B的连接 …… 65
2.2.3　FYD12864－0402B汉字图形点阵液晶显示模块编程示例 …… 66
2.3　触摸屏模块的连接与编程 …… 73
2.3.1　触摸屏模块简介 …… 73
2.3.2　LPC2148最小系统开发板与触摸屏模块的连接 …… 75
2.3.3　触摸屏模块的编程示例 …… 76

第3章　ADC和DAC电路 …… 88

3.1　ADC电路的设计与制作 …… 88
3.1.1　LPC214x的ADC简介 …… 88
3.1.2　LPC214x的ADC编程示例 …… 90
3.2　DAC电路设计与制作 …… 99
3.2.1　LPC214x的DAC简介 …… 99
3.2.2　LPC214x的DAC编程示例 …… 100

第4章　电机控制 …… 102

4.1　LPC214x的定时器/计数器和脉宽调制器 …… 102
4.1.1　定时器/计数器(定时器0和定时器1) …… 102
4.1.2　脉宽调制器 …… 105
4.2　直流电机控制 …… 109
4.2.1　直流电机电枢的调速原理与调速方式 …… 109
4.2.2　直流电机驱动电路设计 …… 110
4.2.3　直流电机与LPC214x的连接 …… 115
4.2.4　直流电机控制编程示例 …… 115
4.3　步进电机控制 …… 118
4.3.1　步进电机的工作原理及方式简介 …… 118
4.3.2　基于“L297＋L298N”的步进电机驱动与控制电路 …… 119

4.3.3　基于“L297＋L298N”的步进电机控制编程示例 …… 123
4.3.4　基于 TA8435H 的步进电机驱动与控制电路 …… 129
4.3.5　基于 TA8435H 的步进电机控制编程示例 …… 134
4.4　舵机控制 …… 138
4.4.1　舵机简介 …… 138
4.4.2　舵机与 LPC214x 的连接 …… 141
4.4.3　舵机控制编程示例 …… 141

第 5 章　传感器电路 …… 144

5.1　光电传感器及其应用 …… 144
5.1.1　光电传感器选型 …… 144
5.1.2　利用反射式光电传感器检测障碍物 …… 145
5.1.3　利用反射式光电传感器检测黑线 …… 147
5.1.4　利用光电传感器检测光源 …… 151
5.2　超声波传感器及其应用 …… 162
5.2.1　超声波传感器的基本特性与选型 …… 162
5.2.2　超声波传感器用于障碍物检测与测距 …… 163
5.2.3　超声波传感器用于障碍物检测与测距编程示例 …… 167
5.3　图像识别传感器及其应用 …… 172
5.3.1　图像识别模组的内部结构 …… 172
5.3.2　图像识别模组的电路 …… 172
5.3.3　图像识别模组的应用 …… 177
5.3.4　SPCA563A 图像识别模块编程示例 …… 179
5.4　色彩传感器及其应用 …… 193
5.4.1　常用的几种色彩传感器的解决方案 …… 193
5.4.2　TCS230 可编程颜色光-频率转换器 …… 195
5.4.3　颜色识别模块的编程示例 …… 198
5.5　电子罗盘及其应用 …… 204
5.5.1　电子罗盘简介 …… 204
5.5.2　BQ－CA80－TTL 电子罗盘与微控制器的连接 …… 205
5.5.3　BQ－CA80－TTL 电子罗盘模块的编程示例 …… 205
5.6　倾角传感器及其应用 …… 209
5.6.1　倾角传感器简介 …… 209
5.6.2　LPC214x 开发板与 MSIN－LD60 倾角传感器的连接 …… 210
5.6.3　MSIN－LD60 倾角传感器编程示例 …… 210
5.7　角度传感器及其应用 …… 215

5.7.1 WDD35D-4 角度传感器简介 …… 215
5.7.2 LPC214x 开发板与 WDD35D-4 角度传感器的连接 …… 216
5.7.3 WDD35D-4 角度传感器编程示例 …… 216

第 6 章 数据存储 …… 220

6.1 E^2PROM 24LC256 …… 220
6.1.1 E^2PROM 24LC256 简介 …… 220
6.1.2 24LC256 的典型应用电路 …… 220
6.1.3 24LC256 读/写操作编程示例 …… 221
6.2 SK-SDMP3 语音模块及其应用 …… 223
6.2.1 SK-SDMP3 模块简介 …… 223
6.2.2 音频功率放大器电路 …… 226
6.2.3 SK-SDMP3 模块的编程示例 …… 228

第 7 章 数据传输 …… 232

7.1 无线数据传输 …… 232
7.1.1 基于 nRF905 的无线收发器电路模块 …… 232
7.1.2 LPC214x 开发板与无线收发器电路模块的连接 …… 236
7.1.3 无线收发器电路模块的编程示例 …… 237
7.2 CAN 总线应用 …… 249
7.2.1 CAN 总线简介 …… 249
7.2.2 在嵌入式处理器上扩展 CAN 总线接口 …… 250
7.2.3 CAN 总线网络结构 …… 251
7.2.4 CAN 总线模块设计 …… 252
7.2.5 CAN 总线网络编程示例 …… 256

第 8 章 系统应用 …… 262

8.1 基于 ARM 微控制器的随动控制系统 …… 262
8.1.1 设计要求 …… 262
8.1.2 总体方案设计 …… 263
8.1.3 系统各模块方案论证与选择 …… 263
8.1.4 理论分析及计算 …… 268
8.1.5 系统主要单元电路设计 …… 270
8.1.6 系统软件设计 …… 277
8.2 音频信号分析仪 …… 302
8.2.1 赛题要求 …… 302

8.2.2　基于单片机和 FPGA 的设计方案 …… 303
8.2.3　基于 LPC214x ARM 微控制器的设计方案实例 …… 305
8.3　正弦波信号发生器 …… 310
8.3.1　AD9850/51 DDS 模块简介 …… 310
8.3.2　LPC214x 开发板与 AD9850/51 DDS 模块的连接 …… 312
8.3.3　AD9850/51 DDS 模块的编程示例 …… 312
8.4　基于 ARM 微控制器的声音导引系统 …… 316
8.4.1　设计要求 …… 316
8.4.2　系统方案设计 …… 318
8.4.3　系统主要单元的选择与论证 …… 319
8.4.4　系统组成 …… 321
8.4.5　理论分析及计算 …… 321
8.4.6　系统电路设计 …… 322
8.4.7　系统软件设计 …… 324

第 9 章　开发环境及 ISP 下载 …… 333

9.1　MDK 集成开发环境 …… 333
9.1.1　MDK 集成开发环境简介 …… 333
9.1.2　工程的编辑 …… 334
9.2　ISP 下载 …… 338

参考文献 …… 340

第 1 章

LPC214x ARM7 微控制器与最小系统

1.1 LPC214x 的特性与封装

1.1.1 LPC214x 的主要特性

LPC214x(LPC2141/2/4/6/8)是基于一个支持实时仿真和嵌入式跟踪的 32/16 位 ARM7TDMI - S CPU 微控制器，带有 32 KB 和 512 KB 嵌入的高速 Flash 存储器。128 位宽的存储器接口和独特的加速结构使 32 位代码能够在最大时钟速率下运行。对代码规模有严格控制的应用，可使用 16 位 Thumb 模式，代码规模降低超过 30%，而性能的损失却很小。

较小的封装和较低的功耗使 LPC214x 特别适合于访问控制和 POS 机等小型应用。由于内置了宽范围的串行通信接口(从 USB 2.0 全速器件、多个通用异步收发器 UART、串行外设接口 SPI、同步串行口控制器 SSP 到 I^2C 总线)和 8～40 KB 的片内 SRAM，所以它们非常适合于通信网关、协议转换器、软 Modem、语音识别、低端成像，以及为这些应用提供大规模的缓冲区和强大的处理功能。多个 32 位定时器、1 个或 2 个 10 位 ADC、10 位 DAC、PWM 通道、45 个高速 GPIO，以及多达 9 个边沿或电平触发的外部中断引脚，使它们特别适用于工业控制和医疗系统。

LPC214x 的主要特性如下：

- 16/32 位 ARM7TDMI - S 微控制器，超小的 LQFP64 封装。
- 8～40 KB 的片内静态 RAM 和 32～512 KB 的片内 Flash 存储器，128 位宽接口/加速器可实现高达 60 MHz 的工作频率。
- 通过片内 boot 装载程序软件实现在系统编程/在应用编程(ISP/IAP)。单个 Flash 扇区或整片擦除时间为 400 ms，256 字节编程时间为 1 ms。
- Embedded ICE RT 和嵌入式跟踪接口通过片内 Real Monitor 软件提供实时调试和高速跟踪指令的执行。
- USB 2.0 全速设备控制器具有 2 KB 的终端 RAM。此外，LPC2146/8 提供 8 KB的片内 RAM，可通过 DMA 访问 USB。
- 1 个或 2 个 10 位 ADC，提供总共 6/14 路模拟输入，每个通道的转换时间低至 2.44 μs。

- 1个10位的DAC,可产生不同的模拟输出(仅用于LPC2142/4/6/8)。
- 2个32位定时器/外部事件计数器(带4路捕获和4路比较通道)、PWM单元(6路输出)和看门狗。
- 低功耗实时时钟(RTC)具有独立的电源和特定的32 kHz时钟输入。
- 多个串行接口,包括2个UART(16C550)、2个高速I^2C总线(400 kbit/s)、SPI,以及具有缓冲作用和数据长度可变功能的SSP。
- 向量中断控制器(VIC)可配置优先级和向量地址。
- 小型的LQFP64封装上包含多达45个通用I/O口(可承受5 V电压)。
- 多达21个可用的外部中断引脚。
- 通过一个可编程的片内PLL(100 μs的设置时间)可实现最大为60 MHz的CPU时钟操作频率。
- 片内集成振荡器与外部晶体的操作频率范围是1～25 MHz。
- 有空闲和掉电两种低功耗模式。
- 可通过个别使能/禁止外围功能和外围时钟分频来优化额外功耗。
- 通过外部中断或掉电检测(BOD)将处理器从掉电模式中唤醒。
- 单电源,具有上电复位(POR)和掉电检测电路,CPU操作电压范围为3.0～3.6 V(即3.3(1±0.1)V),I/O口可承受5 V的电压。
- 温度范围:－40～＋85 ℃。

LPC2141/2/4/6/8的主要特性区别如表1-1所列。

表1-1 LPC2141/2/4/6/8的主要特性区别

器件型号	Flash 存储器	RAM	端点 USB RAM	ADC (全部通道)	DAC
LPC2141FBD64	32 KB	8 KB	2 KB	1(6通道)	—
LPC2142FBD64	64 KB	16 KB	2 KB	1(6通道)	1
LPC2144FBD64	128 KB	16 KB	2 KB	2(14通道)	1
LPC2146FBD64	256 KB	32 KB＋8 KB与USB DMA共用*	2 KB	2(14通道)	1
LPC2148FBD64	512 KB	32 KB＋8 KB与USB DMA共用*	2 KB	2(14通道)	1

* 当USB DMA是额外8 KB RAM的主要用户时,该RAM作为通用RAM可被CPU在任何时候访问用于数据和代码存储。

LPC214x的一些极限参数如下:

- V_{DD}引脚的电源电压最大值为＋3.6 V。
- V_{DDA}引脚的模拟电源电压最大值为4.6 V。
- V_{BAT}引脚的输入电压最大值为4.6 V(用于RTC)。
- V_{REF}引脚的输入电压最大值为4.6 V。
- V_{IA}模拟输入电压最大值为5.1 V。

- V_I 输入电压，可承受 5 V 的 I/O 口最大值为 6.0 V，其他 I/O 口最大值为 V_{DD} +0.5 V。
- T_{stg} 储存温度范围为 −40～+125 ℃。
- I_{DD} DC 电源电流最大值为 100 mA。
- I_{SS} DC 地电流最大值为 100 mA。

注意：使用时不能够超过这些极限参数，否则将造成芯片永久性损坏。

有关 LPC214x 的更多信息，请登录 http://www.zlgmcu.com 查询《PHILIPS 单片 16/32 位微控制器——LPC2141/42/44/46/48 数据手册》。

1.1.2　LPC214x 的封装形式与引脚功能

LPC214x 采用小型的 LQFP64 封装，LPC2144/6/8 的封装形式如图 1-1 所示，LPC214x 的引脚功能如表 1-2 所列。

注意：LPC2141/2 与 LPC2144/6/8 采用相同的封装形式，仅个别引脚功能不同。

表 1-2　LPC214x 的引脚功能

引脚符号	类　型	功　能
P0.0～P0.31	I/O	P0，是一个 32 位 I/O 口，每个位都有独立的方向控制。有 31 个 P0 口可用作通用双向数字 I/O 口，P0.31 只用作输出口。P0 口引脚多数是多功能引脚端，其操作取决于引脚连接模块所选择的功能。引脚 P0.24、P0.26 和 P0.27 不可用。 引脚端不同符号表示如下： P0.0～P0.31——通用输入/输出数字引脚(GPIO)； TXDx——UARTx 的发送器输出； RXDx——UARTx 的接收器输入； RTSx——UARTx 请求发送输出，仅用于 LPC2144/6/8； CTSx——UARTx 的清零发送输入，仅用于 LPC2144/6/8； DSRx——UARTx 的数据设置就绪输入，仅用于 LPC2144/6/8； DTRx——UARTx 的数据终端就绪输出，仅用于 LPC2144/6/8； DCDx——UARTx 数据载波检测输入，仅用于 LPC2144/6/8； RIx——UARTx 铃声指示输入，仅用于 LPC2144/6/8； PWMx——脉宽调制器 x 输出； EINTx——外部中断 x 输入； SCLx——I2Cx 时钟输入/输出，开漏输出(符合 I^2C 规范)； SDAx——I2Cx 数据输入/输出，开漏输出(符合 I^2C 规范)； SCKx——SPIx 串行时钟，主机输出或从机输入的时钟； MISOx——SPIx 主机输入/从机输出，从机到主机的数据传输； MOSIx——SPIx 主机输出/从机输入，主机到从机的数据传输； SSELx——SPIx 从机选择，选择 SPI 接口用作从机； SCKx——SSP 串行时钟，主机输出或从机输入的时钟；

续表 1-2

引脚符号	类 型	功 能
P0.0～P0.31	I/O	MISOx——SSP主机输入/从机输出,从机到主机的数据传输; MOSIx——SSP主机输出/从机输入,主机到从机的数据传输; SSELx——SSP从机选择,选择SSP接口用作从机; CAPx.x——定时器x,捕获输入x; MATx.x——定时器x,匹配输出x; ADx.x——ADCx,输入x; AOUT——DAC输出,仅用于LPC2144/6/8; V_{BUS}——USB总线电源; UP_LED——表示USB良好连接的LED指示器; CONNECT——在软件控制下,信号用来切换外部15 kΩ的电阻。使用Soft Connect USB特性
P1.0～P1.31	I/O	P1口,一个32位双向I/O口,每个位都有独立的方向控制。P1口引脚多数是多功能引脚端,其操作取决于引脚连接模块所选择的功能。P1口的P1.0～P1.15不可用。 引脚端不同符号表示如下: P1.16～P1.31——通用输入/输出数字引脚(GPIO); TRACEPKTx——跟踪包位x,带内部上拉的标准I/O口; TRACESYNC——跟踪同步,带内部上拉的标准I/O口; TRACECLK——跟踪时钟,带内部上拉的标准I/O口; PIPESTATx——流水线状态位x,带内部上拉的标准I/O口; EXTINx——外部触发输入,带内部上拉的标准I/O口; RTCK——返回的测试时钟输出; TDO——JTAG接口测试数据输出; TDI——JTAG接口测试数据输入; TCK——JTAG接口测试时钟; TMS——JTAG接口模式选择; TRST——JTAG接口测试复位
D+	I/O	USB双向D+线
D-	I/O	USB双向D-线
$\overline{\text{RESET}}$	I	外部复位输入,该引脚的低电平将器件复位,并使I/O口和外围功能恢复默认状态,处理器从地址0开始执行。带迟滞的TTL电平,引脚可承受5 V电压
XTAL1	I	振荡器电路和内部时钟发生器的输入
XTAL2	O	振荡放大器的输出
RTXC1	I	RTC振荡电路的输入
RTXC2	O	RTC振荡电路的输出
V_{SS}	I	地,0 V参考点
V_{SSA}	I	模拟地,0 V参考点。标称电压与V_{SS}相同,但应当互相隔离以减少噪声和故障
V_{DD}	I	3.3 V电源,内核和I/O口的电源电压
V_{DDA}	I	模拟3.3 V端口电源,标称电压与V_{DD}相同,但应当互相隔离以减少噪声和故障。该电压也用来向片内ADC和DAC供电

续表 1-2

引脚符号	类　型	功　能
V_{REF}	I	A/D 转换器参考电压，标称电压应少于或等于 V_{DD} 电压，但应当互相隔离以减少噪声和故障。该引脚的电平用作 ADC 和 DAC 的参考电压
V_{BAT}	I	RTC 电源，RTC 的 3.3 V 电源端

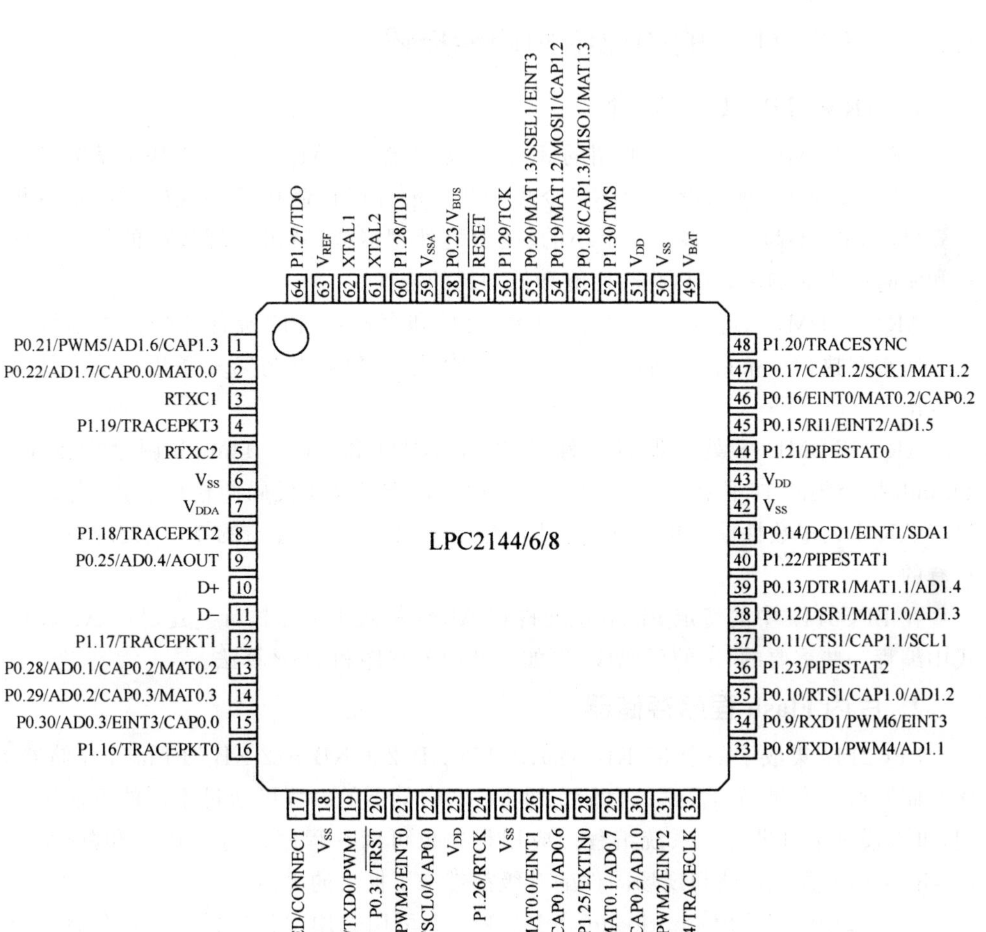

图 1-1　LPC2144/6/8 的封装形式

1.2　LPC214x 的内部结构与功能

1.2.1　LPC214x 的内部结构

LPC214x 内部结构方框图如图 1-2 所示。

1.2.2　LPC214x 的内部结构功能描述

1. ARM7TDMI-S 简介

ARM7TDMI-S 是一个通用的 32 位微处理器,其 ARM 结构是基于精简指令集计算机(RISC)原理而设计的。指令集和相关的译码机制比微编程的复杂指令集计算机(CISC)要简单得多。一个小的、廉价的处理器核,就可实现较高的指令吞吐量和实时的中断响应,并具有高性能和低功耗的特性。

ARM7TDMI-S 使用了流水线技术,处理和存储系统的所有部分都可连续工作。通常在执行一条指令的同时对下一条指令进行译码,并将第三条指令从存储器中取出。

ARM7TDMI-S 处理器具有标准 32 位 ARM 和 16 位 Thumb 两个指令集。Thumb 指令集的 16 位指令长度使其可以达到标准 ARM 代码两倍的密度,却仍然保持 ARM 大多数性能上的优势,这些优势是使用 16 位寄存器的 16 位处理器所不具有的。

在 LPC214x 中的特定 Flash 也允许在 ARM 模式中全速执行。建议在 ARM 模式中编程一些重要的、短的代码段,例如中断服务程序和 DSP 算法。

2. 片内 Flash 程序存储器

LPC214x 集成了一个 32 KB/64 KB/128 KB/256 KB/512 KB 的 Flash 存储器。该存储器可用作代码和数据的存储。对 Flash 存储器的编程可通过不同的方法来实现,可以通过串口进行在系统编程。应用程序也可以在程序运行时擦除和/或编程 Flash,这样为数据存储和现场固件的升级都带来了极大的灵活性。

由于选择使用片内 Bootloader,LPC214x 上可用作用户代码的 Flash 存储器分别为 32 KB/64 KB/128 KB/256 KB/512 KB。

LPC214x Flash 存储器至少含有 100 000 个擦除/写周期,数据至少可保存 20 年。

3. 片内静态 RAM

片内静态 RAM 可用作代码和/或数据存储。SRAM 支持 8 位、16 位和 32 位访问。LPC2141、LPC2142/4 和 LPC2146/8 分别提供 8 KB、16 KB、32 KB 的静态 RAM。在仅为 LPC2146/8 的情况下,主要由 USB 使用的 8 KB SRAM 模块也可用作数据存储、代码存储和执行的通用 RAM。

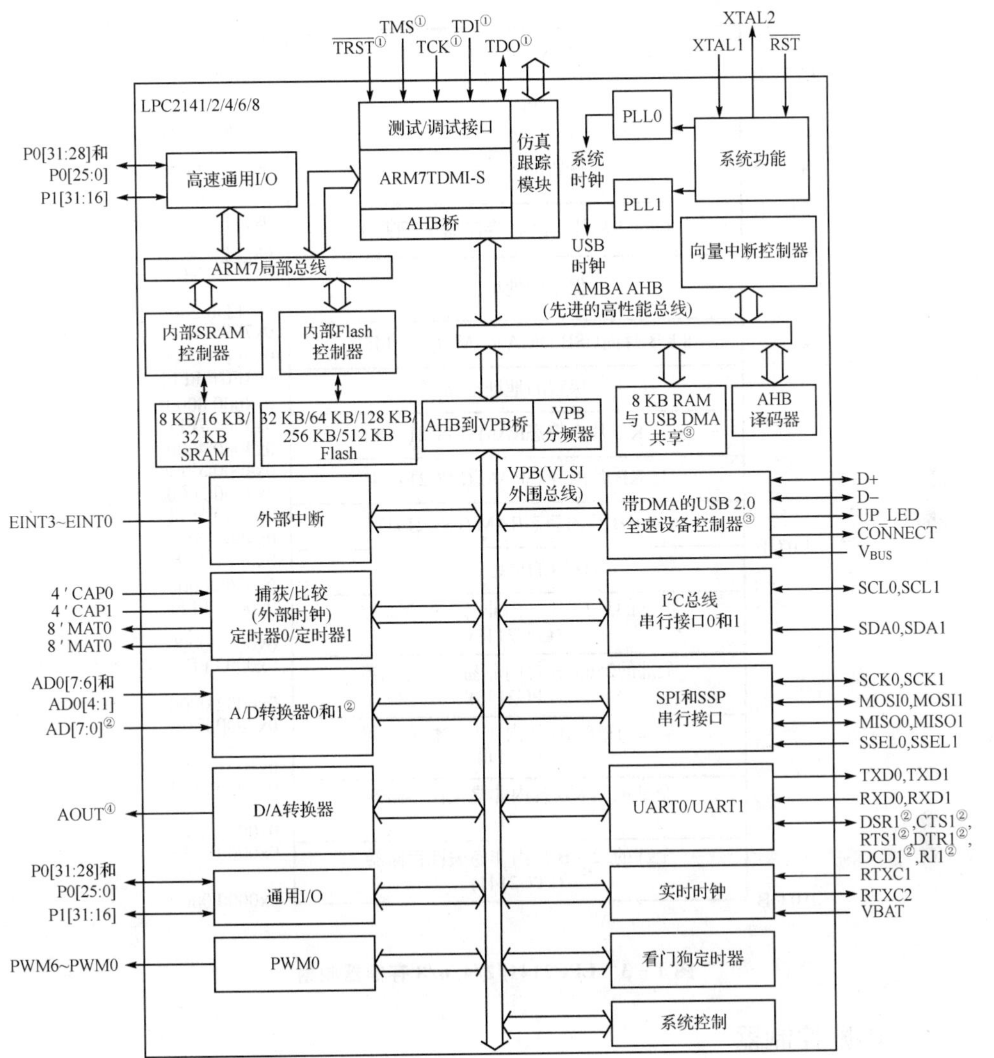

注：① 该引脚与 GPIO 共用。

② 仅在 LPC2144/6/8 中可用。

③ USB DMA 控制器带 8 KB RAM，可作为通用 RAM 访问，DMA 仅在 LPC2146/8 中可用。

④ 仅在 LPC2142/4/6/8 中可用。

图 1-2　LPC214x 内部结构方框图

4. 存储器映射

LPC214x 的存储器映射包含几个不同的区域，如图 1-3 所示。此外，CPU 的中断向量可以重新映射，允许它们位于 Flash 存储器(默认)或者片内静态 RAM 中。

图 1-3　LPC2141/2/4/6/8 存储器映射

5. 中断控制器

向量中断控制器(VIC)接收所有的中断请求(IRQ)输入,并将它们编程分配为 3 类:快速中断请求(FIQ)、向量 IRQ 和非向量 IRQ。可编程分配机制意味着不同外设的中断优先级可以动态分配和调整。

快速中断请求(FIQ)具有最高优先级。如果分配给 FIQ 的请求多于 1 个,VIC 将中断请求"相或"向 ARM 处理器产生 FIQ 信号。当只有一个被分配为 FIQ 时可实现最短的 FIQ 等待时间。如果分配给 FIQ 级的请求多于 1 个,FIQ 服务程序从 VIC 中读出一个字以识别产生中断请求的 FIQ 中断源是哪一个。

向量 IRQ 具有中等优先级。该级别可分配 16 个中断请求。中断请求中的任意一个都可分配到 16 个向量 IRQ slot 中的任意一个,其中 slot0 具有最高优先级,而 slot15 则为最低优先级。

非向量 IRQ 的优先级最低。

VIC 将所有向量和非向量 IRQ 组合后向 ARM 处理器产生 IRQ 信号。IRQ 服务程序可通过读取 VIC 的一个寄存器立即启动并跳转到相应地址。如果有任意一个向量 IRQ 发出请求，VIC 则提供最高优先级请求 IRQ 服务程序的地址；否则提供默认程序的地址，该默认程序由所有非向量 IRQ 共用。默认程序可读取另外一个 VIC 寄存器以确定哪个 IRQ 被激活。

每个外设都有一条中断线连接到 VIC，但可能有几个内部中断标志。单个中断标志也可能代表多于一个的中断源。

6. 引脚连接模块

引脚连接模块允许将微控制器的引脚配置为不同的功能。配置寄存器控制连接引脚和片内外设的多路开关。应当在激活外设以及使能任何相关的中断之前，将外设连接到相应的引脚。任何一个被使能的外设，如果其功能没有映射到相关的引脚，那么对它的激活将被认为是未定义的。

带有自身引脚选择寄存器的引脚控制模块在一个给定的硬件环境中定义了微控制器的功能。

复位后，P0 和 P1 的所有引脚都配置为输入，但以下情况除外：如果调试被使能，则 JTAG 引脚将假定为 JTAG 功能；如果跟踪被使能，则跟踪引脚将假定为跟踪功能。与 I2C0 和I2C1 接口有关的引脚为开漏。

7. 快速通用并行 I/O 口

没有连接到特定外设功能的引脚由 GPIO 寄存器进行控制。引脚可以动态配置为输入或输出。寄存器可以同时对任意输出口进行置位或清零。输出寄存器的值以及引脚的当前状态都可以读出。

LPC2141/2/4/6/8 的 GPIO：GPIO 寄存器被转移到 ARM 局部总线，可以实现高速 I/O 时序；所有 GPIO 寄存器为字节可寻址；整个端口值可用一条指令写入；位电平置位和清零寄存器允许一条指令置位和清零一个端口的任何位数；输出置位和清零可单独控制；所有 I/O 在复位后的默认状态都为输入；可以实现单个位的方向控制。

8. 10 位模/数转换器 ADC

LPC2141/2 含有 1 个 ADC，LPC2144/6/8 含有 2 个 ADC。这些 ADC 为单个 10 位逐次逼近 ADC。其中，ADC0 有 6 个通道时，ADC1 有 8 个通道。LPC2141/2 可用的 ADC 输入总数为 6，LPC2144/6/8 可用的 ADC 输入总数为 14。

ADC 的测量范围为 0 V～V_{REF}（2.0 V≤V_{REF}≤V_{DDA}），每个转换器每秒可执行多于400 000次 10 位采样，每个模拟输入有一个指定的结果寄存器来减少中断开销，具有单路或多路输入的突发转换模式，2 个转换器全局启动命令（仅用于 LPC2142/4/

6/8),可以根据输入引脚的跳变或定时器匹配信号执行转换。

9. 10 位数/模转换器 DAC

LPC2141/2/4/6/8 具有一个 10 位 DAC,DAC 用来产生模拟量输出;另外,其还具有缓冲输出,可用于掉电模式,可选择速率与功耗,最大的 DAC 输出电压为 V_{REF} 电压。

10. USB 2.0 设备控制器

LPC214x 带有 USB 设备控制器,该控制器能与 USB 主机控制器以 12 Mbit/s 的速率进行数据传输。它由寄存器接口、串行接口引擎、端点缓冲存储器和 DMA 控制器组成。串行接口引擎对 USB 数据流进行译码,并将数据写入相应的端点缓冲存储器。完整的 USB 传输状态或错误条件通过状态寄存器来指示,若中断使能则产生中断。

DMA 控制器(仅用于 LPC2146/8)可传输端点缓冲区和 USB RAM 之间的数据。

LPC214x 的 USB 设备控制器完全兼容 USB 2.0 全速规范,支持 32 个物理(16 个逻辑)端点,支持控制、批量、中断和同步端点,运行时调整使用的端点,并且可通过软件来选择端点最大包长度(取决于 USB 最大规格,RAM 信息缓冲区大小取决于使用的端点和最大包的长度),支持 Soft Connect 特性和 Good Link LED 指示器,支持总线供电功能,具有较低的挂起电流,支持所有非控制端点的 DMA 传输(仅用于 LPC2146/8),所有端点都有一个双向的 DMA 通道(仅用于 LPC2146/8),允许 CPU 控制和 DMA 模式之间的动态切换(仅用于 LPC2146/8),实现了批量和同步端点的双缓冲。

11. UART

LPC214x 包含 2 个 UART。除了标准的发送和接收数据线外,LPC2144/6/8 UART1 还提供了一个完全的调制解调器控制握手接口。

LPC214x 的 UART 具有 16 字节接收和发送 FIFO,寄存器位置遵循 550 工业标准,接收器 FIFO 触发点为 1、4、8 和 14 字节,内置分数波特率发生器(可使能微控制器来激活标准波特率如 115 200),发送 FIFO 控制使能实现 2 个 UART 的软件(XON/XOFF)流控制,LPC2144/6/8 的 UART1 带有标准的调制解调器接口信号,在硬件中完全实现自动 CTS/RTS 流控制功能。

12. I^2C 总线串行 I/O 控制器

LPC214x 包含 2 个 I^2C 总线控制器。I^2C 为双向总线,它使用串行时钟线(SCL)和串行数据线(SDA)实现互联芯片的控制。

LPC214x 中的 I^2C 总线遵循标准的 I^2C 总线接口,可配置为主机、从机或主/从机;可编程时钟可实现通用速率控制;主机与从机之间双向数据传输;多主机总线(无中央主机),在同时发送的主机之间进行仲裁可避免总线上串行数据冲突;串行时钟同步使器件在一条串行总线上实现不同位速率的通信;串行时钟同步可作为握手机制使串行传输挂起和恢复;I^2C 总线可用于测试和诊断;支持高达 400 kbit/s 的位速率(高速 I^2C 总线)。

13. SPI 串行 I/O 控制器

LPC214x 包含 1 个 SPI 控制器。SPI 是一个全双工的串行接口,可以设计成在一条给定总线上能够处理多个互连的主机和从机。在数据传输过程中,接口上只能有一个主机和一个从机可以通信。在一次数据传输中,主机总是向从机发送一个数据字节,而从机也总是向主机发送一个数据字节。

LPC214x 的 SPI 控制器遵循串行外设接口(SPI)规范,可实现同步、串行、全双工通信,组合的 SPI 主机和从机,其最大数据位速率是输入时钟速率的 1/8。

14. SSP 串行 I/O 控制器

LPC214x 包含 1 个同步串行口控制器(SSP)。SSP 可以控制 SPI、4 线 SSI 或 Microwire总线。

它可与总线的多个主机和从机互相通信,但是,在一个给定的数据传输过程中,总线只允许一个主机和一个从机通信。SSP 支持全双工传输,具有发送和接收的 8 帧FIFO,允许主机和从机之间传输 4～16 位的数据流帧(即每帧包含4～16 位数据)。

15. 通用定时器/外部事件计数器

LPC214x 的定时器/计数器对外设时钟周期(PCLK)或外部时钟进行计数,LPC214x 的定时器/计数器带可编程 32 位预分频器的 32 位定时器/计数器;外部事件计数器或定时器操作;当输入信号跳变时,4 个 32 位捕获通道可捕获定时器的瞬时值,捕获事件可选择产生中断;4 个 32 位匹配寄存器可实现连续操作,匹配时停止定时器,匹配时复位定时器,并可选择产生中断;每个定时器有 4 个对应于匹配寄存器的外部输出,具有匹配时置低电平,匹配时置高电平,匹配时翻转,匹配时不变的特性。

16. 看门狗定时器

LPC214x 的看门狗定时器的作用是使微控制器在进入错误状态经过一段时间后复位。当看门狗使能时,如果没有在预先确定的时间内“喂”(重装)看门狗,那么它将会产生一次系统复位。

LPC214x 的看门狗定时器带有内部预分频器的可编程 32 位定时器;可选择时间周期从 $T_{PCLK}\times 2^8\times 4$ 到 $T_{PCLK}\times 2^{32}\times 4$,可选值为 $4T_{PCLK}$的倍数。

17. 实时时钟

当选择正常或空闲模式时，实时时钟 RTC 提供一套用于测量时间的计数器。RTC 消耗的功率非常低，适合由电池供电或 CPU 不连续工作（空闲模式）的系统。

LPC214x 的实时时钟可以对时间段进行测量，以实现一个日历和时钟；具有超低功耗设计，支持电池供电系统；可提供秒、分、小时、日、月、年和星期；可使用 RTC 专用的 32 kHz 振荡器输入或 XTAL1 连接的外部晶体/振荡器输入的时钟。可编程基准时钟分频器，允许调节 RTC 以适应不同的晶振频率；采用专用电源引脚连接到电池或 3.3 V 的电压。

18. 脉宽调制器

LPC214x 的脉宽调制器 PWM 基于标准的定时器模块并具有其所有特性。LPC214x 的 PWM 具有 7 个匹配寄存器，可实现 6 个单边沿控制或 3 个双边沿控制 PWM 输出，或这两种类型的混合输出；匹配寄存器允许执行连续操作，匹配时停止定时器，匹配时复位定时器，可选择产生中断的操作；支持单边沿控制和双边沿控制的 PWM 输出；脉冲周期和宽度可以是任何的定时器计数值；双边沿控制的 PWM 输出可编程为正脉冲或负脉冲；匹配寄存器更新与脉冲输出同步，防止产生错误的脉冲；如果不使能 PWM 模式，则可作为一个标准定时器；带可编程 32 位预分频器的 32 位定时器/计数器。

19. 系统控制

(1) 晶　振

LPC214x 的片内集成振荡器支持的晶振范围为 1～25 MHz。晶振输出频率称为 f_{OSC}，而 ARM 处理器时钟频率称为 f_{CLK}。除非连接并运行 PLL，否则在该文档中 f_{OSC} 和 f_{CLK} 的值是相同的。

(2) PLL

LPC214x 的 PLL 可以接收 10～25 MHz 的输入时钟频率。输入频率通过一个电流控制振荡器（CCO）可以倍增为 10～60 MHz。倍增器可以是 1～32 的整数（实际上在该系列微控制器当中，由于 CPU 频率的限制，所以倍增器的值不可能大于 6）。CCO 操作的范围为 156～320 MHz。输出分频器可设置为 2、4、8 或者 16 分频以产生输出时钟。PLL 在芯片复位后关闭并且被旁路，可通过软件使能。程序必须配置并且激活 PLL，等待 PLL 锁定之后再将 PLL 作为时钟源。PLL 设置时间为 100 μs。

(3) 复位和唤醒定时器

LPC214x 有 2 个复位源：RESET 引脚和看门狗复位。RESET 引脚是一个施密特触发输入引脚，带有附加的干扰滤波器。任何复位源所导致的芯片复位都会启动唤醒定时器，使内部芯片复位保持有效直到外部复位撤除，然后振荡器开始运行。

振荡器运行经过固定数目的时钟后片内 Flash 控制器完成其初始化。

当内部复位撤除后,处理器从复位向量地址 0 开始执行。此时所有的处理器和外设寄存器都被初始化为预设的值。

唤醒定时器的用途,是确保振荡器和其他芯片操作所需要的模拟功能,在处理器能够执行指令之前完全正常工作。由于振荡器和其他功能在掉电模式下关闭,因此将处理器从掉电模式中唤醒就要利用唤醒定时器。

唤醒定时器监视晶体振荡器是否可以安全地开始执行代码。当芯片上电时,或某些事件导致芯片退出掉电模式时,振荡器需要一定的时间以产生足够振幅的信号驱动时钟逻辑。时间的长度取决于许多因素,包括 V_{DD}上升速度(上电时)、晶振的类型及电气特性以及其他外部电路(例如,电容)和外部环境下振荡器自身的特性。

(4) 掉电检测器

LPC214x 可以对 V_{DD}引脚电压进行二级检测。如果 V_{DD}引脚电压低于 2.9 V,则掉电检测器(BOD)向向量中断控制器声明一个中断。该信号可通过中断使能,也可由软件通过读取相应的寄存器来监控。

当 V_{DD}引脚的电压低于 2.6 V 时,第二级的低电压检测将产生复位,禁能 LPC214x。

(5) 代码安全

LPC214x 可以控制应用代码是否被调试或被保护(以防盗用)。当片内 Bootloader 在 Flash 中检测到一个有效校验,以及在 Flash 的 0x1FC 地址单元读取 0x8765 4321 时,禁止调试,Flash 代码被保护。一旦调试被禁能,它就只能通过执行芯片擦除来使能。

(6) 外部中断输入

根据可选引脚功能的设定,LPC214x 最多可包含 9 个边沿或电平触发的外部中断输入。外部事件可作为 4 个独立的中断信号来处理。外部中断输入可用于将处理器从掉电状态唤醒。

(7) 存储器映射控制

存储器映射控制可以改变从地址 0x0000 0000 开始的中断向量的映射。向量可以映射到片内 Flash 存储器的底部,也可以映射到片内静态 RAM。这使得在不同存储器空间中运行的代码都能够对中断进行控制。

(8) 功率控制

LPC214x 支持空闲模式和掉电模式两种低功耗模式。

在空闲模式中,指令的执行被暂停,直到产生复位或中断为止。外围功能在空闲模式下继续工作并可产生中断唤醒处理器。空闲模式使处理器自身、存储器系统和相关的控制器以及内部总线不再消耗功率。

在掉电模式中,振荡器被关闭,芯片没有任何的内部时钟。处理器状态和寄存器、外设寄存器和内部 SRAM 的值在掉电模式下保持不变。芯片输出引脚的逻辑电平保持静态。通过复位或特定的不需要时钟还可工作的中断可终止掉电模式并恢复正常操作。由于芯片的所有动态操作都被暂停,掉电模式使芯片消耗的功率降低到几乎为零。

(9) VPB 总线

VPB 分频器决定处理器时钟(C_{CLK})和外设时钟(P_{CLK})之间的关系。VPB 分频器通过 VPB 总线为外设提供所需的时钟(P_{CLK}),使外设可工作在 ARM 处理器选择的速率下。为了实现该功能,VPB 总线频率可以降低为处理器时钟频率的1/2~1/4。VPB 总线在复位后的默认状态是以 1/4 的处理器时钟速率运行。当所有外设都不必在全速率下运行时,VPB 分频器可以降频以降低功耗。由于 VPB 分频器连接到 PLL 的输出,PLL(如果正在运行)在空闲模式时保持有效。

20. 仿真和调试

LPC214x 支持通过 JTAG 串行端口进行仿真和调试。跟踪端口允许跟踪程序的执行。调试和跟踪功能只在 GPIO 的 P1 口复用。这意味着当应用在嵌入式系统内运行时,位于 P0 口的所有通信、定时器和接口外设在开发和调试阶段都可用。

标准的 ARM Embedded ICE 逻辑提供对片内调试的支持。对目标系统进行调试需要一个主机来运行调试软件和 Embedded ICE 协议转换器。Embedded ICE 协议转换器将远程调试协议命令转换成所需要的 JTAG 数据,从而对目标系统上的 ARM 内核进行访问。

嵌入式跟踪宏单元(ETM)对嵌入式处理器内核提供了实时跟踪能力。它向一个跟踪端口输出处理器执行的信息。ETM 直接连接到 ARM 内核而不是 AMBA 系统总线。它将跟踪信息压缩并通过一个窄带跟踪端口输出。外部跟踪端口分析仪在软件调试器的控制下捕获跟踪信息。

Real Monitor 是一个可配置的软件模块,它由 ARM 公司开发,可以提供实时的调试。LPC214x 包含一个编程到片内 Flash 存储器中的 Real Monitor 软件的指定配置。

1.3 LPC214x ARM7 最小系统设计与制作

1.3.1 LPC214x ARM7 CPU PACK 板电路

LPC214x(LPC2141/2/4/6/8)CPU PACK 板电原理图和板电路如图 1-4 和图 1-5所示,通过 J1~J4 这 4 个插座连接 LPC214x(LPC2141/2/4/6/8)的 52 个引脚端到系统板。

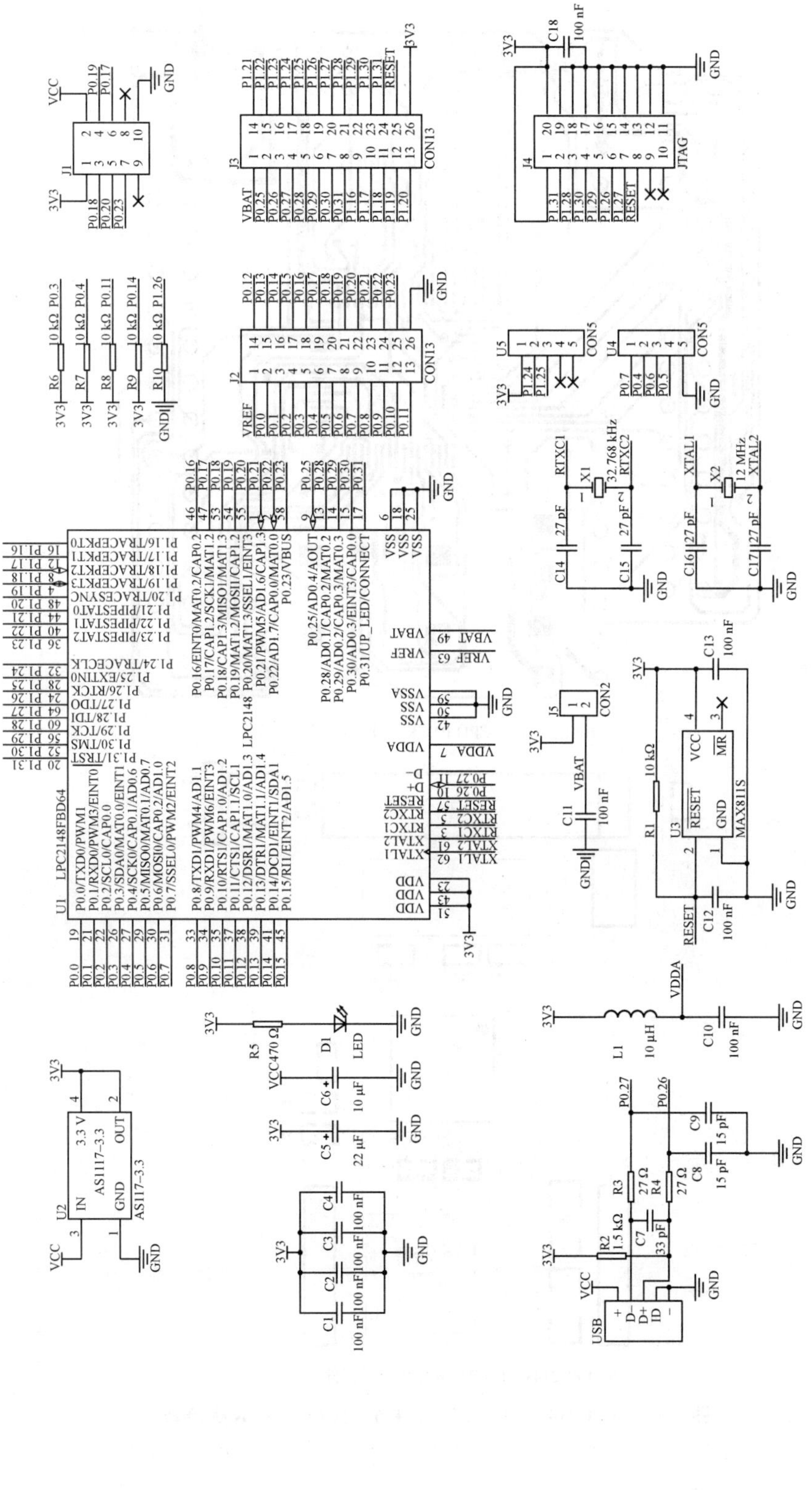

图 1-4　LPC214x(LPC2141/2/4/6/8) CPU PACK板电原理图

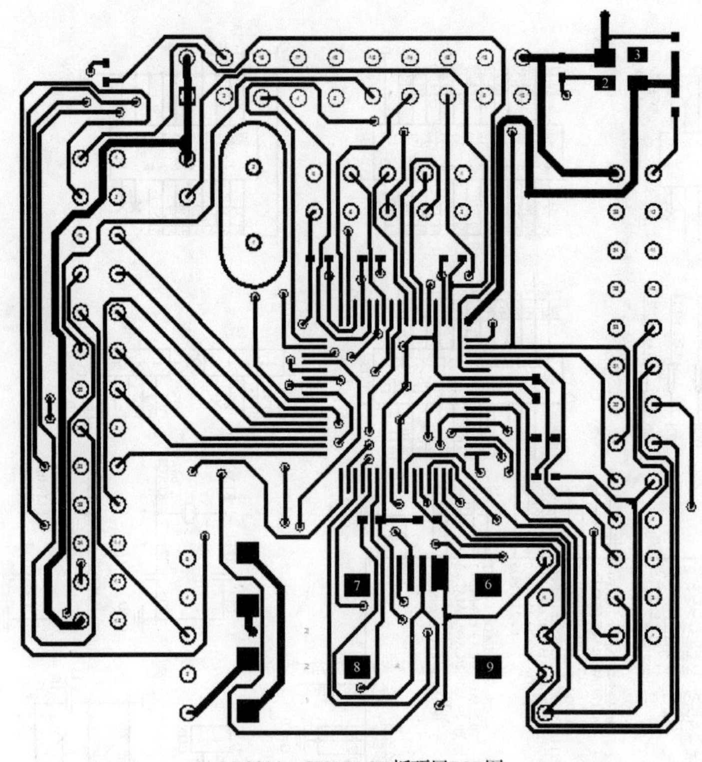

(a) LPC214x CPU PACK板顶层PCB图

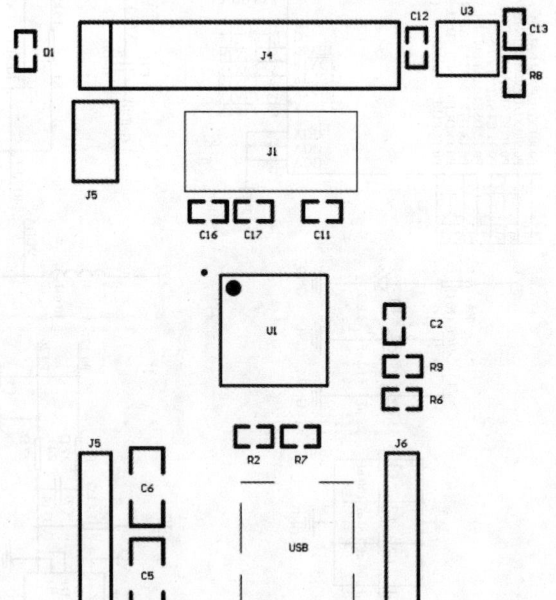

(b) LPC214x CPU PACK板顶层元器件布局图

图 1-5　LPC214x(LPC2141/2/4/6/8)CPU PACK 板电路

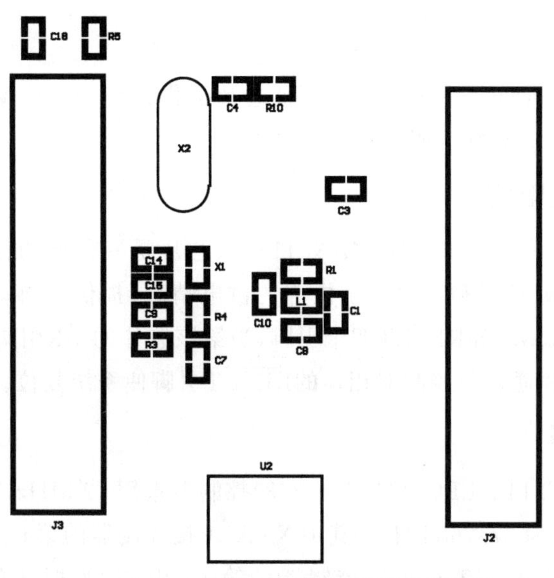

(c) LPC214x CPU PACK板底层元器件布局图

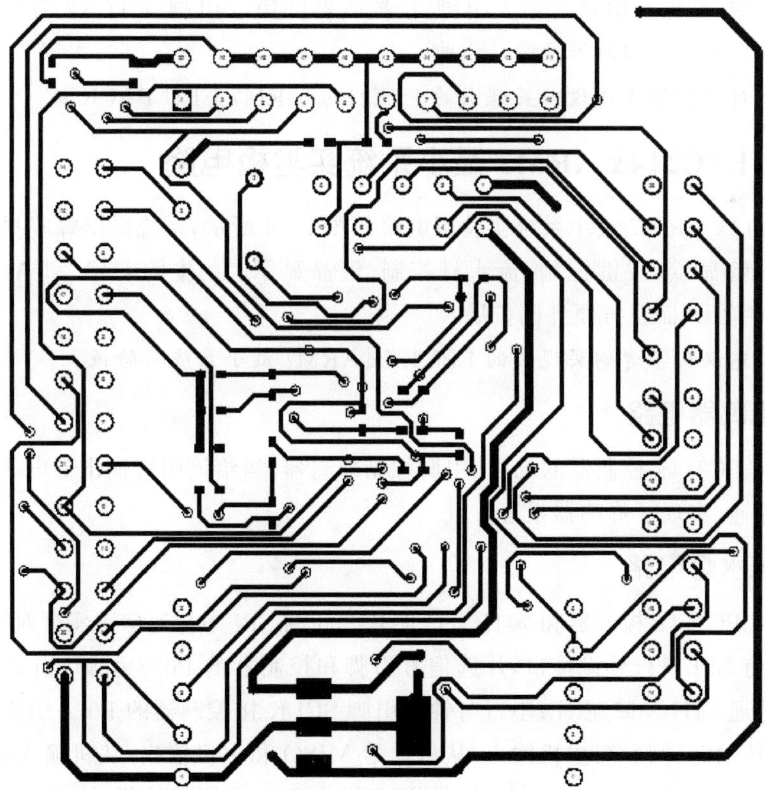

(d) LPC214x CPU PACK板底层PCB图

图 1－5　LPC214x(LPC2141/2/4/6/8)CPU PACK 板电路(续)

1. 电源电路

电源电路采用AS1117-3.3 V稳压器芯片提供系统板3.3 V电源电压。当接通电源时,LED D1点亮则说明电路接通电源。

2. 系统复位电路

系统复位电路采用专用芯片MAX811S,该芯片是一种低功耗微处理器监控电路IC,用于监控微处理器和数字系统内部的供电情况,提供去抖动的手动复位。芯片共有VCC、GND、$\overline{\text{MR}}$和$\overline{\text{RESET}}$四个引脚,当给该芯片的$\overline{\text{MR}}$引脚一个大于144 ms的低脉冲时,该芯片通过与控制器相连的$\overline{\text{RESET}}$引脚使系统复位。

3. 晶振电路

本系统的LPC214x(LPC2141/2/4/6/8)控制器选用12 MHz晶振,内部RTC实时时钟的晶振频率为37.768 kHz。其中XTAL1接控制器的第62引脚,为振荡电路和内部时钟提供输入,XTAL2接控制器的第61引脚,为振荡放大器提供输出;RTCX1为RTC振荡电路的输入引脚,接控制器的第3引脚,RTCX2为RTC振荡电路的输出引脚,接控制器的第5引脚。

注意:对于初学者建议购买现成的LPC214x ARM7 CPU PACK板。

1.3.2 LPC214x ARM7最小系统实验板电路

LPC214x ARM7最小系统实验板电路如图1-6所示,系统提供蜂鸣报警、串行通信、数码管显示、按键控制、流水灯控制、液晶显示、10位的D/A和A/D转换、PWM脉冲输出,而且方便外围扩展。

注意:建议初学者购买现成的LPC214x ARM7最小系统实验板。

1. 蜂鸣器电路

使用LPC214x控制器的P0.7引脚控制蜂鸣器,当给该引脚输出低电平时,三极管8550导通,实现了蜂鸣器报警。

2. 流水灯电路

通过LPC214x控制器和74HC595芯片之间的SPI串行通信来实现对流水灯的控制。具有SPI串行外设接口,片选信号引脚和控制器的P0.8引脚,当P0.8为低电平时,选通74HC595芯片;串行时钟线引脚SCLK接控制器的P0.4引脚,为芯片提供数据传输时钟频率;主机输入/从机输出MISO和主机输出/从机输入MOSI引脚分别接控制器的P0.6和P0.5引脚,用于设定SPI数据传输模式,本系统设为主机控制器输出,从机74HC595输入模式;Q0~Q1共8个并行输出引脚分别和LED4~LED11的8个发光二极管相接;74HC595芯片拥有8位串行输入、并行输出的存储

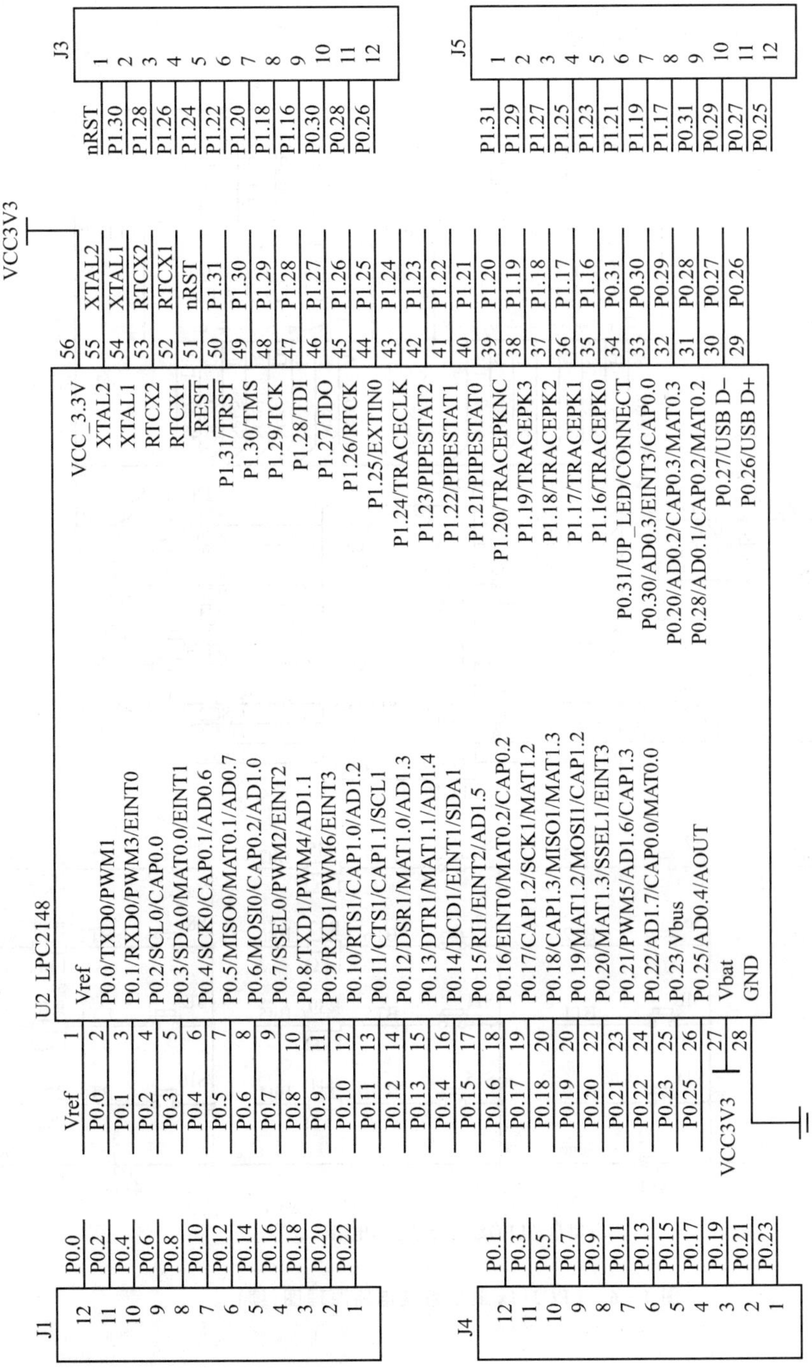

(a) LPC214x CPU PACK板插座

图 1-6 LPC214x最小系统电路原理图

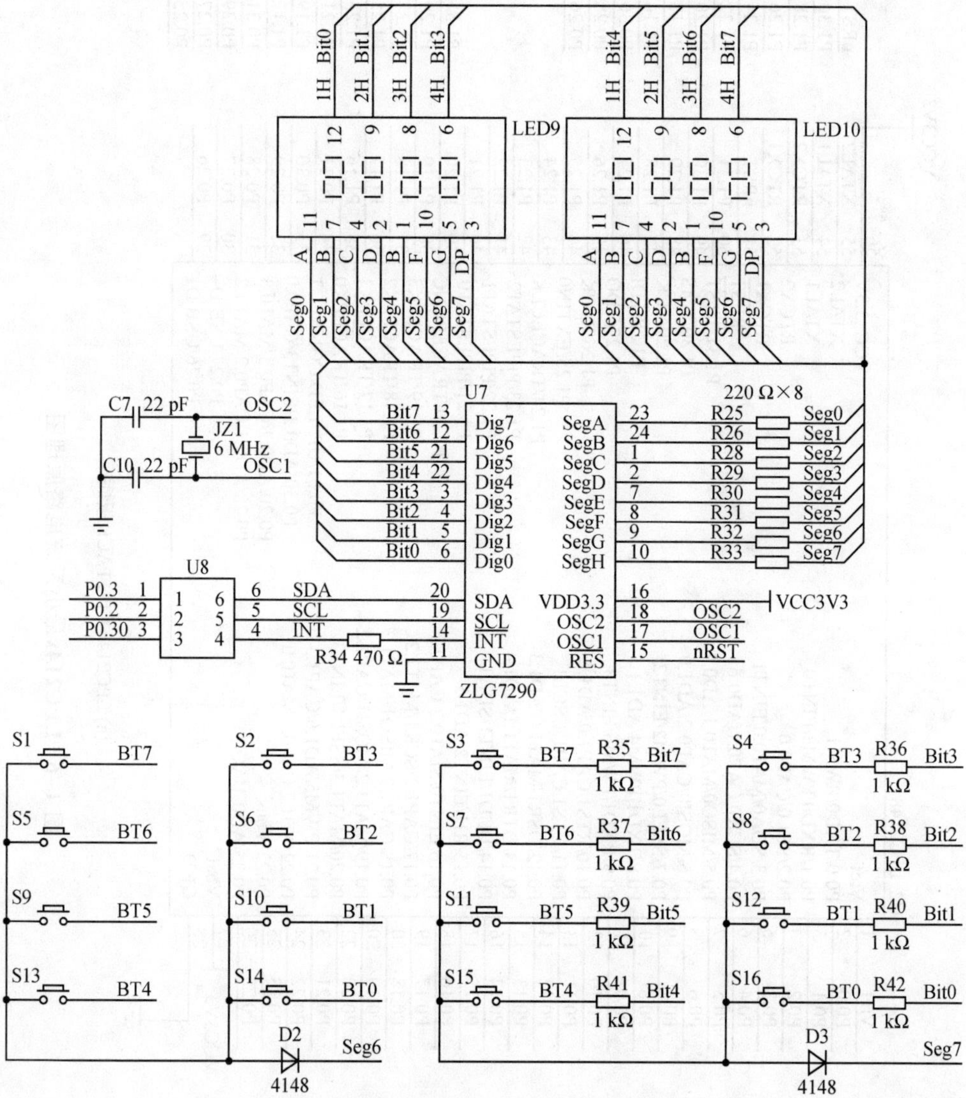

(b) 键盘与LED数码管显示器电路

图 1-6　LPC214x 最小系统电路原理图（续）

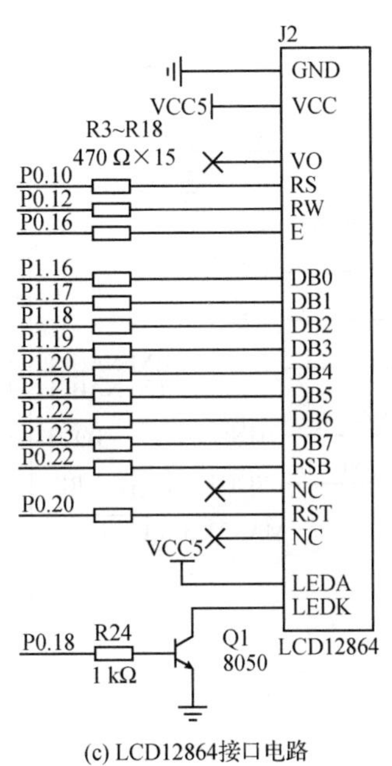

(c) LCD12864接口电路

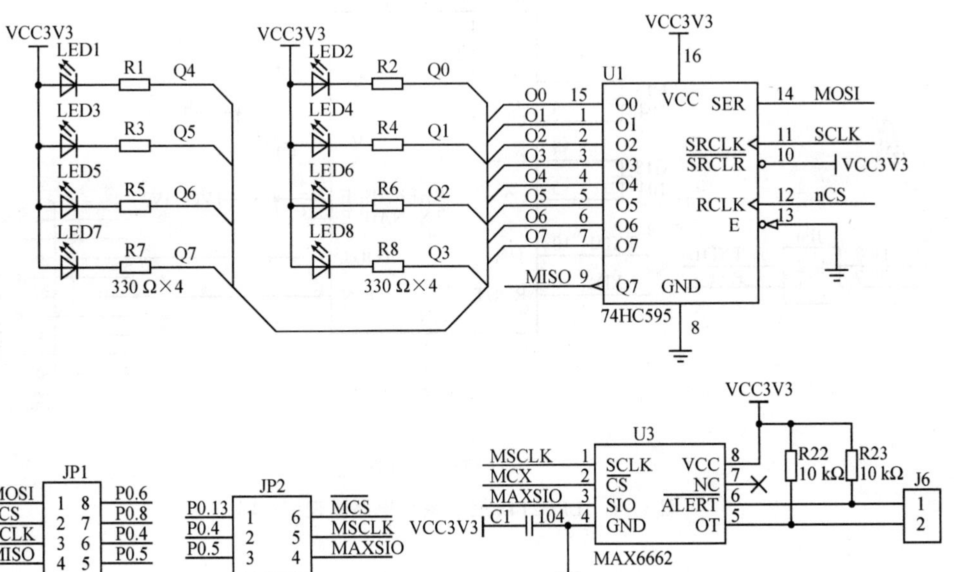

(d) SPI接口与LED

图 1-6　LPC214x 最小系统电路原理图(续)

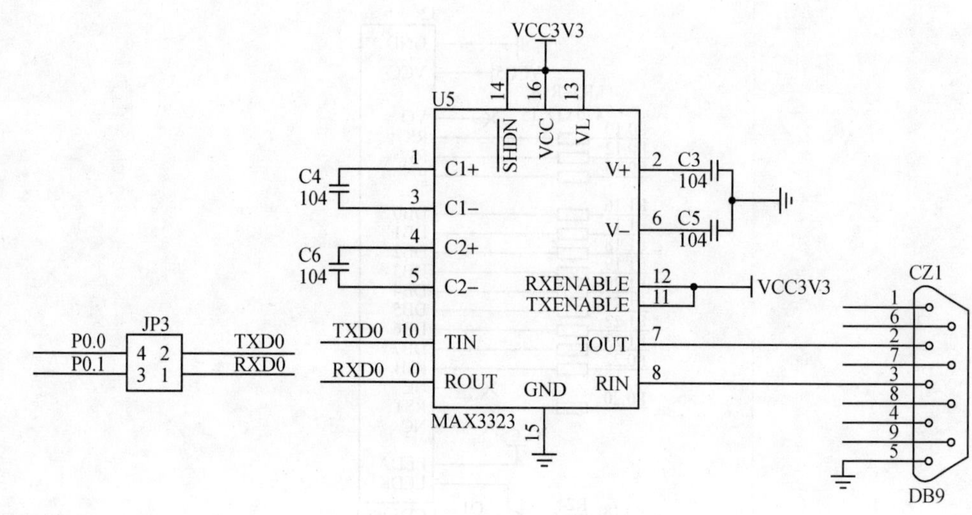

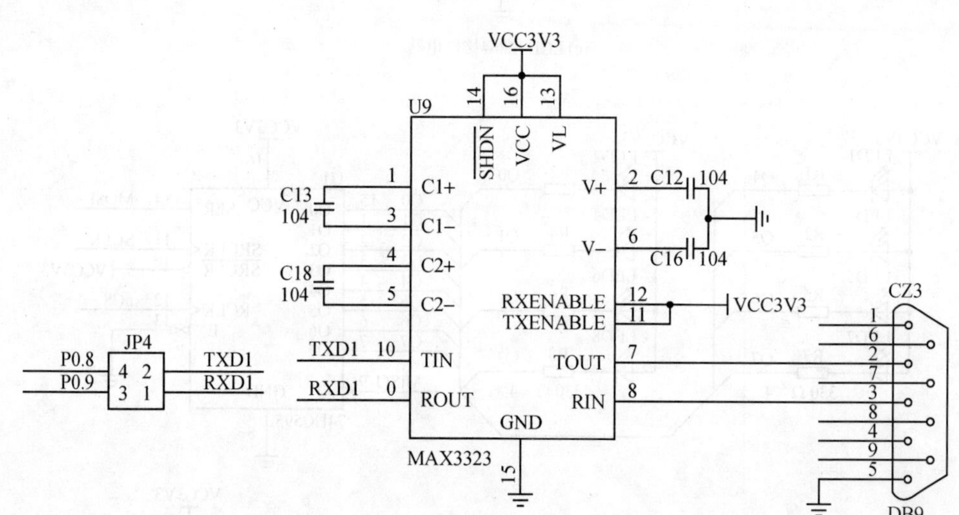

(e) RS232接口电路

图 1－6　LPC214x 最小系统电路原理图(续)

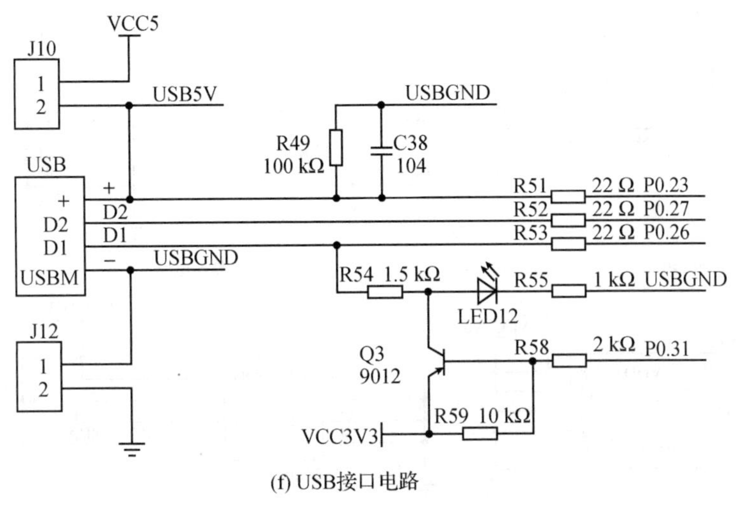

(f) USB接口电路

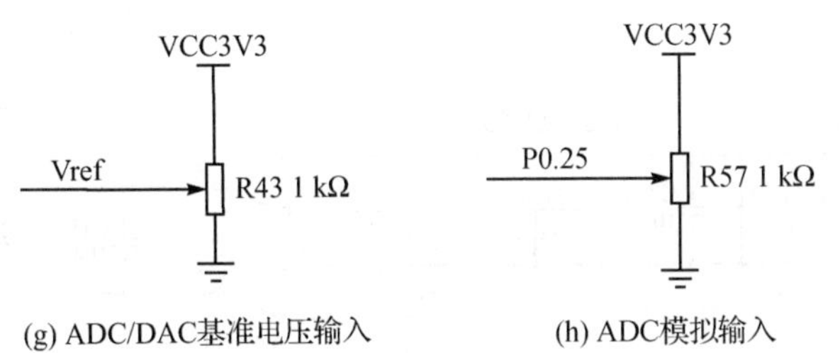

(g) ADC/DAC基准电压输入　　(h) ADC模拟输入

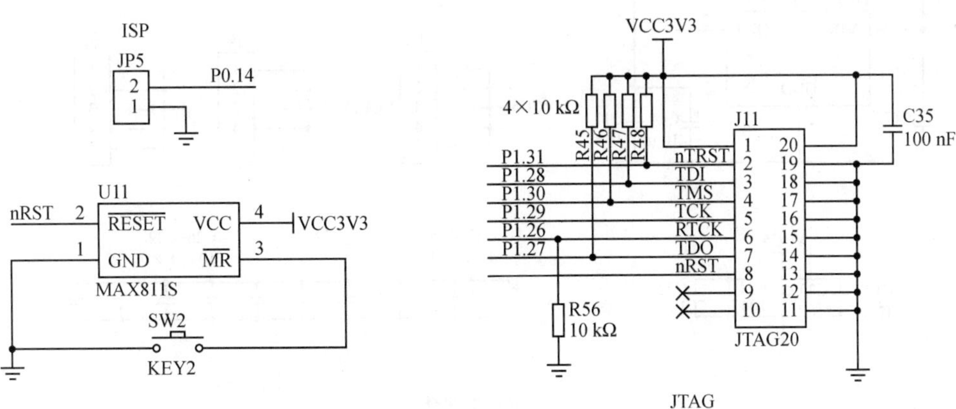

(i) 复位和JTAG接口电路

图 1-6　LPC214x 最小系统电路原理图（续）

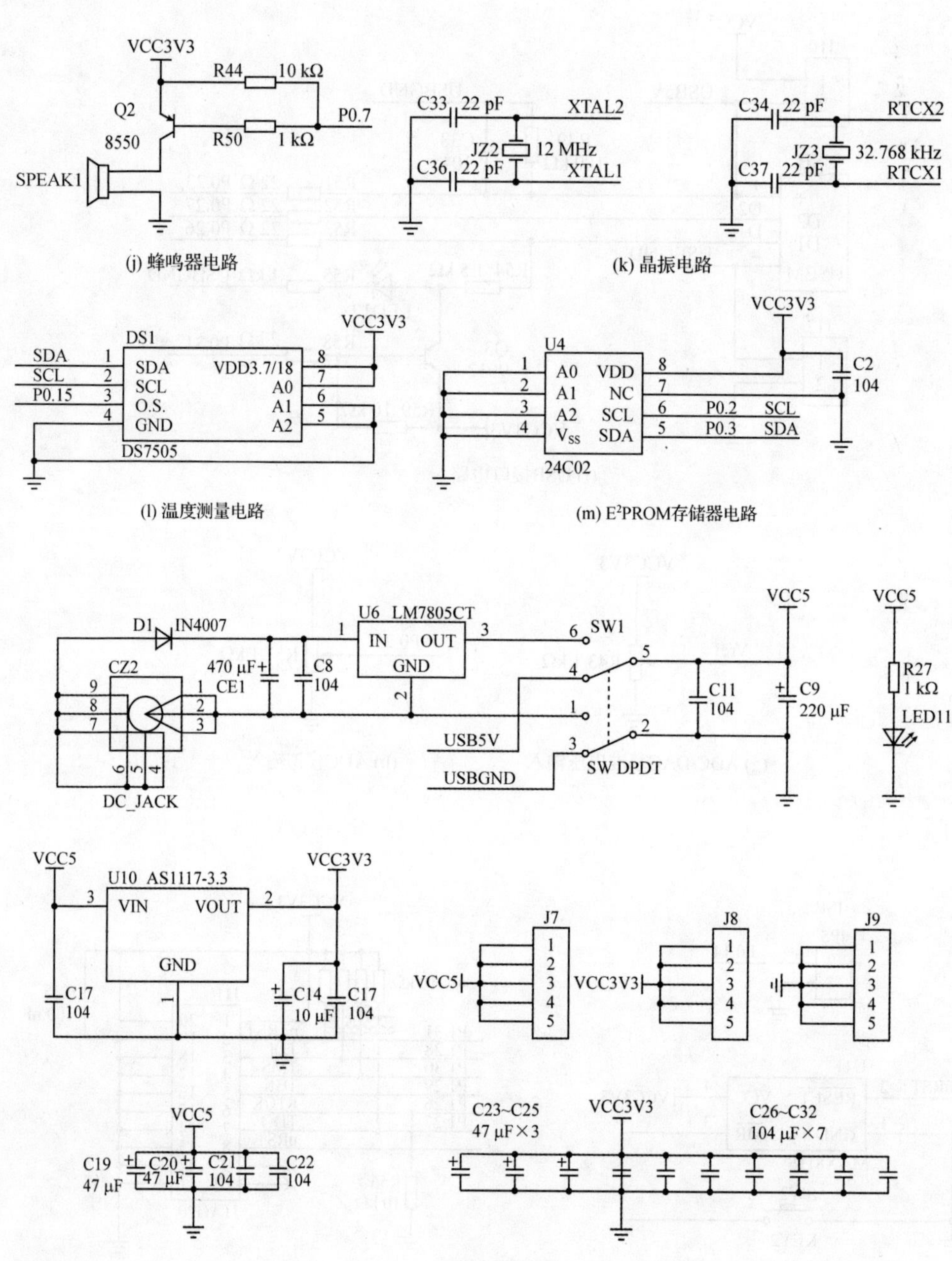

图 1-6 LPC214x 最小系统电路原理图(续)

器，实现将控制器通过 SPI 串行接口输入的一字节数并行地传送给 8 个 LED 灯，控制流水灯显示。SPI 的最大通信速率为 $f_{PCLK}/8$，频率计算公式为

$$f_{SPI} = f_{PCLK}/SPCCR$$

式中，SPCCR 为 SPI 时钟计数寄存器，$f_{PCLK} = f_{CCLK}/VPB = 12$ MHz，设 SPCCR 寄存器值为 12，则 SPI 通信速率为 1 MHz。

3. LED 数码管显示和按键电路

LED 数码管显示和按键电路采用 ZLG7290 作为驱动器，ZLG7290 是一个专用的数码管显示驱动及键盘扫描管理芯片，能够直接驱动 8 位共阴式数码管(或 64 只独立的 LED)，同时还可以扫描管理多达 64 只按键，其中有 8 只按键还可以作为功能键使用，就像计算机键盘上的 Ctrl、Shift、Alt 键一样。

4. 液晶显示电路

LCD12864 液晶显示电路由 LPC214x 控制器引脚直接控制。本系统选用的 12864 型号为 FYD12864－0402B，共 20 个引脚，选择与控制器并行通信，将液晶的使能信号引脚 E 与控制器的 P0.16 连接，当控制器给其低电平时，使能液晶。RS 引脚与控制器的 P0.10 相连，用于选择控制器是对液晶写指令还是写数据，当 P0.10 为高电平时，传送的是数据；当其为低电平时，传送的是指令。R/W 引脚与控制器的 P0.12 相连，用于控制对液晶读数据或写数据。DB0～DB7 为三态数据线，分别连接控制器的 P1.16～P1.23；RST 为复位引脚，接控制器的 P0.22；LEDK 为背光灯控制引脚，接控制器的 P0.18，当其为高电平时，三极管导通，背光灯亮。

1.4 可选择的 ARM 微处理器

1.4.1 ARM 体系结构简介

ARM(Advanced RISC Machines)公司 1990 年成立于英国剑桥，是专门从事基于 RISC 技术芯片设计开发的公司，主要出售芯片设计技术的授权。作为知识产权供应商，其本身不直接从事芯片生产，靠转让设计许可由合作公司生产各具特色的芯片，半导体生产商从 ARM 公司购买其设计的 ARM 微处理器核，根据各自不同的应用领域，加入适当的外围电路，从而形成自己的 ARM 微处理器芯片进入市场。目前，全世界有几十家大的半导体公司都使用 ARM 公司的授权，使得 ARM 技术获得了更多的第三方工具、制造、软件的支持，使整个系统的成本降低，产品更容易进入市场，且更具有竞争力。到目前为止已销售了超过 150 亿枚基于 ARM 的芯片，向 200 多家公司出售了 600 个处理器许可证。ARM 微处理器已经深入到工业控制、无

线通信、网络应用、消费类电子产品、成像和安全产品等各个领域。

采用 RISC 架构的 ARM 微处理器一般具有如下特点:

- 支持 Thumb(16 位)/ARM(32 位)双指令集,能很好地兼容 8 位/16 位器件;Thumb 指令集比通常的 8 位和 16 位 CISC/RISC 处理器具有更好的代码密度。
- 指令执行采用 3 级流水线/5 级流水线技术。
- 带有指令 Cache 和数据 Cache,使用大量寄存器,使指令执行速度更快;大多数数据操作都在寄存器中完成;寻址方式灵活简单,执行效率高;指令长度固定(在 ARM 状态下是 32 位,在 Thumb 状态下是 16 位)。
- 支持大端格式和小端格式两种方法存储字数据。
- 支持 Byte(字节,8 位)、Halfword(半字,16 位)和 Word(字,32 位)三种数据类型。
- 支持用户、快中断、中断、管理、中止、系统和未定义 7 种处理器模式,除了用户模式外,其余模式均为特权模式。
- 处理器芯片上都嵌入了在线仿真 ICE-RT 逻辑,便于通过 JTAG 来仿真调试 ARM 体系结构芯片,可以避免使用昂贵的在线仿真器。另外,在处理器核中还可以嵌入跟踪宏单元 ETM,用于监控内部总线,实时跟踪指令和数据的执行。
- 具有片上总线 AMBA(Advanced Micro-controller Bus Architecture)。AMBA定义了 3 组总线:先进的高性能总线 AHB(Advanced High performance Bus);先进的系统总线 ASB(Advanced System Bus);先进的外围总线 APB(Advanced Peripheral Bus)。通过 AMBA 可以方便地扩充各种处理器及 I/O,可以把 DSP、其他处理器和 I/O(如 UART、定时器和接口等)都集成在一块芯片中。
- 采用存储器映像 I/O 的方式,即把 I/O 端口地址作为特殊的存储器地址。
- 具有协处理器接口。ARM 允许接 16 个协处理器,如 CP15 用于系统控制,CP14 用于调试控制器。
- 采用了降低的电源电压,可工作在 3.0 V 以下;减少门的翻转次数,当某个功能电路不需要时,禁止门翻转;减少门的数目,即降低芯片的集成度;通过降低时钟频率等一些措施降低功耗。
- 体积小、成本低、性能高。

一个典型的 ARM 体系结构方框图如图 1-7 所示,包含有 32 位 ALU、31 个 32 位通用寄存器及 6 个状态寄存器、32×8 位乘法器、32×32 位桶形移位寄存器、指令译码和控制逻辑、指令流水线,以及数据/地址寄存器等。

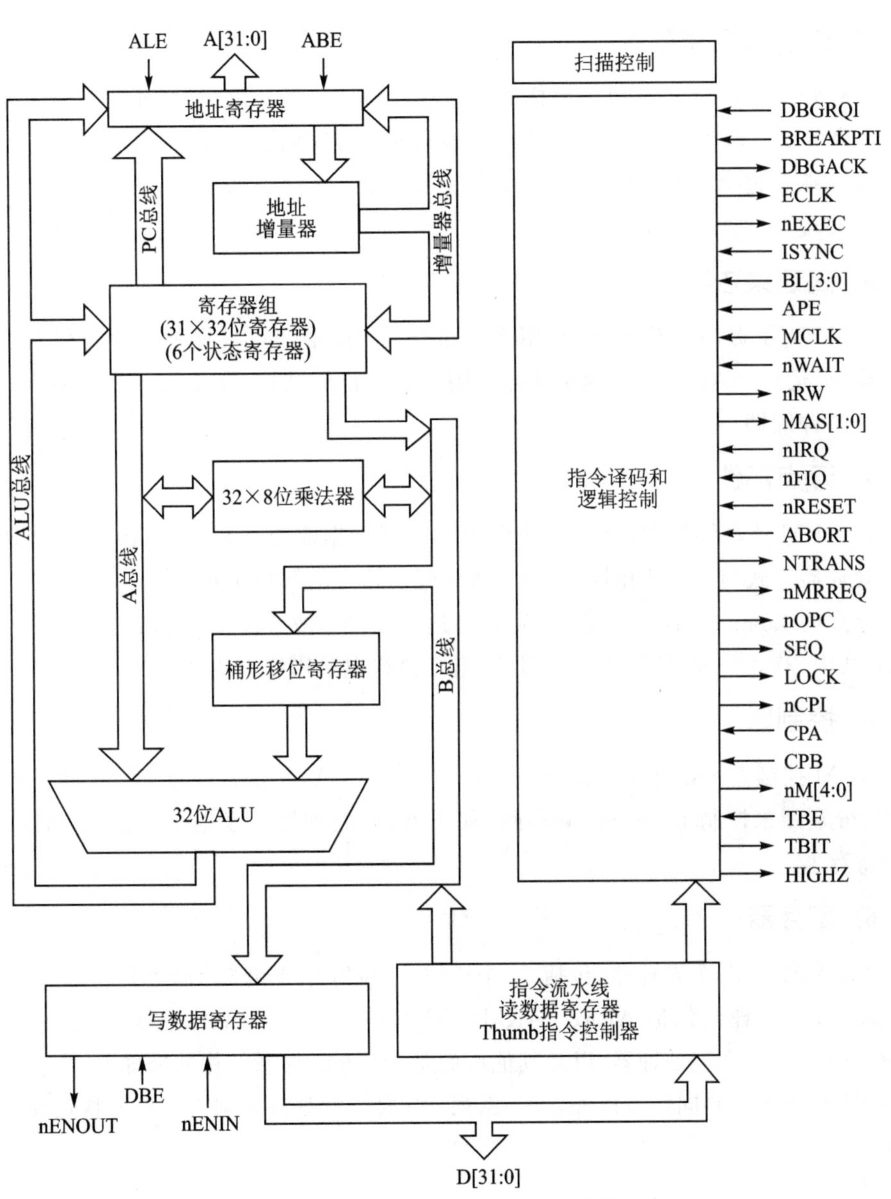

图 1-7　ARM 体系结构方框图

1. ALU

ARM 体系结构的 ALU 与常用的 ALU 逻辑结构基本相同，由两个操作数锁存器、加法器、逻辑功能、结果及零检测逻辑构成。ALU 的最小数据通路周期包含寄存器读时间、移位器延迟、ALU 延迟、寄存器写建立时间、双相时钟间非重叠时间等几部分。

2. 桶形移位寄存器

ARM 采用了 32×32 位桶形移位寄存器,左移/右移 n 位、环移 n 位和算术右移 n 位等都可以一次完成,可以有效地减少移位的延迟时间。在桶形移位寄存器中,所有的输入端通过交叉开关(crossbar)与所有的输出端相连。交叉开关采用 NMOS 晶体管来实现。

3. 高速乘法器

ARM 为了提高运算速度,采用 2 位乘法。2 位乘法可根据乘数的 2 位来实现"加-移位"运算。ARM 的高速乘法器采用 32×8 位的结构,完成 32×2 位乘法只需要 5 个时钟周期。

4. 浮点部件

在 ARM 体系结构中,浮点部件作为选件可根据需要选用,FPA10 浮点加速器以协处理器方式与 ARM 相连,并通过协处理器指令的解释来执行。

浮点的 Load/Store 指令使用频度要达到 67%,故 FPA10 内部也采用 Load/Store 结构,有 8 个 80 位浮点寄存器组,指令执行也采用流水线结构。

5. 控制器

ARM 控制器采用硬接线的可编程逻辑阵列 PLA,其输入端有 14 根、输出端有 40 根,分别用来控制 Load/Store 多路、乘法器、协处理器以及地址寄存器、ALU 和移位寄存器。

6. 寄存器

ARM 内含 37 个寄存器,包括 31 个通用 32 位寄存器和 6 个状态寄存器。

ARM 微处理器包括 ARM7、ARM9、ARM11、Cortex - A、Cortex - R、Cortex - M、SecurCore 等系列处理器,以及其他厂商基于 ARM 体系结构的处理器,除了具有 ARM 体系结构的共同特点以外,每一系列的 ARM 微处理器都有各自的特点和应用领域。

1.4.2 ARM7 微处理器

ARM7 微处理器自 1994 年推出以来,一直都深受用户欢迎,它帮助 ARM 体系结构在数字领域确立了领先地位。ARM7 微处理器是目前世界上使用范围最广的 32 位嵌入式微处理器,具有 170 多个芯片授权使用方,已销售了 100 多亿片。ARM7 微处理器为众多关注成本和功耗的嵌入式应用提供了有力的支持。虽然现在 ARM7 微处理器仍用于某些简单的 32 位设备,但是对于新的设计,不建议使用 ARM7 微处理器(如 ARM7TDMI - S 和 ARM7EJ - S)。目前,一些最新的 ARM 微处理器(例如 Cortex-M0 和 Cortex-M3 处理器)在技术和性能上比 ARM7 微处理器有了显著改进。

Cortex-M0 和 Cortex-M3 处理器是目前嵌入式市场中 ARM7TDMI-S 用户选择的优秀的替代产品,可以使用户以更低的成本获得更多功能,实现代码重用,提高能效和增强连接性,并且为将来的嵌入式应用提供支持。从 ARM7TDMI-S 升级到 Cortex-M0 和 Cortex-M3 的好处如表 1-3 所列。

表 1-3 从 ARM7TDMI-S 升级到 Cortex-M0 和 Cortex-M3 的好处

功 能	ARM7TDMI	Cortex-M0/M3	升级的好处
中断控制器	无标准中断控制器	集成的嵌套矢量中断控制器(NVIC)	灵活而强大的中断处理
ISR 条目	非确定性 ISR 条目	H/W 入栈可以确保确定性 ISR 条目	完全确定的中断处理
功耗管理	无内置电源管理	基于架构的睡眠模式支持	极低的功耗模式
需要汇编语言代码	需要汇编器代码(对于 ISR 等)	不需要汇编器代码	简化了软件开发过程,用 C 语言编写所有代码,降低了项目成本,缩短了时间
指令集性能与代码大小	为更好地平衡性能-代码大小,需要 ARM 与 Thumb 代码交互操作	Thumb-2 以 Thumb 代码密度提供 ARM 性能	简化了程序员模型,代码密度更高,简化了软件开发过程
易于将应用从一台设备移植到另一台设备	缺少标准化,制约了应用移植	NVIC、SysTick 与内存映射定义兼容的 CMSIS	标准化支持 IP 重用,上市速度更快

有关将 ARM7TDMI-S 编写的软件移植到 Cortex-M3 微处理器的建议,请登录 http://www.arm.com/zh/products/processors/classic/arm7/index.php,查询 *ARM Cortex-M3 Processor Software Development for ARM7TDMI Processor Programmers*(《面向 ARM7TDMI 微处理器程序员的 ARM Cortex-M3 微处理器软件开发》)白皮书。

1.4.3 ARM9 微处理器

ARM9 微处理器是迄今最受欢迎的 ARM 微处理器,有 250 多个芯片授权使用方和 100 多个 ARM926EJ-S 微处理器授权使用方,已销售了 50 多亿片。目前,

ARM9 微处理器系列包括:ARM926EJ-S、ARM946E-S 和 ARM968E-S 三种微处理器。ARM9 微处理器为微控制器、DSP 和 Java 应用提供单处理器解决方案,可以为要求苛刻、成本敏感的嵌入式应用提供可靠的高性能和灵活性。丰富的 DSP 扩展使 SoC 设计不再需要单独的 DSP。此外,PPA 也特别适合各种应用。

ARM926EJ-S 处理器具有一个采用 Jazelle 技术的增强型 32 位 RISC CPU,灵活的大小指令和数据高速缓存,紧密耦合内存(TCM)接口和内存管理单元(MMU)。此外它还提供单独指令和数据 AMBA AHB 接口,适合基于多层 AHB 的系统。ARM926EJ-S 处理器可执行 ARMv5TEJ 指令集,其中包括功能得到增强的 16×32 位乘法器,可进行单周期 MAC 运算;16 位定点 DSP 指令,可增强多个信号处理应用程序的性能并支持 Thumb 技术,保持与 ARM7TDMI 处理器的二进制兼容。ARM926EJ-S 处理器为入门级微处理器,支持完整版操作系统,如 Linux、Windows CE 和 Symbian。ARM926EJ-S 处理器是最流行的 ARM 微处理器之一,是众多应用的理想之选。

ARM946E-S 可合成处理器非常适合各种嵌入式应用。它能够提供灵活的指令和数据高速缓存、指令和数据紧密耦合内存(TCM)接口、内存保护单元以及 AMBA AHB 接口。该微处理器执行 ARMv5TE 指令集,并包括一个增强型 16×32 位乘法器,可进行单周期 MAC 运算。它还可执行 16 位定点 DSP 指令,以改进信号处理算法和应用。

ARM968E-S 是一种主要针对实时、低功耗和数据密集型应用的 32 位 RISC 处理器,是功耗最低的 ARM9 微处理器。ARM968E-S 处理器具有如下特点:可直接单独连接指令和数据紧密耦合内存(TCM),并且大小可变;提供专用 AMBA AHB-lite 辅助设备直接内存访问(DMA)端口和双存储数据 TCM,使处理器和 DMA 控制器可共享对 TCM 的访问权限;保持与 ARM7TDMI 处理器的二进制兼容。

1.4.4 ARM11 微处理器

ARM11 微处理器系列目前包括:ARM1136J(F)-S、ARM1156T2(F)-S、ARM1176JZ(F)-S、ARM11MPCore 四种微处理器。ARM11 微处理器广泛用于消费类、家庭和嵌入式应用领域,是许多智能手机的首选。ARM11 微处理器的功耗非常低。ARM11 微处理器软件可以与之前的所有 ARM 处理器兼容,并引入了用于多媒体处理的 32 位 SIMD,用于提高操作系统上下文切换性能的物理标记高速缓存,用于强制实施硬件安全性的 TrustZone,以及针对实时应用的紧密耦合内存。它在多媒体、操作系统和浏览器性能方面比 ARM926EJ-S 处理器有显著改进。其可与 Mali-200 组合,共同为 UI 提供 Open GL ES 2.0 支持。

ARM1136 处理器包含带多媒体扩展的 ARMv6 指令集、Thumb 代码压缩技术以及可选的浮点协处理器。ARM1136 是一个成熟的内核，作为一种应用程序处理器广泛应用在手机和消费类应用场合中。ARM1136 处理器在整体性能、多媒体编解码器和操作系统性能方面，比 ARM926EJ－S 处理器有显著改进。ARM1136 处理器在软件方面与早期 ARM 内核兼容，并与最新的 Cortex－A 系列微处理器的 ARM 和 Thumb 指令兼容。ARM1136 不支持 Cortex－A 系列的 Neon 或 Thumb－2 指令。对于新的应用程序处理器和 SoC 设计，建议使用 Cortex－A5 微处理器，因为它比 ARM1136 或 ARM1176 的面积更小，但提供的性能更高，而且功耗更低。ARM1136 处理器内部结构示意图如图 1－8 所示。

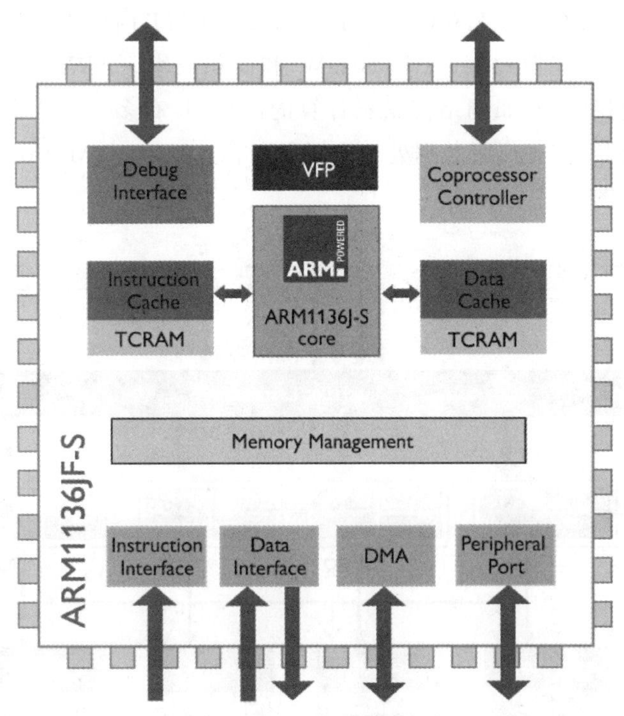

图 1－8　ARM1136 处理器内部结构示意图

ARM1156 处理器对 ARM11 性能进行了优化，以适合高可靠性和实时嵌入式应用。ARM1156T2－S 和 ARM1156T2F－S 处理器基于 ARMv6 指令集体系结构，并借助 Cortex 微处理器系列中的 Thumb－2，增强功能得到了扩展。ARM1156 处理器使用九阶段整数管道，合并了同类最佳分支预测技术，可提供 ARM11 类处理器的最高指令吞吐量。

ARM1176 处理器可提供多媒体和浏览器功能、安全计算环境，在低成本设计的情况下性能高达 1 GHz。ARM1176JZ－S 处理器采用针对安全应用领域的 ARMTrustZone 技术，以及用于执行高效嵌入式 Java 的 ARM Jazelle 技术。可选的紧密耦合内存可以简化 ARM9 微处理器的移植和实时设计。同时,AMBA 3 AXI 接口提高了内存总线性能。DVFS 可以实现功耗优化。ARM1176 处理器适合智能手机、数字电视、电子阅读器的应用。

ARM11 MPCore 多核处理器实现了 ARM11 微体系结构,并引入了基于单个 RTL,从 1 个内核到 4 个内核的多核可扩展性,从而使具有单个宏的简单系统设计可以集成高达单个内核 4 倍的性能。ARM11 MPCore 处理器使用内置 SCU,实现高效一致性,此外还受到具有 ARM SMP 功能的众多操作系统的支持。该处理器使用 PIPT 高速缓存扩展 ARMv6 体系结构,可以高效支持 16～64 KB L1 高速缓存。ARM11 MPCore 每个内核 2.0 Coremarks/MHz,与 MT 内核比较,具有可预测的单线程性能,具有高的能源效率(DMIPS/mW)和面积效率。ARM11 MPCore 非常适合计算密集型应用场合,如网络流量和计算的混合,具有多个进程的操作系统 GUI 环境,特别适合编写多工作者数据处理应用程序等。ARM11 MPCore 多核处理器内部结构示意图如图 1-9 所示。

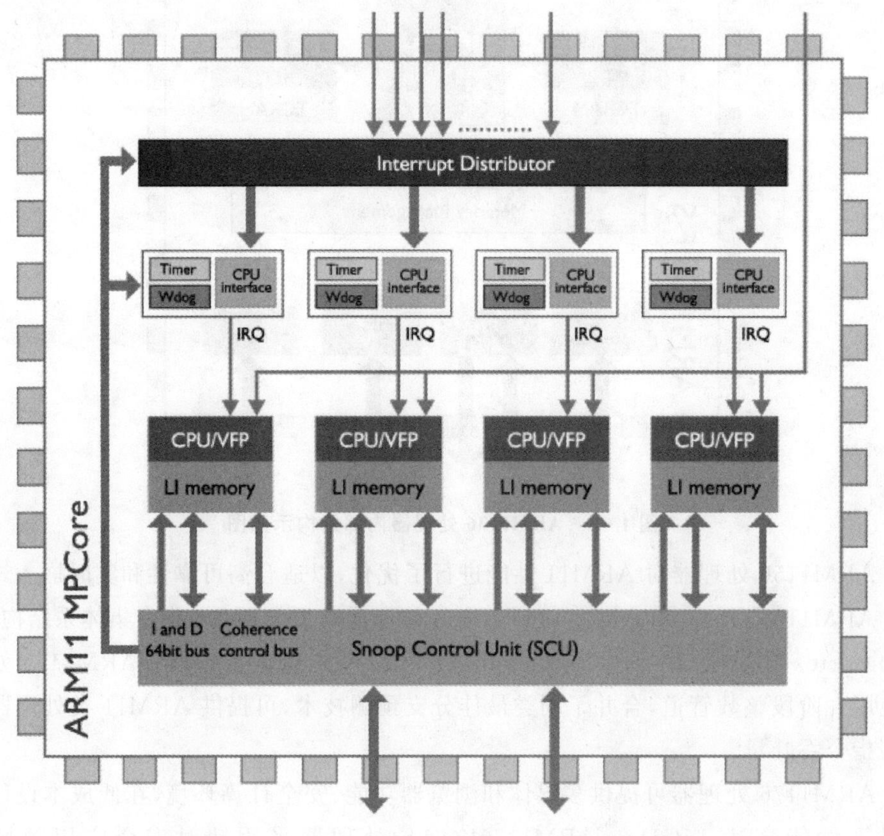

图 1-9　ARM11 MPCore 多核处理器内部结构示意图

1.4.5　Cortex－A 微处理器

Cortex－A 微处理器采用 ARMv7－A 体系结构，支持所有操作系统，如：Linux 完整分配（Android、Chrome、Ubuntu 和 Debian）、Linux 第三方（MontaVista、QNX、Wind River）、Symbian、Windows CE，以及支持需要使用内存管理单元的其他操作系统。指令集支持：ARM、Thumb－2、Thumb、Jazelle、DSP。支持 TrustZone 安全扩展。支持高级单精度和双精度浮点运算。具有高性能 NEON 引擎，广泛支持多媒体编解码器。

Cortex－A 微处理器除了具有与上一代经典 ARM 和 Thumb 体系结构的二进制兼容性外，Cortex－A 微处理器采用的 Thumb－2 提供了最佳代码大小和性能，TrustZone 安全扩展提供了可信计算，Jazelle 技术提高了执行环境（如 Java、Net、MSIL、Python 和 Perl）的速度。

Cortex－A 微处理器系列目前包括：Cortex－A5、Cortex－A8、Cortex－A9 和 Cortex－A15 处理器，这些处理器都共享共同的体系结构和功能集。尽管这些处理器都支持同样卓越的基础功能和完整的软件兼容性，但它们提供了显著不同的特性，可确保其完全符合未来的高级嵌入式解决方案的要求。Cortex－A5、Cortex－A9 和 Cortex－A15 处理器都支持 ARM 的第二代多核技术，可以实现单核到四核，支持面向性能的应用领域，支持对称和非对称的操作系统。

Cortex－A5 处理器是能效最高、成本最低的微处理器之一。Cortex－A5 可以工作在 400～800 MHz 的频率下，提供的性能超过 1 200 DMIPS。Cortex－A5 处理器可为现有的 ARM926EJ－S 和 ARM1176JZ－S 处理器设计提供很有价值的迁移途径。它可以获得比 ARM1176JZ－S 更好的性能，比 ARM926EJ－S 更好的功效和能效以及 100％的 Cortex－A 兼容性。Cortex－A5 多核处理器内部结构示意图如图 1－10所示。

Cortex－A8 单核解决方案，可提供经济有效的高性能，在 600 MHz～1 GHz 的频率下，提供的性能超过 2 000 DMIPS。Cortex－A8 处理器可以满足需要在 300 mW以下运行的移动设备的功率优化要求，以及需要 2 000 Dhrystone MIPS 的消费类应用领域的性能优化要求。Cortex－A8 与 ARM926、ARM1136 和 ARM1176 处理器的二进制兼容。

Cortex－A9 是目前性能最高的 ARM 微处理器之一，可以实现 ARMv7 体系结构的丰富功能。Cortex－A9 微体系结构既可用于可伸缩的多核处理器（Cortex－A9 MPCore 多核处理器），也可用于传统的微处理器（Cortex－A9 单核处理器）。可伸缩的多核处理器和单核处理器支持 16 KB、32 KB 或 64 KB 4 路关联的 L1 高速缓存配置，对于可选的 L2 高速缓存控制器，最多支持 8 MB 的 L2 高速缓存配置，它们具

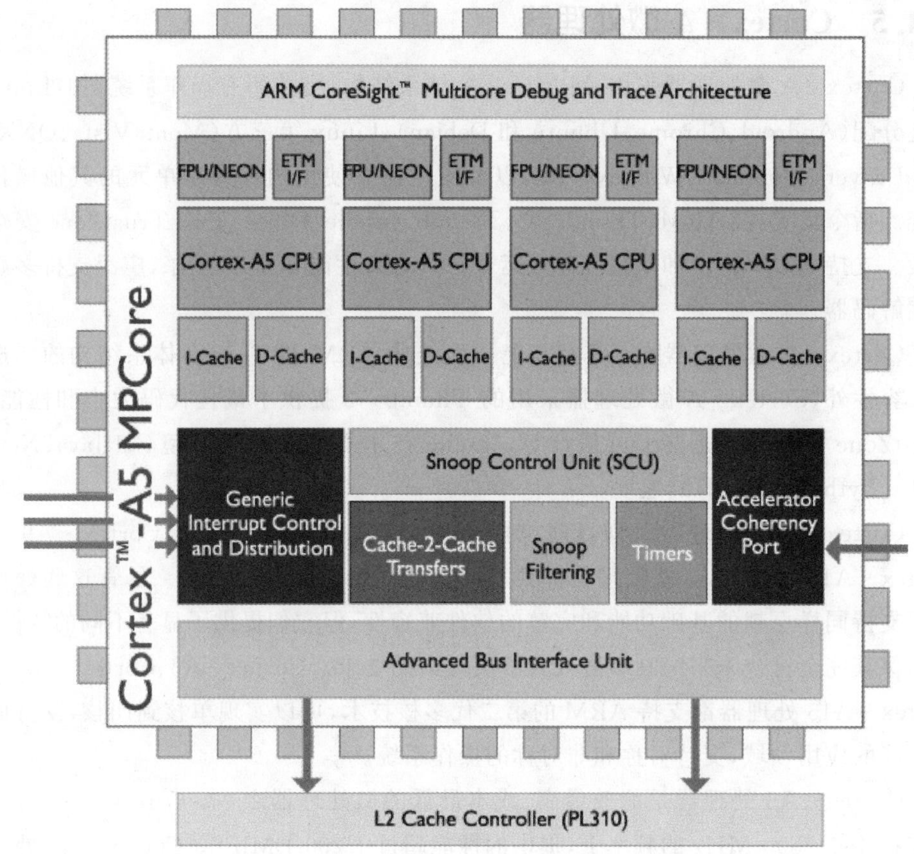

图 1-10　Cortex-A5 多核处理器内部结构示意图

有极高的灵活性,适用于特定应用领域和市场。Cortex-A9 处理器提供了高性能和高能效,使其成为在低功耗或散热受限的成本敏感型设备中的理想解决方案。Cortex-A9可提供 800 MHz～2 GHz 的标准频率,每个内核可提供 5 000 DMIPS 的性能。

Cortex-A15 可为新一代移动基础结构应用和要求苛刻的无线基础结构应用提供性能最高的解决方案。Cortex-A15 MPCore 处理器是 Cortex-A 系列微处理器的最新成员,在应用方面与其他所有的 Cortex-A 微处理器完全兼容。支持的开发平台和软件体系包括 Android、Adobe Flash Player、Java Platform Standard Edition (Java SE)、JavaFX、Linux、Microsoft Windows Embedded、Symbian 和 Ubuntu,以及 700 多个 ARM Connected Community 成员所提供的应用软件、硬件和软件开发工具、中间件以及 SoC 设计服务。Cortex-A15 MPCore 处理器具有无序超标量管道,带有紧密耦合的低延迟 2 级高速缓存,该高速缓存的大小最高可达 4 MB。浮点和

NEON多媒体性能方面的改进，能够为 Web 基础结构应用提供高性能计算。在高级基础结构应用中，Cortex－A15 的运行速度最高可达 2.5 GHz，可以支持在降低功耗、散热和降低成本预算方面的设计要求。Cortex－A15 多核处理器内部结构示意图如图 1－11所示。

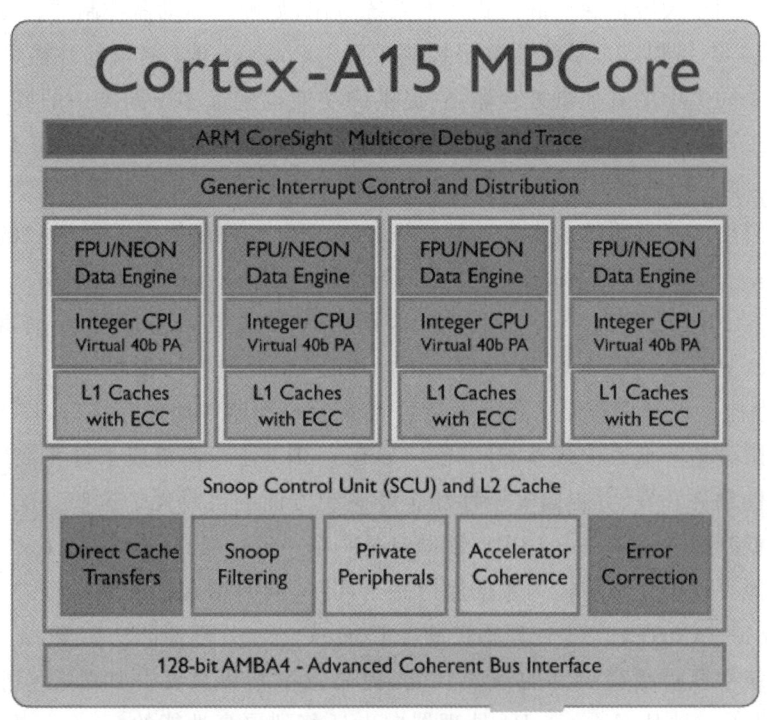

图 1－11　Cortex－A15 多核处理器内部结构示意图

1.4.6　Cortex－R 微处理器

Cortex－R 微处理器为具有严格的实时响应限制的嵌入式系统提供高性能计算解决方案。Cortex－R 微处理器目前包含：Cortex－R4、Cortex－R5 和 Cortex－R7 三种处理器。Cortex－R 微处理器保持与经典 ARM 微处理器(如 ARM7TDMI－S、ARM946E－S、ARM968E－S 和 ARM1156T2－S)的二进制兼容性，因此可确保应用的可移植性。

Cortex－R 微处理器具有 Thumb－2 指令的 ARMv7－R 架构，可在不牺牲性能的情况下实现高代码密度。其具有高性能、高时钟频率、深管道化的微架构；采用指令预取、分支预测和超标量执行等性能增强技术；快速且确定的中断响应；硬件除法器、浮点单元(FPU)可选项；可用于 DSP 和多媒体处理的增强指令集；内存保护单元(MPU)的用户和授权软件操作模式；指令和数据高速缓存控制器的哈佛架构；用于获得快速响应代码和数据的微处理器本地的紧密耦合内存(TCM);高性能 64 位 AMBA 3 AXI 总线接口;奇偶校验检测和 ECC,用于 1 级内存系统和总线的软错误和硬错误检测/更正等功能。

Cortex-R4 处理器是第一个基于 ARMv7-R 体系结构的深层嵌入式实时微处理器。它专用于大容量深层嵌入式片上系统应用,如硬盘驱动器控制器、无线基带处理器、消费类产品和汽车系统的电子控制单元。Cortex-R4 提供的性能、实时响应性大大高于同类中的其他微处理器,它提供的功能也远远多于同类中的其他微处理器。Cortex-R4 可以实现近 1 GHz 的频率运行,此时它可提供 1 500 Dhrystone MIPS 的性能。该处理器提供高度灵活且有效的双周期本地内存接口,使 SoC 设计者能最大限度地降低系统成本和功耗。Cortex-R4 使用高性能实时 SoC 的标准,取代了许多基于 ARM9 和 ARM11 微处理器的设计。

Cortex-R5 处理器于 2010 年推出,该处理器基于 ARMv7-R 体系结构。Cortex-R5 处理器可以实现近 1 GHz 的频率运行,此时它可提供 1 500 Dhrystone MIPS 的性能。该处理器提供高度灵活且有效的双周期本地内存接口,使 SoC 设计者能最大限度地降低系统成本和功耗。Cortex-R5 处理器集成了许多高级系统级功能来帮助软件开发,并提高安全性和企业系统方面的可靠性。这些功能中包括一个全新的低延迟外设端口(LLPP),该端口是一个一致性接口,允许 Cortex-R5 高速缓存与智能外设中正在传输的数据保持完全同步,同时增强扩展到所有处理器接口的 ECC 支持。Cortex-R5 处理器扩展了 Cortex-R4 处理器的功能集,支持在可靠的实时系统中获得更高级别的系统性能,提高效率和可靠性并加强错误管理。因此,它提供了一种从 Cortex-R4 处理器向上迁移到更高性能的 Cortex-R7 处理器的简单迁移途径。

Cortex-R7 处理器为深层嵌入式应用提供高性能的双核、实时解决方案。Cortex-R7 处理器通过引入新技术(包括无序执行和动态寄存器重命名),并与改进的分支预测、超标量执行功能和用于除法和其他功能的更快的硬件支持相结合,提供了比其他 Cortex-R 系列微处理器高得多的性能。Cortex-R7 处理器是性能最高的 Cortex-R 系列微处理器。它使用高性能实时 SoC 的标准。Cortex-R7 处理器的设计重点在于提升能效、实时响应、高级功能和简化系统设计。Cortex-R7 处理器可实现超过 1 GHz 的频率运行,此时它可提供 2 700 Dhrystone MIPS 的性能。该处理器提供支持紧密耦合内存(TCM)的本地共享内存和外设端口的灵活的本地内存系统,使 SoC 设计人员在受限制的芯片资源内达到高标准的硬实时要求。Cortex-R7 多核处理器内部结构示意图如图 1-12 所示。

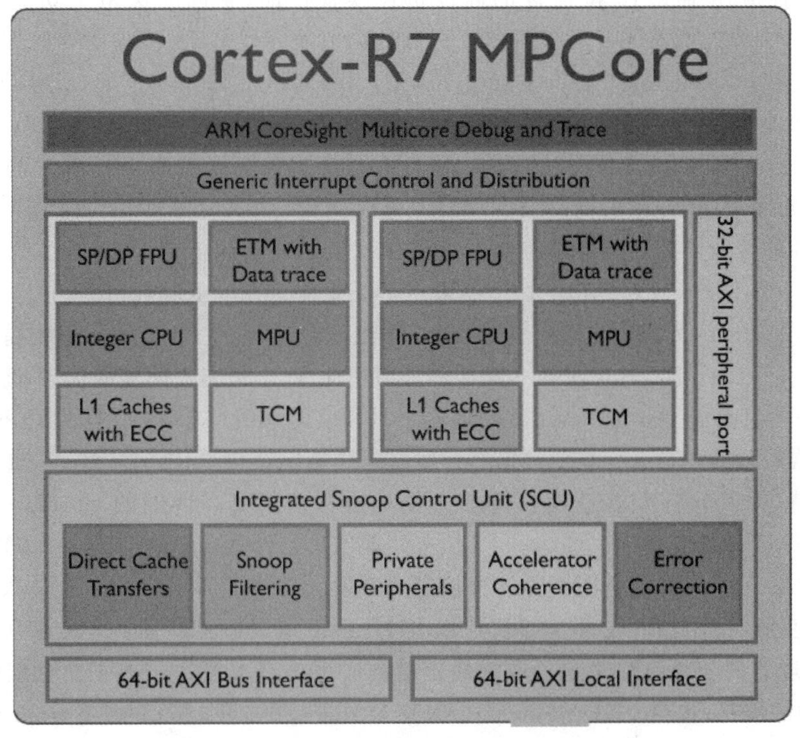

图 1－12　Cortex－R7 多核处理器内部结构示意图

1.4.7　Cortex－M 微处理器

Cortex－M 微处理器是可向上兼容的高能效、易于使用的微处理器，它针对成本和功耗敏感的 MCU 和终端应用（如智能测量、人机接口设备、汽车和工业控制系统、大型家用电器、消费类产品和医疗器械）。Cortex－M 微处理器以更低的时钟频率或更短的活动时段运行，支持基于架构的睡眠模式，比 8/16 位器件的工作方式更智能，睡眠时间更长。Cortex－M 微处理器具有高密度指令集，比 8/16 位器件每字节完成更多的操作，具有更小的 RAM、ROM 或闪存要求，能够以更低的功耗实现更丰富的功能。Cortex－M 微处理器目前已经许可给 40 个以上的 ARM 合作伙伴，包括：NXP Semiconductors、STMicroelectronics、Texas Instruments 和 Toshiba 等厂商。Cortex－M 微处理器目前包含：Cortex－M0、Cortex－M0＋、Cortex－M1、Cortex－M3、Cortex－M4 系列处理器。

Cortex－M0 处理器是现有体积最小的、能耗最低的和能效最高的 ARM 微处理器，在 12 K 门阵列内能耗仅为 85 μW/MHz，指令只有 56 个，可供选择的具有完全确定性的指令和中断计时使得计算响应时间十分容易。该处理器能耗极低并且所需的代码量极少，这使得开发人员能够以 8 位的器件实现 32 位器件的性能。Cortex－

M0 是各种应用中 8/16 位器件的高性价比换代产品,它同时保留了功能丰富的 Cortex－M3 微处理器的开发工具和二进制的向上兼容性。Cortex－M0 处理器支持低能耗连接,如 Bluetooth Low Energy(BLE)、IEEE 802.15 和 Z－wave,可以有效地预处理和传输数据。

Cortex－M0＋处理器是 2012 年推出的一款拥有全球最高功耗效率的微处理器,是针对家用电器、白色商品、医疗监控、电子测量、照明设备以及功耗与汽车控制器件等各种智能传感器与智能控制系统应用提供的一款超低功耗、低成本微控制器(MCU)。32 位 Cortex－M0＋处理器采用了低成本 90 nm 低功耗(LP)工艺,耗电量仅 9 μA/MHz,约为目前主流 8 位或 16 位微处理器功耗的三分之一,却能提供更高的性能。Cortex－M0＋处理器不仅延续了易用性、C 语言编程模型的优势,而且能够二进制兼容已有的 Cortex－M0 微处理器工具和实时系统(RTOS)。作为 Cortex－M 微处理器系列的一员,Cortex－M0＋处理器同样能够获得 ARM Cortex－M 系统的全面支持,而其软件兼容性使其能够方便地被移植到更高性能的 Cortex－M3 或 Cortex－M4 微处理器上。Cortex－M0＋处理器具备已整合 Keil μVision IDE、调试器和 ARM 汇编工具的 ARM Keil 微控制器开发套件的全面支持。作为全球公认的最受欢迎的微控制器开发环境,MDK 以及 ULINK 调试适配器系列均支持 Cortex－M0＋处理器的全新追踪功能。这款处理器同时也拥有大量第三方工具和实时系统(RTOS)的支持,包括 Code Sourcery、Code Red、Express Logic、IAR Systems、Mentor Graphics、Micrium 和 SEGGER。

Cortex－M1 处理器是第一个专门为能够在 FPGA 中实现设计的 ARM 微处理器。Cortex－M1 处理器满足 FPGA 应用的高质量、标准微处理器架构的需要,支持包括 Actel、Altera 和 Xilinx 所有主要的 FPGA 器件,并包括对领先的 FPGA 综合工具的支持,允许设计者为每个项目选择最佳实现。Cortex－M1 处理器使 OEM 能够通过在跨 FPGA、ASIC 和 ASSP 的多个项目之间合理地利用软件和工具投资来节省大量成本,此外还能够通过使用行业标准微处理器实现更大的供应商独立性。Cortex－M1 处理器为 FPGA 用户带来了一系列 ARM Connected Community 工具和操作系统,并提供与 ASIC 优化的微处理器(如 Cortex－M3 处理器)的软件兼容性。

Cortex－M3 处理器是行业领先的 32 位微处理器,适用于具有高确定性的实时应用。Cortex－M3 处理器具有高性能和低动态能耗,功耗为 12.5 DMIPS/mW。Cortex－M3 处理器执行 Thumb－2 指令集以获得最佳性能和代码大小,包括硬件除法、单周期乘法和位字段操作。Cortex－M3 NVIC 在设计时是高度可配置的,最多可提供 240 个具有单独优先级、动态重设优先级功能和集成系统时钟的系统中断。基于 Cortex－M3 的器件可以有效处理多个 I/O 通道和协议标准,如 USB OTG(On－The－Go)。

Cortex－M4 处理器是由 ARM 专门开发的最新嵌入式微处理器,具有高效的信号处理功能，并且与 Cortex－M 微处理器系列的低功耗、低成本和易于使用的优点组合，旨在满足专门面向电动机控制、汽车、电源管理、嵌入式音频以及控制和信号处理混合的控制系统的需要。Cortex－M4 通过一系列出色的软件工具和 Cortex 微控制器软件接口标准（CMSIS）使信号处理算法开发变得十分容易。Cortex－M4 处理器内部结构示意图如图 1－13 所示。

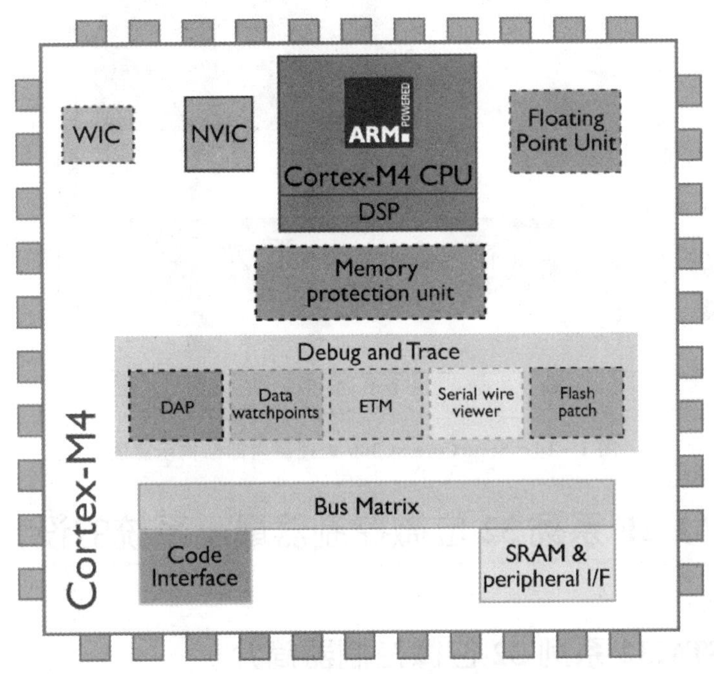

图 1－13　Cortex－M4 处理器内部结构示意图

1.4.8　SecurCore 微处理器

SecurCore 微处理器包含：SecurCore SC100、SecurCore SC110、SecurCore SC200、SecurCore SC210 和 SecurCore SC300 几种类型。它提供了完善的 32 位 RISC 技术的安全解决方案。

SecurCore 微处理器除了具有 ARM 体系结构的各种主要特点外，在系统安全方面：带有灵活的保护单元，以确保操作系统和应用数据的安全；采用软内核技术，防止外部对其进行扫描探测；可集成用户自己的安全特性和其他协微处理器。

SecurCore 微处理器主要应用于如电子商务、电子政务、电子银行业务、网络和认证系统等一些对安全性要求较高的应用产品及应用系统。大多数智能卡芯片供应商和 OEM 都使用 SecurCore 体系结构。

SecurCore SC300 处理器内部结构示意图如图 1－14 所示。

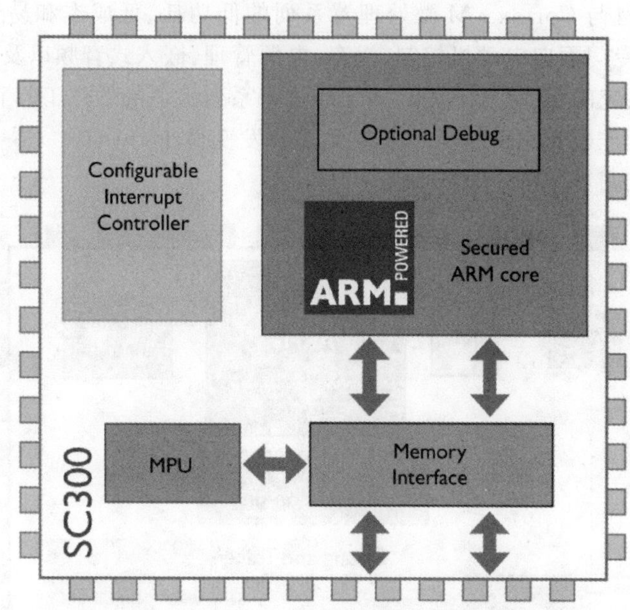

图 1-14 SecurCore SC300 处理器内部结构示意图

1.5 STM32F 系列 32 位微控制器最小系统的设计与制作

1.5.1 STM32 系列 32 位微控制器简介

STMicroelectronics 为用户提供了一系列具有高性能、高兼容、易开发、低功耗、低工作电压、实时、数字信号处理的 32 位闪存微控制器产品。其开发的基于 ARM Cortex-M 处理器的 STM32 32 位闪存微控制器产品系列如图 1-15 所示。

1. STM32 L1 系列超低功耗微控制器

STM32 L1 系列超低功耗微控制器,基于 Cortex-M3 内核,工作频率为 32 MHz,在性能、特性、存储器容量和封装引脚数量方面扩展了超低功耗产品系列,最低功耗模式电流消耗为 0.27 μA,动态运行模式为 230 μA/MHz。STM32 L1 分为两个不同的产品线(STM32L151 和 STM32L152),集合了 STM32F 和 STM8L 的优化功能,是需要高性能的同时特别关注功耗的应用领域的最佳选择。

2. STM32 F1 微控制器系列

STM32 F1 微控制器系列最大化地集成了高性能与一流外设以及低功耗、低电压的工作特性,该系列包含五个产品线,它们之间引脚、外设和软件相互兼容:

- 超值型系列 STM32F100 24 MHz,具有马达控制和 CEC 功能。

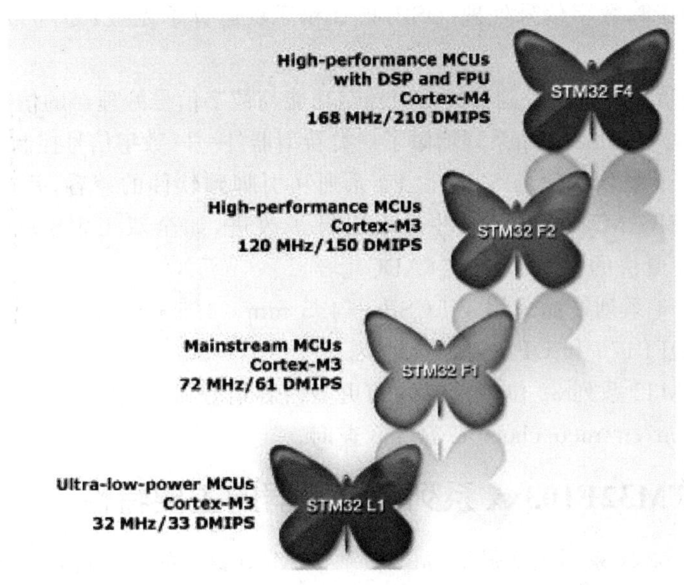

图 1-15　STM32 32 位闪存微控制器产品系列示意图

- 基本型系列 STM32F101 36 MHz,具有高达 1 MB 的片上闪存。
- USB 基本型系列 STM32F102 48 MHz,具有全速 USB 模块。
- 增强型系列 STM32F103 72 MHz,具有高达 1 MB 的片上闪存,兼具马达控制、USB 和 CAN 模块。
- 互联型系列 STM32F105/107 72 MHz(最高主频),具有以太网 MAC、CAN 以及 USB 2.0 OTG 功能。

3. 高性能 STM32 F2 微控制器系列

内置 ARM Cortex-M3 内核的 STM32 F2 系列采用了创新的自适应实时内存加速器(ART 加速器)和多层总线矩阵技术。ST 的加速技术在保持极低水平的 188 A/MHz动态电流消耗的同时,还使得这些 MCU 实现了 120 MHz FCPU,等同于零等待状态执行能力,相当于 150 DMIPS。

STM32 F2 微控制器系列具有:高达 1 MB 闪存、128 KB 以太网 MAC 的 SRAM、USB 2.0 HS OTG、摄像头接口、硬件加密支持和外部存储器接口技术的高度集成化。采用 LQFP64、LQFP100、LQFP144、WLCSP64(<4 mm×4 mm)、UFBGA176和 LQFP176 封装。

4. 具有数字信号处理指令和浮点运算单元的 STM32 F4 高性能微控制器系列

STM32 F4 系列是具有数字信号处理指令和浮点运算单元的高性能微控制器产品,具有意法半导体公司特有的 ART 加速器,在工作频率为 168 MHz 时处理性能达到 210 DMIPS,数字信号处理(DSP)指令和浮点运算单元(FPU)扩大了产品的应用范围。

STM32 F4 系列是微控制器的实时控制功能与数字信号处理器的信号处理功能的完美结合体,为 STM32 产品系列增添了一类新型器件——数字信号控制器(DSC)。

STM32 F4 系列保持与 STM32 F2 系列从引脚到软件的兼容,并提供更多静态随机存储器(SRAM),同时对一些外设进行了改进,如全双工 I^2S 总线、实时时钟(RTC)和速度更快的模/数转换器(ADC)。

STM32 F4 系列产品采用 WLCSP(＜4.5 mm×4.5 mm)、LQFP64、LQFP100、LQFP144、LQFP176 和 UFBGA176 封装。

有关 STM32 系列 32 位微控制器的更多内容请登录 http://www.stmicroelectronics.com.cn/cn/mcu/class/1734.jsp 查询。

1.5.2　STM32F103xx 系列微控制器的主要特性

STM32F103xx 系列是增强型的 32 位基于 ARM 核心的微控制器,具有如下特性:

① 内核为 ARM 32 位的 Cortex-M3 CPU。

- 最高 72 MHz 工作频率,在存储器的 0 等待周期访问时可达 1.25 DMIPS/MHz (Dhrystone 2.1);
- 单周期乘法和硬件除法。

② 存储器。

- 256～512 KB 的闪存程序存储器;
- 64 KB 的 SRAM;
- 带 4 个片选的静态存储器控制器,支持 CF 卡、SRAM、PSRAM、NOR 和 NAND 存储器;
- 并行 LCD 接口,兼容 8080/6800 模式。

③ 时钟、复位和电源管理。

- 2.0～3.6 V 供电和 I/O 引脚;
- 上电/断电复位(POR/PDR)、可编程电压监测器(PVD);
- 4～16 MHz 晶体振荡器;
- 内嵌经出厂调校的 8 MHz 的 RC 振荡器;
- 内嵌带校准的 40 kHz 的 RC 振荡器;
- 带校准功能的 32 kHz 的 RTC 振荡器。

④ 低功耗。

- 睡眠、停机和待机模式;
- VBAT 为 RTC 和后备寄存器供电。

⑤ 3 个 12 位模/数转换器,1 μs 转换时间(多达 21 个输入通道)。

- 转换范围:0～3.6 V;

- 三倍采样和保持功能；
- 温度传感器。

⑥ 2通道12位D/A转换器。

⑦ DMA为12通道DMA控制器。

- 支持的外设：定时器、ADC、DAC、SDIO、I^2S、SPI、I^2C和USART。

⑧ 调试模式。

- 串行单线调试(SWD)和JTAG接口；
- Cortex-M3内嵌跟踪模块(ETM)。

⑨ 112个快速I/O端口。

- 51/80/112个多功能双向的I/O口，所有I/O口可以映像到16个外部中断；几乎所有端口均可容忍5 V信号。

⑩ 11个定时器。

- 4个16位定时器，每个定时器有多达4个用于输入捕获/输出比较/PWM或脉冲计数的通道和增量编码器输入；
- 2个16位带死区控制和紧急刹车，用于电机控制的PWM高级控制定时器；
- 2个看门狗定时器(独立的和窗口型的)；
- 1个系统时间定时器，24位自减型计数器；
- 2个16位基本定时器，用于驱动DAC。

⑪ 13个通信接口。

- 2个I^2C接口(支持SMBus/PMBus)；
- 5个USART接口(支持ISO7816、LIN、IrDA接口和调制解调控制)；
- 3个SPI接口(18 Mbit/s)，2个可复用为I^2S接口；
- 1个CAN接口(2.0B主动)；
- 1个USB 2.0全速接口；
- 1个SDIO接口。

⑫ CRC计算单元，96位的芯片唯一代码。

⑬ ECOPACK封装。

- LFBGA144，144球窄间距球阵列，10 mm×10 mm，0.8 mm间距封装；
- LFBGA100，100球窄间距球阵列封装；
- WLCSP，64球，4.466 mm× 4.395 mm，0.500 mm间距，晶圆级芯片封装；
- LQFP144，20 mm×20 mm，144脚方形扁平封装；
- LQFP100，100脚方形扁平封装；
- LQFP64，64脚方形扁平封装。

⑭ 器件型号：STM32F103xC、STM32F103xD、STM32F103xE。器件型号(订货)代码信息请登录http://www.st.com，参考《STM32F103xx系列数据手册》。

1.5.3 STM32F103xx 系列微控制器的内部结构

STM32F103xx 系列微控制器内部结构方框图如图 1-16 所示。

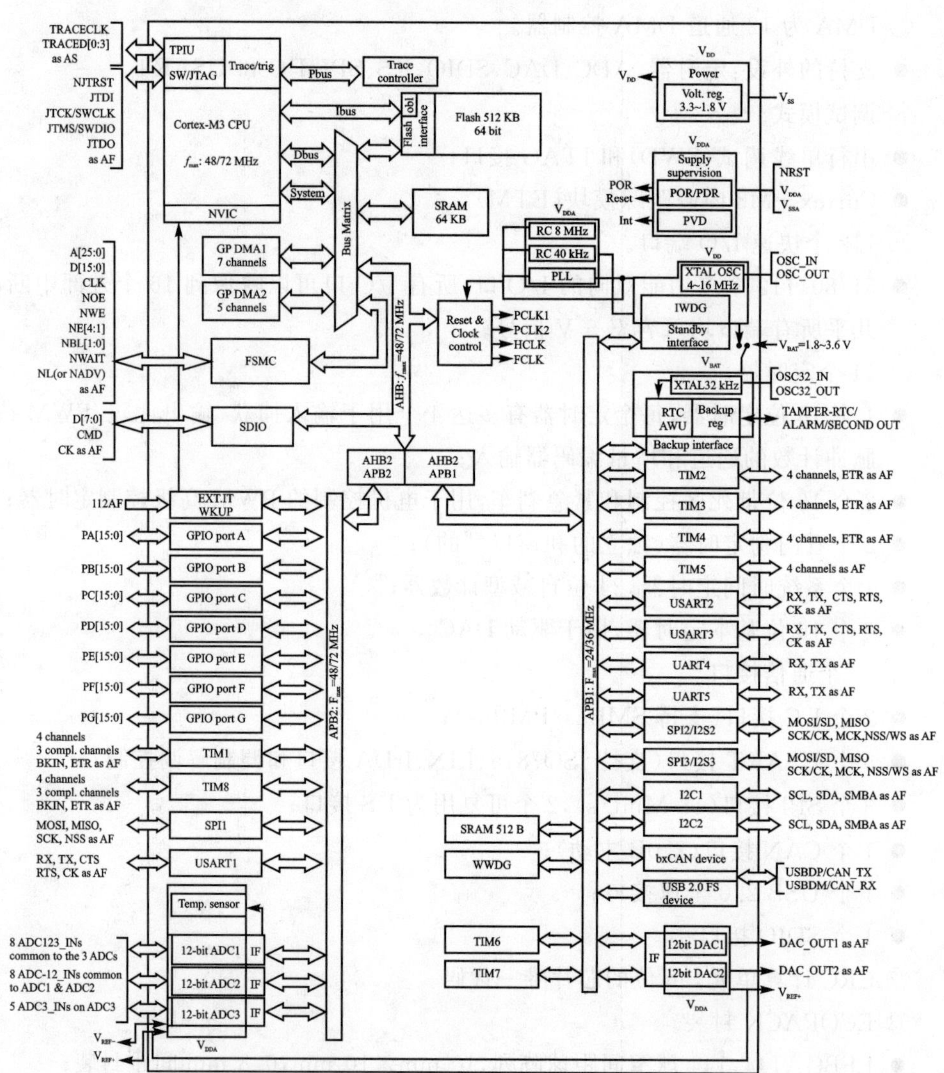

图 1-16 STM32F103xx 系列微控制器内部结构方框图

有关 STM32F 系列 32 位微控制器内部结构的更多内容请登录 http://www.st.com 查询：*ST Microelectronics. RM0008 Reference manual STM32F101xx, STM32F102xx, STM32F103xx, STM32F105xx and STM32F107xx advanced ARM-based 32-bit MCUs.* 或者《ST Microelectronics. STM32F101xx，STM32F102xx，STM32F103xx，STM32F105xx 和 STM32F107xx，ARM 内核 32 位高性能微控制器

参考手册》。

1.5.4　STM32F 系列 32 位微控制器系统板简介

本系统板采用 STM32F103VET6 作为主控制器，系统板包含的资源如下：

① 2 MB 的 Flash（W25X16），SPI 接口。

② 3 V 电压基准芯片 LT6655－3。

③ DS1302 时钟芯片。

④ T－Flash 卡接口。

⑤ nRF24L01 无线模块接口。

⑥ 引出 3.2 in TFT 液晶接口、Jlink 仿真调试接口及所有 I/O 口。

1. STM32F103VET6 微控制器

STM32F103VET6 采用 100 引脚 LQFP 封装，具有 512 KB 闪存存储器，引脚端分布图如图 1－17 所示，各引脚端定义可登录 http://www.st.com 参考《ST Microelectronics. STM32F101xx，STM32F102xx，STM32F103xx，STM32F105xx 和 STM32F107xx，ARM 内核 32 位高性能微控制器参考手册》。

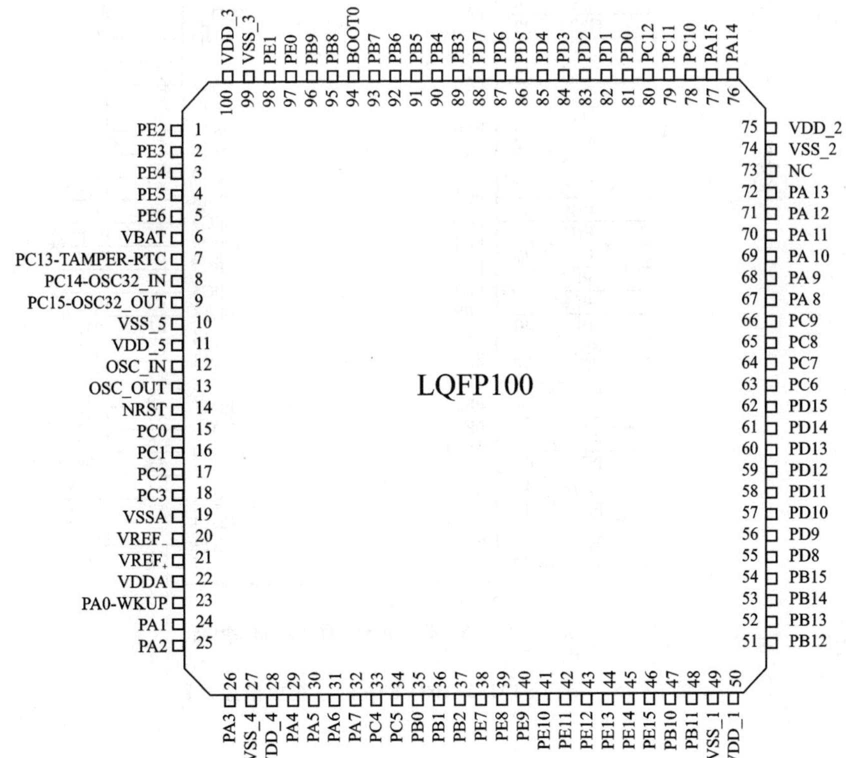

图 1－17　STM32F103VET6 引脚端分布图

采用 STM32F103VET6 作为主控制器的系统板电原理图如图 1-18 所示。

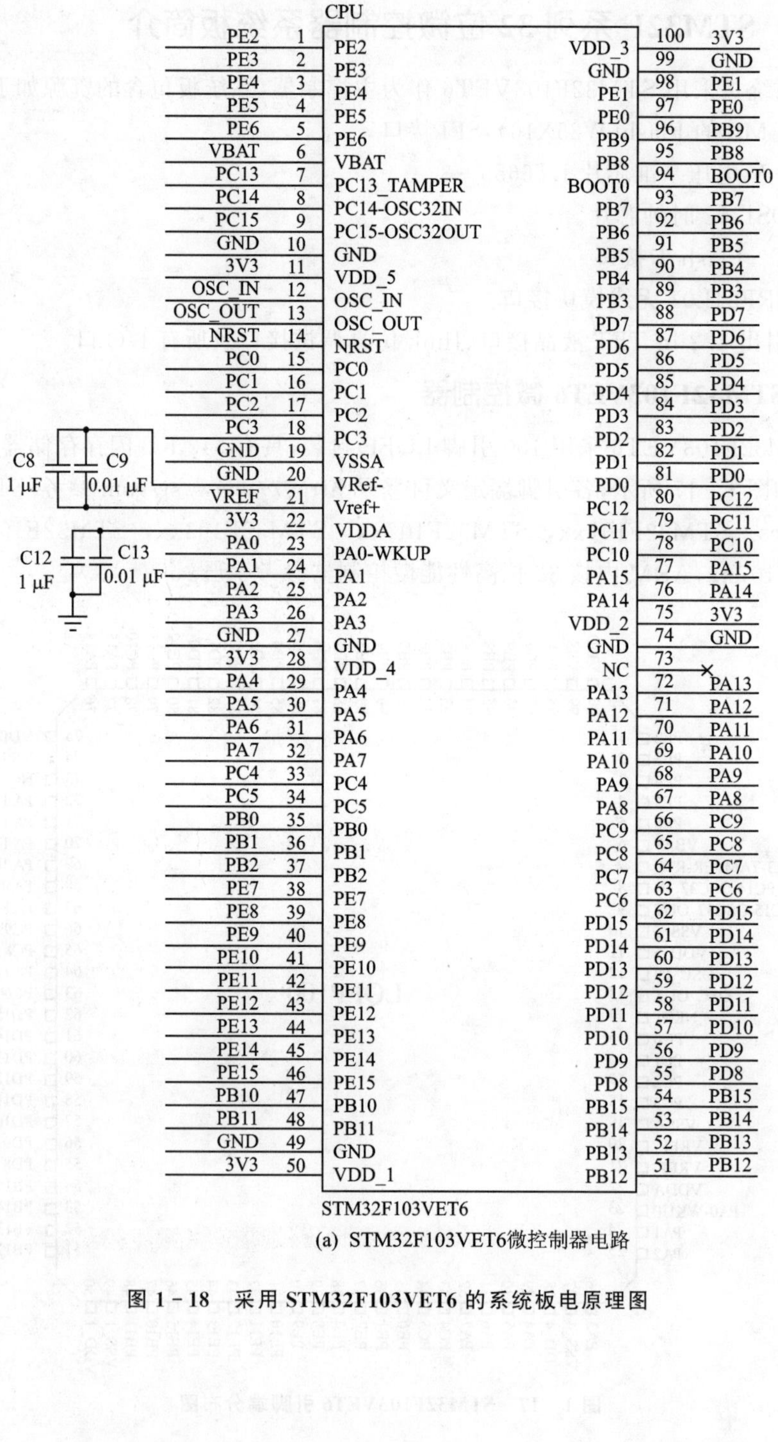

(a) STM32F103VET6微控制器电路

图 1-18　采用 STM32F103VET6 的系统板电原理图

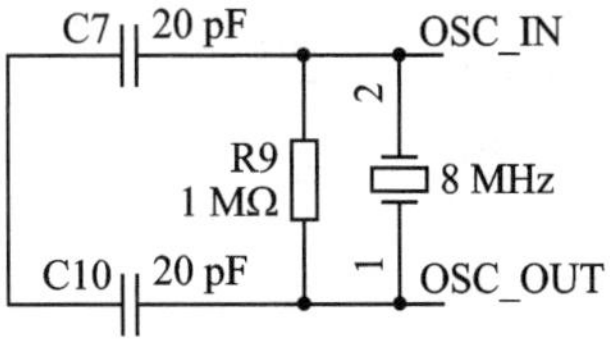

(b) 晶振电路

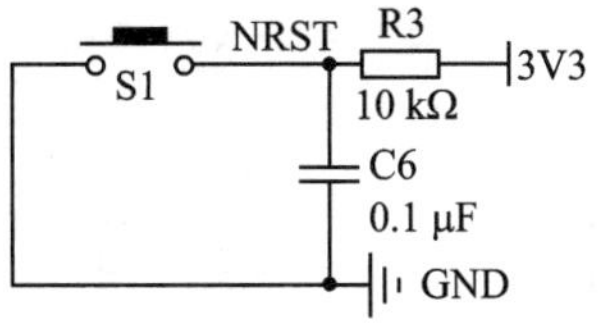

(c) 复位电路

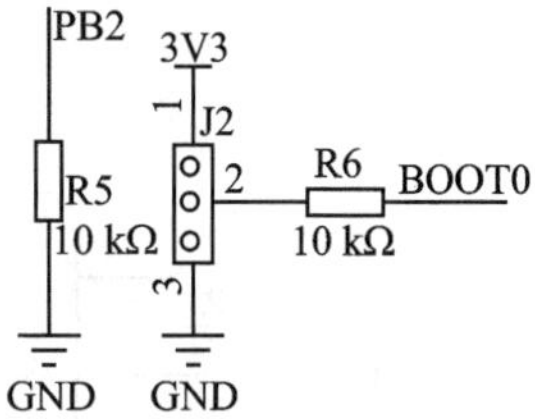

(d) Boot启动电路

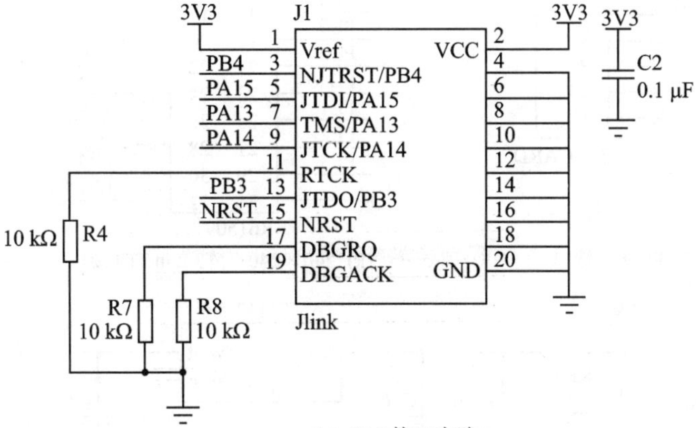

(e) Jlink接口电路

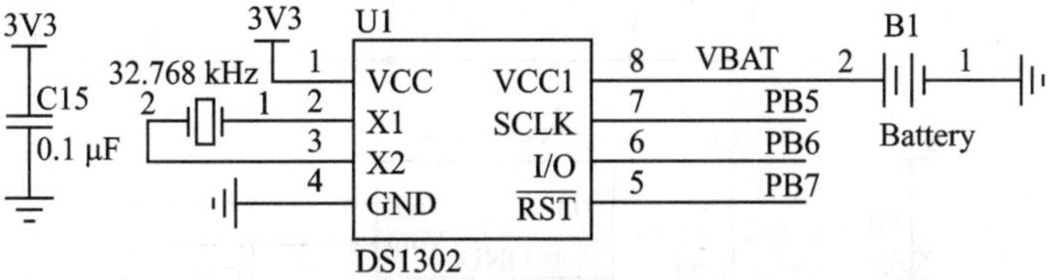

(f) DS1302时钟电路

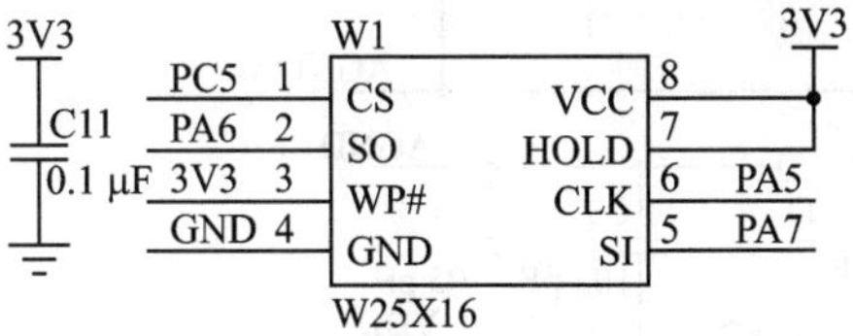

(g) 2 MB的Flash

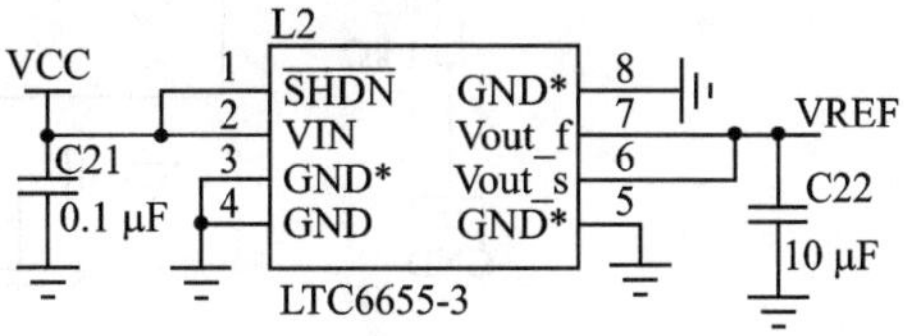

(h) ADC 3 V电压基准

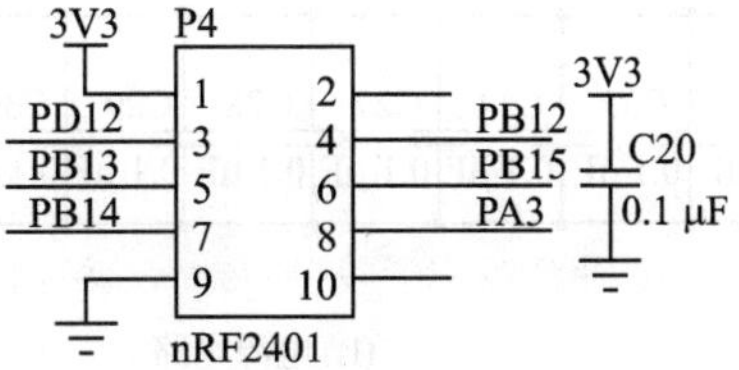

(i) nRF24L01接口

图 1-18　采用 STM32F103VET6 的系统板电原理图(续)

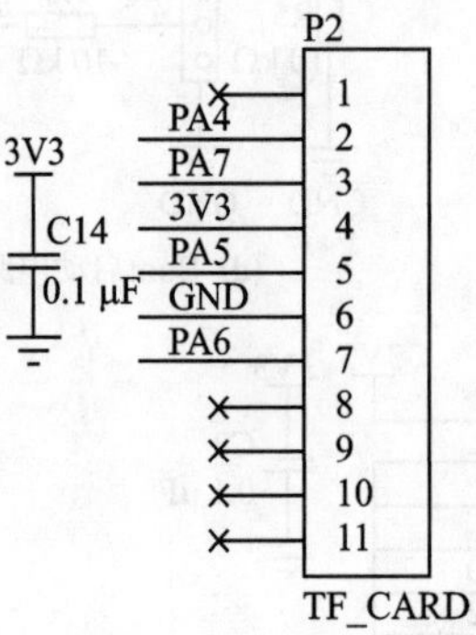

(j) T-Flash卡接口

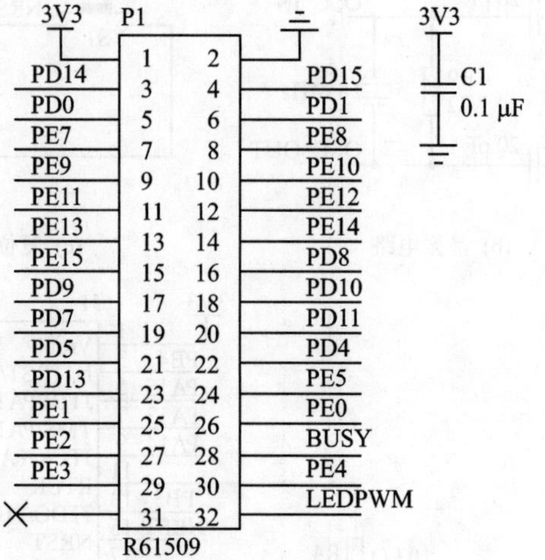

(k) 240×240点阵3.2 in TFT接口

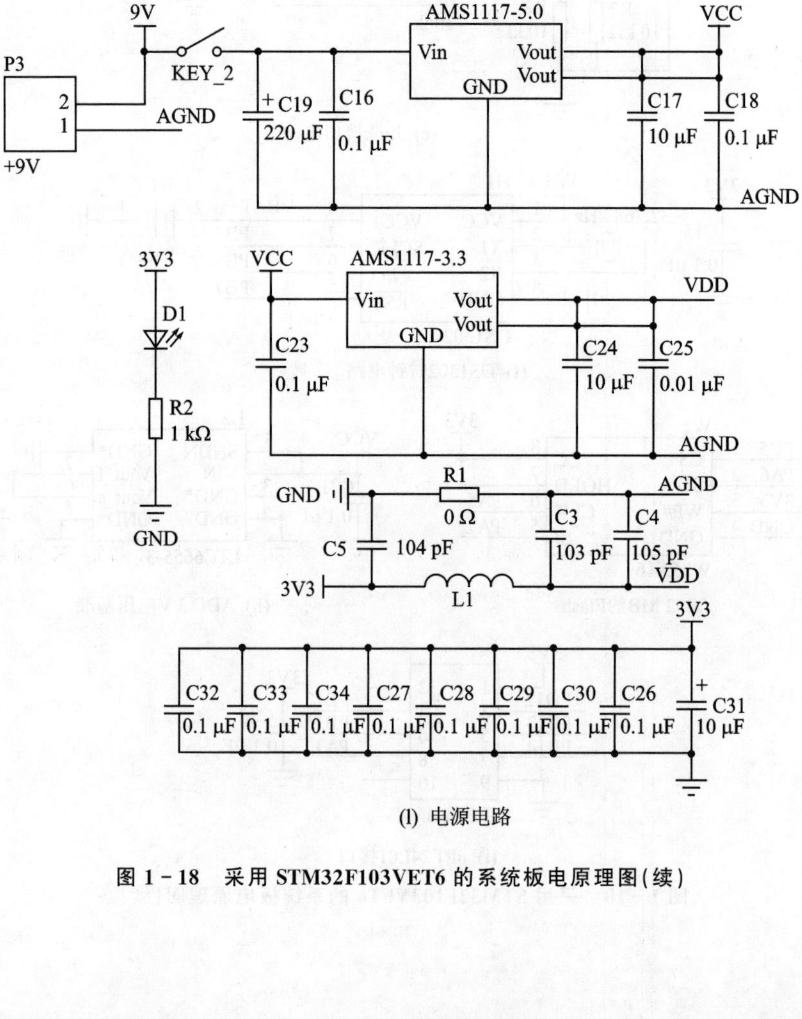

(l) 电源电路

图1-18　采用STM32F103VET6的系统板电原理图（续）

2. 2 MB 的 Flash W25X16

W25X16 是一个 2 MB 的 Flash 存储器，支持标准的 SPI 接口，支持 JEDEC 工业标准，传输速率最大为 75 MHz。四线制：串行时钟引脚 CLK，芯片选择引脚 $\overline{CS}$，串行数据输出引脚 DO，串行数据输入/输出引脚 DIO。另外芯片还具有保持引脚 $\overline{HOLD}$，写保护引脚 $\overline{WP}$，以及可编程写保护位等特性。

W25X16 应用电路如图 1-18(g)所示。

3. 3.3 V 电压基准芯片 LTC6655-3

LTC6655 是一个精准带隙电压基准，提供 1.25 V、2.048 V、2.5 V、3 V、3.3 V、4.096 V、5 V 完整的系列电压基准；具有卓越的低噪声（峰-峰值为 2.5×10^{-7}(0.1～10 Hz)）和低漂移（2×10^{-6}/℃）性能，非常适合于仪表和测试设备所要求的高分辨率测量；具有低压差 500 mV、宽电源范围 13.2 V、负载调整率<1×10^{-5}/mA、吸收电流和供应电流±5 mA、低功率停机模式（<20 μA）；采用 8 引脚 MSOP 封装，工作温度范围－40～125 ℃，可以确保其适合汽车和工业应用。

LTC6655-3 应用电路如图 1-18(h)所示。

4. DS1302 时钟芯片

DS1302 是一个涓流充电时钟芯片，内含有一个实时时钟/日历和 31 B 的静态 RAM。通过简单的串行接口与微控制器进行通信。实时时钟/日历电路提供秒、分、时、日、日期、月、年的信息。每月的天数和闰年的天数可自动调整。时钟操作可通过 AM/PM 指示决定采用 24 或 12 小时格式。

DS1302 与微控制器之间能简单地采用同步串行的方式进行通信，仅需用到三条接口线：RES 复位、I/O 数据线和 SCLK 串行时钟。时钟和 RAM 的读/写数据以 1 B 或多达 31 B 的字符组方式通信。DS1302 工作时功耗很低，保持数据和时钟信息时功率小于 1 mW。DS1302 工作电压为 2.0～5.5 V，工作电流小于 300 nA(2.0 V 时)。

DS1302 采用 8 引脚 DIP 封装或 8 引脚 SOIC 封装，其中：X1、X2 为 32.768 kHz 晶振引脚端，GND 为接地引脚端，RST 为复位引脚端，I/O 为数据输入/输出引脚端，SCLK 为串行时钟引脚端，VCC1、VCC 为电源供电引脚端。

DS1302 应用电路如图 1-18(f)所示。

5. Jlink 仿真调试接口

Jlink 仿真调试接口可以与 IAR Jlink 仿真器等接口实现与用于 ARM 处理器的小型 USB-JTAG/SWD 调试工具的连接，以及直接与 IAR Embedded Workbench for ARM 等集成开发环境的无缝连接。

6. T-Flash 卡接口

Tran Flash 卡，简称 T-Flash 卡或 TF 卡，是 Motorola 与 SanDisk 共同推出的

一种记忆卡规格，它采用了最新的封装技术，并配合 SanDisk 最新的 NAND MLC 技术及控制器技术，尺寸大小为 11 mm×15 mm×1 mm，约等于半张 SIM 卡。

T-Flash 卡产品采用 SD 架构设计而成，SD 协会于 2004 年年底正式将其更名为 Micro SD，已成为 SD Card 产品中的一员，附有 SD 转接器，可兼容任何 SD 读卡器，T-Flash 卡可经 SD 卡转换器后当 SD 卡使用。T-Flash 卡是市面上最小的闪存卡，适用于支持 SD 协议的 PDA、数码相机、MP3、PC 等多媒体应用。

T-Flash 卡接口电路如图 1-18(j)所示。

7. 3.2 in TFT 液晶显示器接口

3.2 in TFT 液晶显示器接口电路如图 1-18(k)所示。

8. nRF24L01 无线模块接口

nRF24L01 无线模块接口如图 1-18(i)所示。

1.5.5 STM32F 系统板电原理图和 PCB 图

在对 STM32F 最小系统的 PCB(印制电路板)进行设计时应注意以下几点：

① 出于技术的考虑，最好使用带有专门独立的接地层(VSS)和专门独立的供电层(VDD)的多层印制电路板，这样能提供较好的耦合性能和屏蔽效果。很多应用中，当受经济条件的限制不能使用多层印制电路板时，就需要保证有一个好的接地和供电的结构形式。

② 为了减少 PCB 上的交叉耦合，设计板图时就需要根据不同器件对 EMI 的影响，设计正确的器件位置，把大电流电路、低电压电路以及数字器件等不同的电路分开。

③ 每个模块(噪声电路、敏感度低的电路、数字电路)都应该单独接地，所有的地最终都应在一个点上连到一起。尽量避免或者减小回路的区域。为了减少供电回路的区域，电源应该尽量靠近地线。这是因为，供电回路就像个天线，可能成为 EMI 的发射器和接收器。PCB 上没有器件的区域，需要填充为地，以提供好的屏蔽效果(特别是对单层 PCB，尤其如此)。

④ STM32F 上每个电源引脚应该并联去耦合的滤波陶瓷电容(100 nF)和电解电容(10 μF)。这些电容应该尽可能地靠近电源/地引脚，或者在 PCB 另一层，处于电源/地引脚之下。典型值一般为 10～100 nF，具体的电容值取决于实际应用的需要。所有的引脚都需要适当地连接到电源。这些连接包括焊盘、连线和过孔，它们的阻抗应该尽量小。通常采用增加连线宽度的办法，包括在多层 PCB 中使用单独的供电层以减小阻抗。

⑤ 对于受暂时的干扰会影响运行结果的信号(比如中断或者握手抖动信号，而不是 LED 命令之类的信号)，其信号线周围铺地，缩短走线距离，消除邻近的噪声和敏感的连线都可以提高 EMC 性能。对于数字信号，为有效区别两种逻辑状态，必须

能够达到最佳可能的信号特性余量。

⑥ 所有微控制器都为各种应用而设计,而通常的应用都不会用到所有的微控制器资源。为了提高 EMC 性能,不用的时钟、计数器或者 I/O 引脚,需要做相应的处理,比如,I/O 端口应该被设置为"0"或"1"(对用不到的 I/O 引脚上拉或者下拉);没有用到的模块应该禁止或者"冻结"。

有关 STM32F 系列 32 位微控制器系统电路和 PCB(印制电路板)设计时应注意的更多内容请登录 http://www.st.com,查询《ST Microelectronics. AN2586 应用笔记 STM32F10xxx 硬件开发使用入门》和 *ST Microelectronics. AN2867 Application note Oscillator design guide for ST microcontrollers*。

采用 STM32F103VET6 的系统板 PCB 图和元器件布局图如图 1-19 所示。

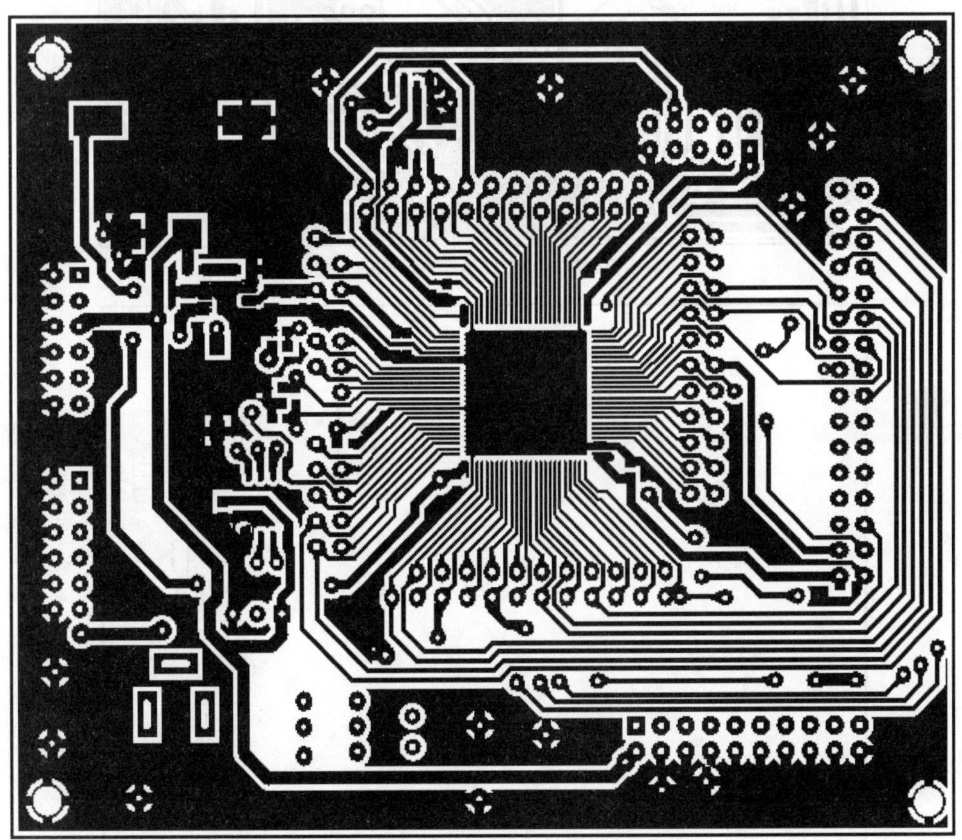

(a) STM32F103VET6系统PCB顶层布线图

图 1-19　采用 STM320F103VET6 的系统板 PCB 图和元器件布局图

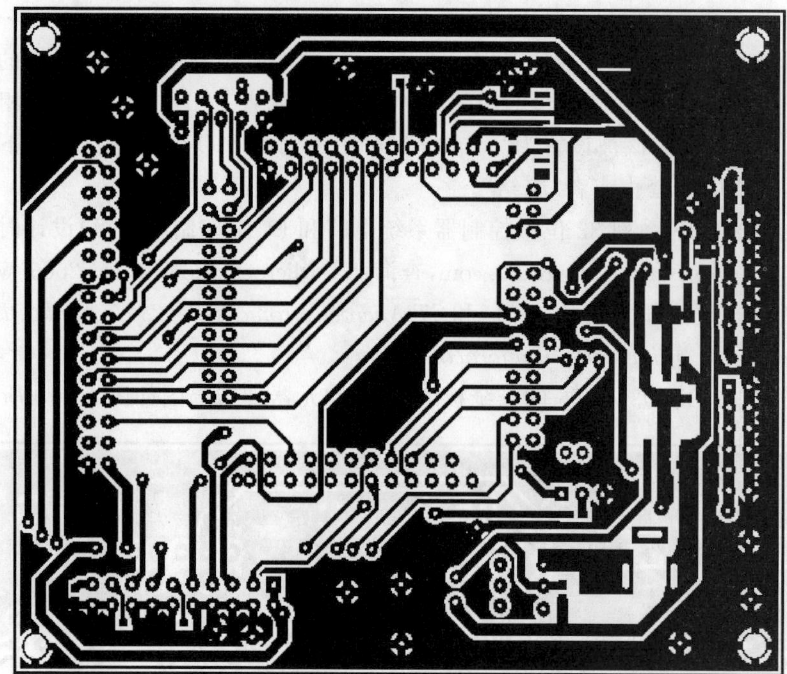

(b) STM32F103VET6系统PCB底层布线图

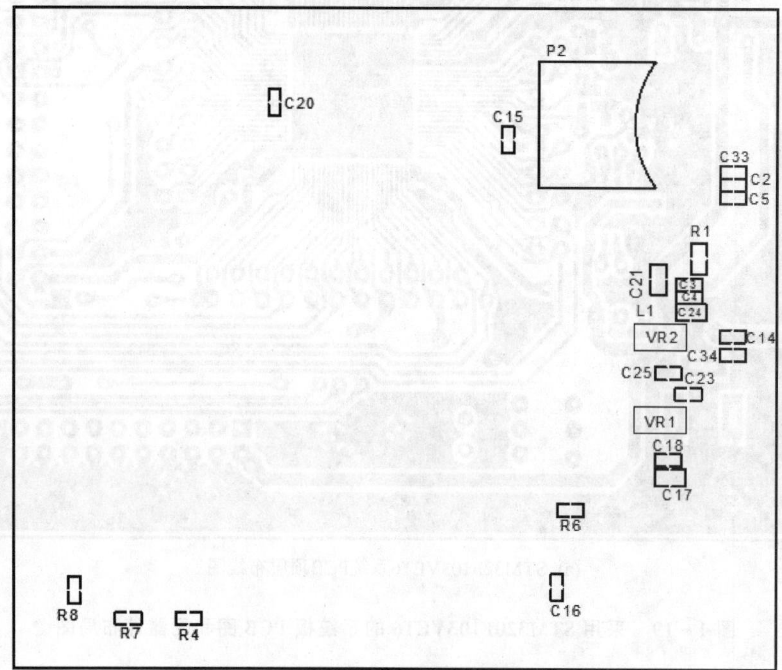

(c) STM32F103VET6系统底层元器件布局图

图 1-19 采用 STM320F103VET6 的系统板 PCB 图和元器件布局图(续)

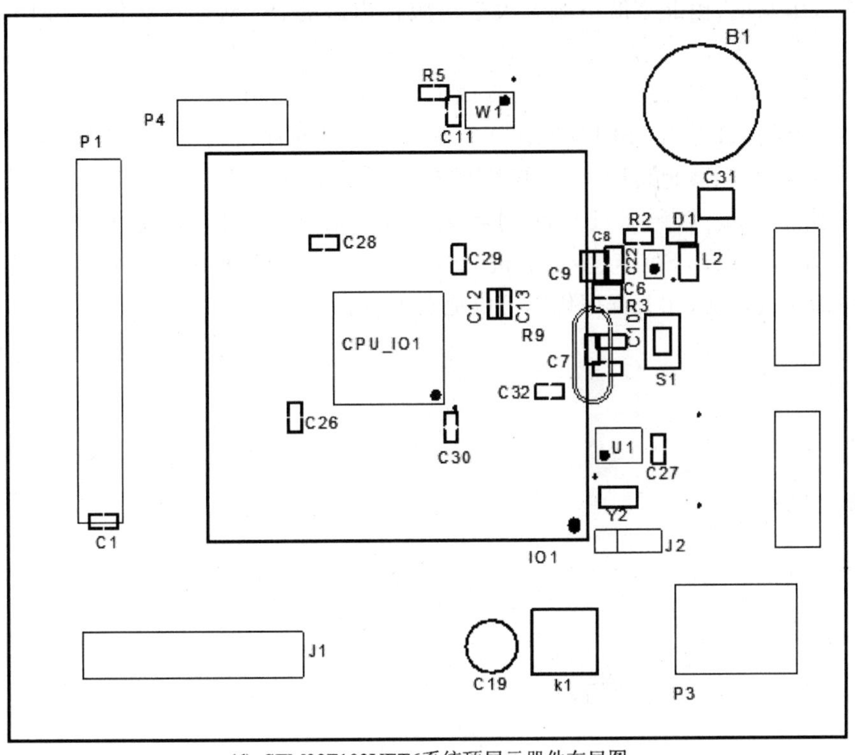

(d) STM32F103VET6系统顶层元器件布局图

图1-19 采用STM320F103VET6的系统板PCB图和元器件布局图(续)

1.5.6 STM32F系统板的应用设计与实践

LPC214x(LPC2141/2/4/6/8)是基于一个支持实时仿真和嵌入式跟踪的32/16位ARM7TDMI-S CPU的微控制器。

STM32F103xx系列是增强型的32位基于ARM核心的微控制器,其内核采用的是ARM 32位的Cortex-M3 CPU。

Cortex-M0和Cortex-M3处理器是目前嵌入式市场中ARM7TDMI-S用户选择的优秀的替代产品,可以使用户以更低的成本获得更多功能,实现代码重用,提高能效和增强连接性,并为未来的嵌入式应用提供支持。

有关将ARM7TDMI-S编写的软件移植到Cortex-M3微处理器的建议,请登录http://www.arm.com/zh/products/processors/classic/arm7/index.php,查询*ARM Cortex-M3 Processor Software Development for ARM7TDMI Processor Programmers*(《面向ARM7TDMI微处理器程序员的ARM Cortex-M3微处理器软件开发》)白皮书。

有关STM32F 32位微控制器的应用设计与实践的更多内容,也可以参考由黄智伟等人编写的,由北京航空航天大学出版社出版的《STM32F 32 位 ARM 微控制器应用设计与实践》一书。该书共分 14 章,系统介绍了 STM32F 系列 32 位微控制器最小系统设计,工程建立、软件仿真调试与程序下载,GPIO、USART、ADC、DAC、定时器、看门狗、FSMC、SPI、I^2C、CAN、SDIO 接口的使用与编程,以及 LCD、触摸屏、Flash 存储器、颜色传感器、光强检测传感器、图像传感器、加速度传感器、角度位移传感器、音频编/解码器、RFID、射频无线收发器、数字调频无线电接收机、DDS、CAN 收发器、Micro SD 卡、步进电机、交流调压等模块的使用与编程。该书所有示例程序都通过验证,相关程序代码可以免费下载。

第 2 章

显示器电路

2.1 键盘及 LED 数码管显示器电路的设计与制作

2.1.1 ZLG7290B 的主要特性

键盘及 LED 显示器电路采用 ZLG7290B 实现。ZLG7290B 是广州周立功单片机发展有限公司自行设计的数码管显示驱动及键盘扫描管理芯片，能够直接驱动 8 位共阴式数码管（1 in以下）或 64 只独立的 LED；能够管理多达 64 只按键，自动消除抖动，其中有 8 只可以作为功能键使用；段电流可达 20 mA，位电流可达 100 mA 以上；利用功率电路可以方便地驱动 1 in以上的大型数码管；具有闪烁、段点亮、段熄灭、功能键、连击键计数等强大功能；提供具有 10 种数字和 21 种字母的译码显示功能，或者直接向显示缓存写入显示数据；不接数码管而仅使用键盘管理功能时，工作电流可降至 1 mA；与微控制器之间采用 I^2C 串行总线接口，只需两根信号线，节省 I/O资源；工作电压范围为 3.3～5.5 V；工作温度范围为－40～＋85 ℃；该芯片为工业级芯片，抗干扰能力强，在工业测控中已有大量应用。

1. ZLG7290B 的引脚功能

ZLG7290B 采用 DIP－24（窄体）或者 SOP－24 封装，其引脚功能如表 2－1 所列。

表 2－1 ZLG7290B 的引脚功能

引脚号	符 号	属 性	功 能
13,12,21,22,3～6	Dig7～Dig0	输入/输出	LED 显示位驱动及键盘扫描线
10～7,2,1,24,23	SegH～SegA	输入/输出	LED 显示段驱动及键盘扫描线
20	SDA	输入/输出	I^2C 总线接口数据/地址线
19	SCL	输入/输出	I^2C 总线接口时钟线
14	$\overline{INT}$	输出	中断输出端，低电平有效
15	$\overline{RES}$	输入	复位输入端，低电平有效

续表 2-1

引脚号	符　号	属　性	功　能
17	OSC1	输入	连接晶体以产生内部时钟
18	OSC2	输出	
16	VCC	电源	电源正端(3.3～5.5 V)
11	GND	电源	电源地

2. ZLG7290B的工作原理

ZLG7290B是一种采用 I^2C 总线接口的键盘及LED驱动管理器件，需外接6 MHz的晶振。使用时ZLG7290B的从地址为70H，器件内部通过 I^2C 总线访问的寄存器地址范围为00H～17H，任一寄存器都可按字节直接读/写，并支持自动增址功能和地址翻转功能。

(1) 驱动数码管显示

使用ZLG7290B驱动数码管显示有两种方法，第一种方法是向命令缓冲区(07H～08H)写入复合指令，向07H写入命令并选通相应的数码管，向08H写入所要显示的数据，这种方法每次只能写入一个字节的数据，多字节数据的输出可在程序中用循环写入的方法实现；第二种方法是向显示缓存寄存器(10H～17H)写入所要显示的数据的段码，段码的编码规则从高位到低位依次为a b c d e f g d p，这种方法每次可写入1～8个字节数据。

(2) 读取按键

使用ZLG7290B读取按键时，读普通键的入口地址和读功能键的入口地址不同，读普通按键的地址为01H，读功能键的地址为03H。读普通键返回按键的编号，读功能键返回的不是按键编号，需要程序对返回值进行翻译，转换成功能键的编号。ZLG7290B具有连击次数计数器，通过读取该寄存器的值可区别单击键和连击键，判断连击次数还可以检测被按时间；连击次数寄存器只为普通键计数，不为功能键计数。此外，ZLG7290B的功能键寄存器，实现了2个以上按键同时按下，来扩展按键数目或实现特殊功能，类似于PC的Shift、Ctrl、Alt键。

3. 与微控制器连接

ZLG7290B通过 I^2C 接口与微控制器进行串口通信，I^2C 总线接口传输速率可达32 kbit/s。ZLG7290B的 I^2C 总线通信接口主要由SDA、SCL和 $\overline{INT}$ 这3个引脚组成。SCL线用来传递时钟信号；SDA线负责传输数据；SDA和SCL与LPC2148相连时，需加3.3～10 kΩ的上拉电阻。$\overline{INT}$ 负责传递键盘中断信号，与LPC2148相连时需串联一个470 Ω电阻。ZLG7290B与LPC2148进行 I^2C 通信的原理如图2-1所示。

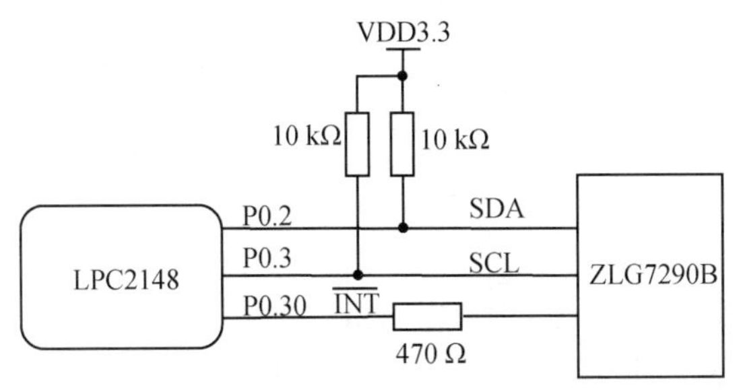

图 2-1　ZLG7290B 与 LPC2148 进行 I²C 通信的原理图

2.1.2　ZLG7290B 的应用电路

一个采用 ZLG7290B 构成的 8 位 LED 显示器和 16 键的应用电路和 PCB 图如图 2-2 和图 2-3 所示。

在图 2-2 中，U1 就是 ZLG7290B。为了使电源更加稳定，一般要在 VDD3.3 到 GND 之间接入 47～470 μF 的电解电容。J1 和 J2 是 ZLG7290B 与微控制器的接口，按照 I²C 总线协议的要求，信号线 SCL 和 SDA 上必须要分别加上拉电阻，其典型值是 10 kΩ。晶振 Y1 通常取值 6 MHz，调节电容 C3 和 C4 通常取值在 22 pF 左右。复位信号是低电平有效，直接通过拉低 RST 引脚的方法进行复位。数码管采用共阴极，不能直接使用共阳极。数码管在工作时要消耗较大的电流，R1～R8 是 LED 的限流电阻，典型值是 270 Ω。如果要增大数码管的亮度，可以适当减小电阻值，最低为 200 Ω。

键盘采用 16 只按键，键盘电阻 R10～R17 的典型值是 3.3 kΩ，这里选择的是 1 kΩ。数码管扫描线和键盘扫描线是共用的，所以二极管 D1 和 D2 是必须有的，有了它们就可以防止按键干扰数码管显示情况的发生。

2.1.3　ZLG7290B 应用中应注意的一些问题

1. ZLG7290B 一定要放在控制面板上

ZLG7290B 可广泛应用于仪器仪表、工业控制器、条形显示器、控制面板等领域。在实际应用中，控制面板和主机板往往是分离的，它们之间有几十厘米的距离，要用长的排线相连。键盘和数码管一般都位于控制面板上，主控制器则在主机板上。

注意：在设计时 ZLG7290B 一定要跟着控制面板走，而不要放在主机板上。

ZLG7290B 采用动态扫描法驱动数码管显示。为了防止显示出现闪烁，采用了比较高的扫描频率。扫描键盘同样也采用频率较高的信号。如果 ZLG7290B 放在主机板上，则这些扫描信号势必要走长线，而高频信号最忌讳走长线了，这容易导致显示混乱、按键失灵等故障。如果 ZLG7290B 放在控制面板上,由于走的是短线,就不易出现上述问题了。不必担心 ZLG7290B 与主控制器之间通信的 I^2C 总线会有问题。因为 I^2C 总线的通信速率是由主控制器控制的,可以做得低一些,所以允许走长线。

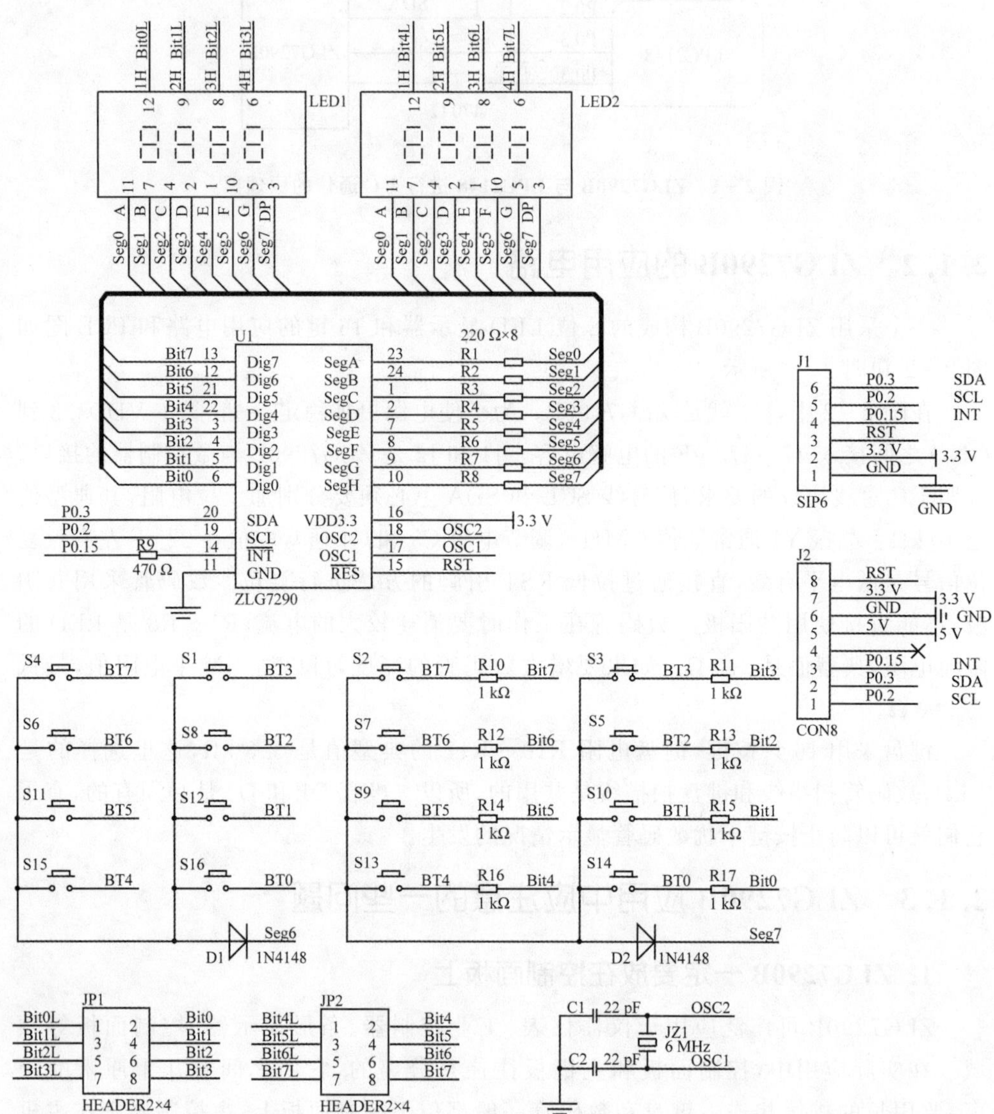

图 2-2 ZLG7290B 8 位 LED 显示器和 16 键应用电路

2. 复位引脚可以由主控制器直接控制

在工业控制应用中,为了增强抗干扰能力,建议采用独立的稳定直流电源给 ZLG7290B 供电,VCC 与 GND 之间的电容也要相应加大。另外复位引脚最好由主

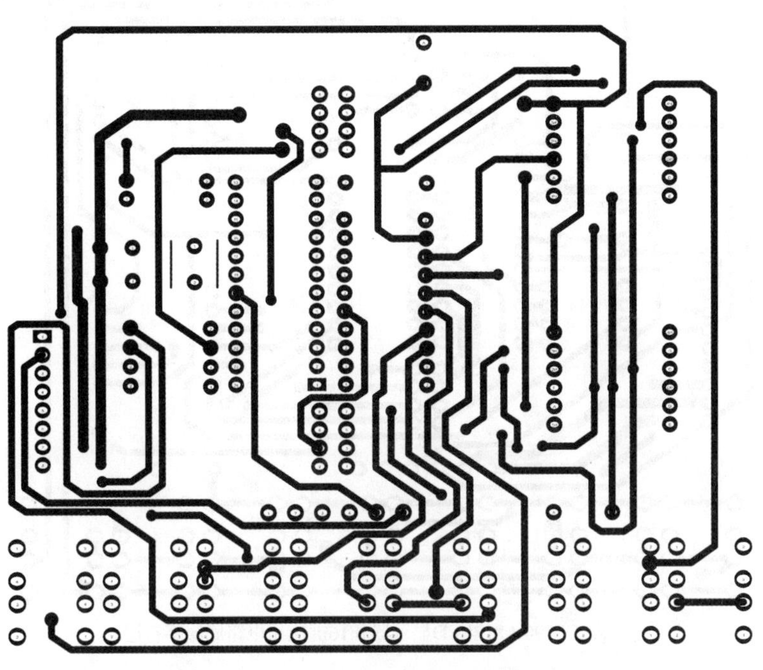

(a) ZLG7290B 8位LED显示器和16键应用电路顶层PCB

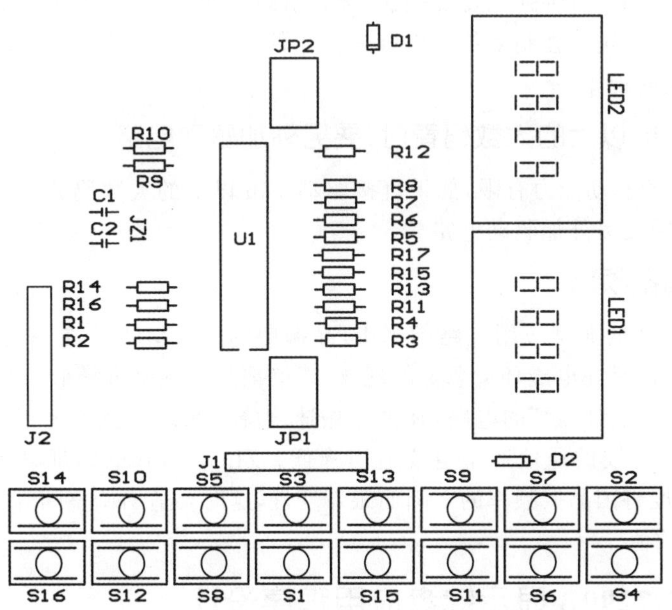

(b) ZLG7290B 8位LED显示器和16键应用电路顶层字符层

图 2-3　ZLG7290B 8 位 LED 显示器和 16 键应用电路 PCB 图

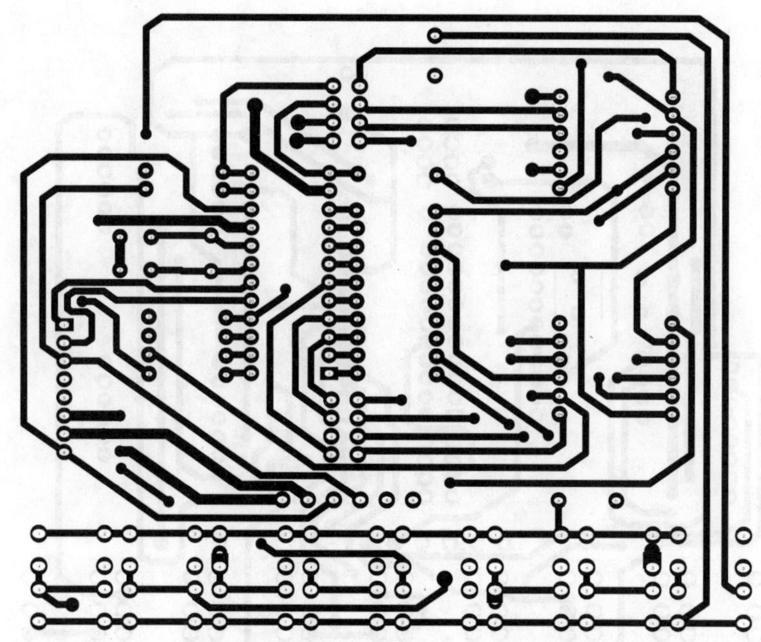

(c) ZLG7290B 8位LED显示器和16键应用电路底层字符层

图 2－3　ZLG7290B 8 位 LED 显示器和 16 键应用电路 PCB 图(续)

控制器来控制，每隔几分钟强制复位一次，复位脉冲宽度可以在 20 ms 左右，一闪而过，肉眼很难察觉。定时强制复位可以有效防止偶尔由于电磁干扰而产生的显示不正常和按键失灵的现象。

3. 驱动 1 in 以上的大数码管时，要另外加驱动电路

ZLG7290B 的驱动能力有限，如果直接驱动 1 in 以上的大数码管则可能会导致显示亮度不够，需要另外加驱动电路。

4. 降低晶振频率

在 ZLG7290B 的典型应用电路图当中，晶振用的是 4 MHz。在一般情况下，能够稳定地工作。但是在电磁环境恶劣的现场，建议把晶振频率再降低一些，降到 1～3 MHz。许多本来“有问题”的电路，在把晶振速度降下来之后就完全正常了。晶振频率降低后，I^2C 总线的通信速率也要适当降低。ZLG7290B 的闪烁显示功能将受到影响，闪烁速度将因晶振频率的下降而跟着变慢，这时要适当调整闪烁控制寄存器 FlashOnOff 的数值。

2.1.4　ZLG7290B 显示键盘应用程序设计

1. ZLG7290B 读/写流程图

下面给出了 LPC2148 使用 I^2C 主模式对 ZLG7290B 进行显示和键盘操作的示

例程序。初始化 I²C 总线后,测试所有数码管显示“87654321”,然后接收键盘输入,根据按键值控制相应显示位闪烁。用户可在键值判断下加入自己的程序,实现其他功能。如采用查询方式处理按键,其处理效率低、反应慢,在实际应用中建议采用键盘中断方式。本示例程序采用中断方式读取键值,读/写流程如图 2-4 所示。

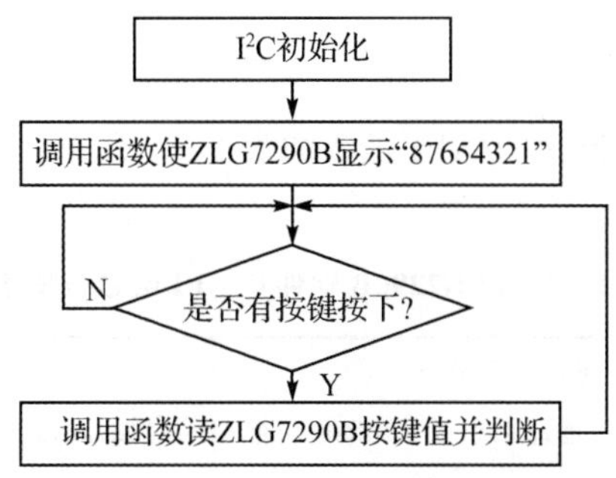

图 2-4　ZLG7290B 读/写流程图

2. ZLG7290B 软件包操作

在软件上操作 ZLG7290B 需要两个软件包,一个是 I2CINT. c,另一个是 zlg7290. c;这两个软件包和用户程序之间的关系如图 2-5 所示,前者是对 I²C 总线操作的软件包,后者封装的是操作 ZLG7290B 的 API 函数。可在周立功公司网站上 http://www.zlgmcu.com 下载这两个软件包。

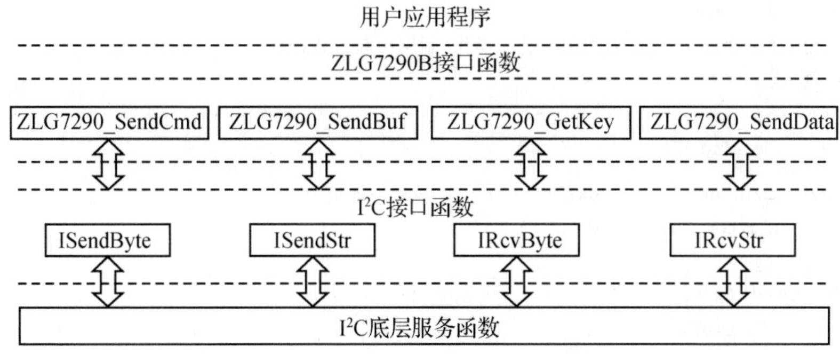

图 2-5　ZLG7290B 软件包结构图

启动 ADS 1.2,使用 ARM Executable Image for lpc2103 工程模板创建一个工程。使用过程中,需要 Startup. s 修改系统模式,堆栈设置为 0x5f,开启中断,还需要将 I²C 软件包文件(I2CINT. c、I2CINT. h)和 ZLG7290 软件包文件(zlg7290. c、zlg7290. h)包含进工程,并且在 config. h 中将 I2CINT. h 和 zlg7290. h 包括进去,如程序清单 2.1 所示。

程序清单 2.1　config.h 中添加 I2CINT.h 和 zlg7290.h

```
/**********************************************************************/
/*      应用程序配置      */
/**********************************************************************/
/* 以下根据需要改动      */
#include "I2CINT.h"
#include "ZLG7290.h"
```

ZLG7290B 软件包所包含的 API 函数如表 2-2 所列。相关函数参数说明请参考软件包说明。

表 2-2　ZLG7290B 软件包 API 函数一览表

函数功能描述	函数名称
向 ZLG7290B 地址 0x07、0x08 发送命令	ZLG7290_SendCmd
向 ZLG7290B 发送显示缓冲区数据	ZLG7290_SendBuf
向某个地址发送单个字节数据	ZLG7290_SendData
读取 ZLG7290B 键值	ZLG7290_GetKey

详细代码见文件目录下 ZLG7290 文件夹的实验例程，部分函数代码如程序清单 2.2所示。

程序清单 2.2　ZLG7290 键盘显示的示例程序

```
/**********************************************************************
 * 文件名:I2CTEST.C
 * 功  能:使用硬件 I²C 对 ZLG7290B 进行操作,利用中断方式操作
 * 说  明:无
**********************************************************************/
#include   "config.h"
#define   ZLG7290          0x70      /* 定义器件地址 */
#define   Glitter_COM     0x70
uint8   table[] = {
0x3f,0x06,0x5b,0x4f,
0x66,0x6d,0x7d,0x07,
0x7f,0x6f,0x77,0x7c,
0x39,0x5e,0x79,0x71,0};
/**********************************************************************
 * 名称:DelayNS()
 * 功能:长软件延时
 * 入口参数:dly  延时参数,值越大,延时越久
 * 出口参数:无
**********************************************************************/
void  DelayNS(uint32   dly)
{  uint32   i;
   for( ; dly>0; dly-- )
```

```
        for(i = 0; i<5000; i++);
}
/****************************************************************
* 名称:I2C_Init()
* 功能:主模式 I2C 初始化,包括初始化其中断为向量 IRQ 中断
* 入口参数:fi2c  初始化 I2C 总线速率,最大值为 400K
* 出口参数:无
****************************************************************/
void  I2C_Init(uint32 fi2c)
{  if(fi2c>400000) fi2c = 400000;
   PINSEL0 = (PINSEL0&0xFFFFFF0F) | 0x50;      // 设置 I2C 控制口有效
   I2C0SCLH = (Fpclk/fi2c + 1) / 2;            // 设置 I2C 时钟为 fi2c
   I2C0SCLL = (Fpclk/fi2c) / 2;
   I2C0CONCLR = 0x2C;
   I2C0CONSET = 0x40;                          // 使能主 I2C
   /* 设置 I2C 中断允许 */
   VICIntSelect = 0x00000000;                  // 设置所有通道为 IRQ 中断
   VICVectCntl0 = 0x29;                        // I2C 通道分配到 IRQ slot 0,即优先级最高
   VICVectAddr0 = (int)IRQ_I2C;                // 设置 I2C 中断向量地址
   VICIntEnable = 0x0200;                      // 使能 I2C 中断
}
/****************************************************************
* 名称:main()
* 功能:对 ZLG7290 进行操作
* 说明:在 STARTUP.S 文件中使能 IRQ 中断(清零 CPSR 中的 I 位);
*      在 CONFIG.H 文件中包含 I2CINT.H、ZLG7290.H
****************************************************************/
int  main(void)
{  uint8  disp_buf[8];
   uint8  key;
   uint8  i;
   PINSEL0 = 0x00000000;                       // 设置引脚连接,使用 I2C 口
   PINSEL1 = 0x00000000;
   DelayNS(10);
   I2C_Init(30000);                            // I2C 配置及端口初始化 I2C 初始化要延时 150
   DelayNS(150);
/* 设置数码管显示 87654321 */
   ISendStr(0x70,0x10,&table[4],1);DelayNS(15);
   ISendStr(0x70,0x11,&table[5],1);DelayNS(50);
   ISendStr(0x70,0x12,&table[6],1);DelayNS(50);
   ISendStr(0x70,0x13,&table[7],1);DelayNS(50);
   ISendStr(0x70,0x14,&table[0],1);DelayNS(50);
   ISendStr(0x70,0x15,&table[1],1);DelayNS(50);
   ISendStr(0x70,0x16,&table[2],1);DelayNS(50);
   ISendStr(0x70,0x17,&table[3],1);DelayNS(50);
```

```
/* 读取按键,设置键值对应的显示位闪烁 */
while(1)
{
    DelayNS(1);
    key = 0;
    IRcvStr(ZLG7290, 0x01, disp_buf, 2);          //读取 ZLG7290B 的 0x01 的值
    if(0 == disp_buf[1])                          //若不等于 0,则表示连击
    {
        key = disp_buf[0];
    }
  /* 将所按下的键所对应的数码管设置为闪烁 */
    switch(key)
    {
        case  1:
        case  9:
              ZLG7290_SendCmd(Glitter_COM, 0x10);
              break;
        case  2:
        case  10:
              ZLG7290_SendCmd(Glitter_COM, 0x20);
              break;
        case  3:
        case  11:
              ZLG7290_SendCmd(Glitter_COM, 0x40);
              break;
        case  4:
        case  12:
              ZLG7290_SendCmd(Glitter_COM, 0x80);
              break;
        case  5:
        case  13:
              ZLG7290_SendCmd(Glitter_COM, 0x01);
              break;
        case  6:
        case  14:
              ZLG7290_SendCmd(Glitter_COM, 0x02);
              break;
        case  7:
        case  15:
              ZLG7290_SendCmd(Glitter_COM, 0x04);
              break;
        case  8:
        case  16:
              ZLG7290_SendCmd(Glitter_COM, 0x08);
              break;
        default:
              break;
```

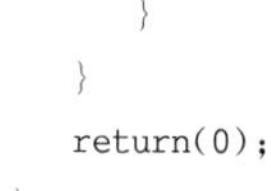

```
        }
    }
    return(0);
}
```

2.2 液晶显示器模块的连接与编程

2.2.1 FYD12864-0402B汉字图形点阵液晶显示模块简介

FYD12864-0402B是一种具有4位/8位并行、2线或3线串行的多种接口方式，内部含有国家一级、二级简体中文字库的点阵图形液晶显示模块；其分辨率为128×64，内置8 192个16×16点阵的汉字和128个16×8点阵的ASCII字符集。LCD显示类型为STN、半透和正显，显示内容128列×64行，显示颜色为黄绿色，电源电压范围为3.0～5.5 V(内置升压电路，无需负压)，配置侧部高亮白色LED背光，功耗仅为普通LED的1/5～1/10。外观尺寸为93 mm×70 mm×12.5 mm，视域尺寸为73 mm×39 mm。

FYD12864-0402B液晶显示器模块的引脚功能如表2-3所列，其与控制器之间的通信有并行和串行两种连接方法。

FYD12864-0402B液晶显示器模块有基本指令集和扩充指令集。更多的内容请参考《FYD12864-0402B液晶显示器模块数据手册》。

表2-3 FYD12864-0402B液晶显示器模块的引脚功能

引脚号	符 号	功 能	引脚号	符 号	功 能
1	GND	模块的电源地	7～14	DB0～DB7	并行数据0～数据7
2	VDD	模块的电源正端	15	PSB	并/串行接口选择：H—并行；L—串行
3	V0	LCD驱动电压输入端			
4	RS(CS)	并行的指令/数据选择信号；串行的片选信号	16	NC	空脚
			17	$\overline{RET}$	复位，低电平有效
5	R/W(SID)	并行的读/写选择信号；串行的数据口	18	NC	空脚
			19	BLA	背光源正极(LED+5 V)
6	E(CLK)	并行的使能信号；串行的同步时钟	20	BLK	背光源负极(LED-0 V)

2.2.2 LPC2148最小系统开发板与FYD12864-0402B的连接

LPC2148最小系统开发板与FYD12864-0402B的串行接口电路如图2-6所示，其连接方式如表2-4所列。在LPC2148最小系统开发板上液晶接口使用的是插槽连接，故在需要使用液晶时，只需将液晶的插针插进插槽即可。

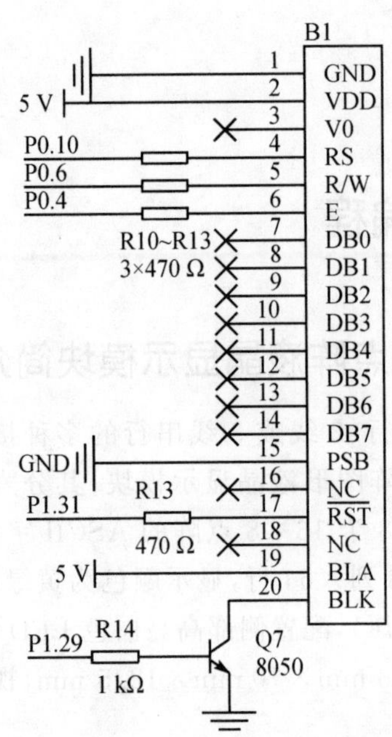

图 2－6　LPC2148 最小系统开发板与 FYD12864－0402B 的接口电路

表 2－4　LPC2148 最小系统开发板与 FYD12864－0402B 的连接方式

FYD12864－0402B 液晶显示器的串行接口	LPC2148 最小系统开发板
VDD	＋5 V
GND	GND
$\overline{\text{RST}}$	P1.31
E(CLK)	P0.4
R/W(SID)	P0.6
RS(CS)	P0.10

2.2.3　FYD12864－0402B 汉字图形点阵液晶显示模块编程示例

启动 ADS 1.2，使用 ARM Executable Image for lpc2148 工程模板创建一个工程。程序流程图如图 2－7 所示。

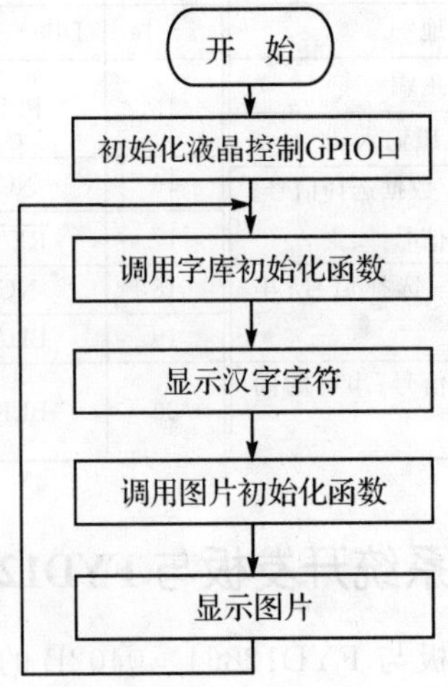

图 2－7　LPC2148 最小系统开发板驱动液晶模块程序流程图

本示例程序利用 LPC2148 最小系统开发板的输出控制 FYD12864－0402B 显

示。本示例程序相关代码和注释如程序清单 2.3 所示。

程序清单 2.3　液晶显示编程示例

```
/*****************************************************************
 *  文件名:LCD12864_test.C
 *  功能:LPC2148 串行模式控制 LCD12864 软件包
*****************************************************************/
#include "config.h"
#define     RS      1<<10           // P0.10
#define     SID     1<<6            // P0.6
#define     E       1<<4            // P0.4
#define     PSB     1<<2            // P0.2,并行或串行,选择低电平串行模式
#define     RST     1<<31           // P1.31,复位脚
#define     LEDAK   1<<29           // P1.29,背光控制
unsigned char  pic1[] =
{
/* --  调入了一幅图像    -- */
/* --  宽度×高度 = 128×64   -- */
0x00,0x00,0x00,0x00,0x00,0x00,0x00,0x00,0x00,0x00,0x00,0x00,0x00,0x00,0x00,0x00,
…………
(注:图片编码数据省略)
0x00,0x00,0x00,0x00,0x00,0x00,0x00,0x00,0x00,0x00,0x00,0x00,0x00,0x00,0x00,0x00,
};
unsigned char  IC_DAT1[64] = "LPC2148 控制 LCD12864 串行模式的示例程序";
unsigned char  IC_DAT2[64] = "可显示汉字、数字、点阵图片等";
void  TransferCom(unsigned char data0);
void  TransferData(unsigned char data1);
void  delayms(unsigned int n);
void  DisplayGraphic(unsigned char  *adder);
void  delay(unsigned int m);
void  lcd_mesg(unsigned char  *adder1);
/*****************************************************************
 *  名称:delayms()
 *  功能:延时(10×n) ms 程序。
 *  入口参数:n  延时 n×10 ms
 *  出口参数:无
*****************************************************************/
void  delayms(unsigned int n)
{
    unsigned int i,j;
    for(i = 0;i<n;i++)
        for(j = 0;j<2000;j++);
}
/*****************************************************************
 *  名称:delay()
 *  功能:延时程序
```

```
 * 入口参数:dly  延时时长控制变量
 * 出口参数:无
*********************************************************************/
void delay(unsigned int m)                          // 延时程序
{
    unsigned int i,j;
    for(i = 0;i<m;i++)
        for(j = 0;j<50;j++);
}
/*********************************************************************
 * 名称:initinal()
 * 功能:LCD12864 字库初始化
 * 入口参数:无
 * 出口参数:无
*********************************************************************/
void initinal(void)
{
   delay(40);                                       // 大于 40 ms 的延时程序
   IO0SET = PSB;                                    // 设置为串行工作模式
   delay(1);                                        // 延时
   IO1CLR = RST;                                    // 复位
   delay(1);                                        // 延时
   IO1SET = RST;                                    // 复位置高
   delay(10);
   TransferCom(0x30);                               // 8BIT 设置,RE = 0,G = 0,图片显示关
   delay(100);                                      // 大于 100 μs 的延时程序
   TransferCom(0x0C);                               // D = 1,显示开
   delay(100);                                      // 大于 100 μs 的延时程序
   TransferCom(0x01);                               // 清屏
   delay(10);                                       // 大于 10 ms 的延时程序
   TransferCom(0x06);                               // 模式设置,光标从右向左加 1 位移动
   delay(100);                                      // 大于 100 μs 的延时程序
}
/*********************************************************************
 * 名称:initina2()
 * 功能:图片初始化,LCD 显示图片(扩展)初始化程序
 * 入口参数:无
 * 出口参数:无
*********************************************************************/
void  initina2(void)
{
   delay(40);                                       // 大于 40 μs 的延时程序
   IO0SET = PSB;                                    // 设置为串行工作模式
   delay(1);                                        // 延时
   IO1CLR = RST;                                    // 复位
```

```
   delay(1);                                    // 延时
   IO1SET = RST;                                // 复位置高
   delay(10);
   TransferCom(0x36);                           // RE = 1
   delay(100);                                  // 大于 100 μs 的延时程序
   TransferCom(0x36);                           // RE = 1
   delay(37);                                   // 大于 37 μs 的延时程序
   TransferCom(0x3E);                           // DL = 8BITS,RE = 1,G = 1
   delay(100);                                  // 大于 100 μs 的延时程序
   TransferCom(0x01);
   delay(100);                                  // 大于 100 μs 的延时程序
}
/*******************************************************************
* 名称:lcd_mesg()
* 功能:在 LCD12864 上显示汉字
* 入口参数:* adder1  显示的数据
* 出口参数:无
*******************************************************************/
void  lcd_mesg(unsigned char  * adder1)
{
    unsigned char i;
    TransferCom(0x80);                          // 设置显示地址
    delay(100);
    for(i = 0;i<32;i ++ )
    {
          TransferData( * adder1);
           adder1 ++ ;
    }
    TransferCom(0x90);                          // 设置显示地址
    delay(100);
    for(i = 16;i<64;i ++ )
    {
          TransferData( * adder1);
          adder1 ++ ;
    }
}

/*******************************************************************
* 名称:SendByte()
* 功能:发送 1 字节数据
* 入口参数:Dbyte  发送的数据
* 出口参数:无
*******************************************************************/
void  SendByte(unsigned char Dbyte)
{
    unsigned char i;
```

```
    for(i = 0;i<8;i ++ )
    {
        IO0CLR = E;
        if((Dbyte&0x80) == 0X80)
            IO0SET = SID;
        else
            IO0CLR = SID;
            Dbyte = Dbyte<<1;                        // 左移一位
            IO0SET = E;
            IO0CLR = E;
    }
}
/*********************************************************************
*  名称:TransferCom()
*  功能:向 LCD12864 发送命令
*  入口参数:data0   要发送的命令
*  出口参数:无
*********************************************************************/
void  TransferCom(unsigned char data0)
{
    IO0SET = RS;
    SendByte(0xf8);                              // 11111,RW(0),RS(1),0
    SendByte(0xf0&data0);                        // 高 4 位
    SendByte(0xf0&data0<<4);                     // 低 4 位
    IO0CLR = RS;
}
/*********************************************************************
*  名称:TransferData()
*  功能:向 LCD12864 发送数据
*  入口参数:data1   要发送的数据
*  出口参数:无
*********************************************************************/
void  TransferData(unsigned char data1)
{
    IO0SET = RS;
    SendByte(0xfa);                              // 11111,RW(0),RS(1),0
    SendByte(0xf0&data1);                        // 高 4 位
    SendByte(0xf0&data1<<4);                     // 低 4 位
    IO0CLR = RS;
}
/*********************************************************************
*  名称:DisplayGraphic()
*  功能:在 LCD12864 上显示图片
*  入口参数:* adder   要显示的图片数据
*  出口参数:无
*********************************************************************/
void DisplayGraphic(unsigned char * adder)
```

```
{
    int i,j;
    /* 显示上半屏内容设置 */
    for(i = 0;i<32;i++)
    {
        TransferCom((0x80 + i));                 // 垂直地址
        TransferCom(0x80);                       // 水平地址
        for(j = 0;j<16;j++)
        {
            TransferData( * adder);
            adder++;
        }
    }
    /* 显示下半屏内容设置 */
    for(i = 0;i<32;i++)
    {
        TransferCom((0x80 + i));                 // 垂直地址
        TransferCom(0x88);                       // 水平地址
        for(j = 0;j<16;j++)
        {
            TransferData( * adder);
            adder++;
        }
    }
}
/*****************************************************************
* 名称:LCD12864_init()
* 功能:LCD12864 端口初始化函数
* 入口参数:无
* 出口参数:无
*****************************************************************/
void LCD12864_init(void)
{

    IO0DIR |= (E | SID | RS);                    // 设置为输出
    IO0CLR =  (E | SID | RS);

    IO1DIR |= RST;

    IO1CLR = RST;                                // 复位
    delay(1);                                    // 延时
    IO1SET = RST;                                // 复位置高

    IO1DIR |= LEDAK;
    IO1SET = LEDAK;                              // 点亮 LCD 背光灯
}
/*****************************************************************
* 名称:LCD_BackLight_ON()
* 功能:液晶背光灯开
```

```
 *  入口参数:无
 *  出口参数:无
*********************************************************************/
void LCD_BackLight_ON(void)
{
    IO1SET = LEDAK;                              // 点亮 LCD 背光灯
}
/*********************************************************************
 *  名称:LCD_BackLight_OFF()
 *  功能:液晶背光灯关
 *  入口参数:无
 *  出口参数:无
*********************************************************************/
void LCD_BackLight_OFF(void)
{
    IO1CLR = LEDAK;                              // 关闭 LCD 背光灯
}
/*********************************************************************
 *  名称:main()
 *  功能:LCD12864 汉字和图片显示程序主函数
*********************************************************************/
int main(void)
{
    PINSEL0 = 0;
    PINSEL1 = 0;
    PINSEL2 &= ~(0x00000006);                    // 设置所有 I/O 口为普通 GPIO 口
    LCD12864_init();                             // 液晶端口初始化
    while(1)
    {
        initinal();
        lcd_mesg(IC_DAT1);                       // 显示汉字界面 1
        delayms(1000);
        initinal();
        lcd_mesg(IC_DAT2);                       // 显示汉字界面 2
        delayms(1000);
        initina2();
        DisplayGraphic(pic1);                    // 显示图片 1
        delayms(1000);
    }
}
/*********************************************************************
**                              结束                                **
*********************************************************************/
```

2.3　触摸屏模块的连接与编程

2.3.1　触摸屏模块简介

触摸屏模块（HMI）选择北京迪文科技有限公司生产的 DMT32240S035_01WT，其分辨率为 320×240，工作温度范围为－20～＋70 ℃；工作电压范围为 5～28 V，功耗为 1 W；12 V 时，背光最亮和背光熄灭时的工作电流分别为 90 mA 和 50 mA。该模块共有 33 MB 字库空间，可存放 60 个字库，支持多语言、多字体、字体大小可变的文本显示，还支持用户自行设计的字库；96 MB 的图片存储空间，最多可存储 384 幅全屏图片，支持 USB 高速图片下载更新，图形功能完善；用户最大串口访问存储器空间为 32 MB，与图片存储器空间重叠；支持触摸屏和键盘，并具有触摸屏漂移处理技术；同时还内嵌拼音输入法、数据排序等简单算法处理。

触摸屏模块采用异步、全双工串口（UART），数据传送采用 10 位：1 个起始位，8 个数据位（低位在前传送，LSB），1 个停止位。

- 上电时，如果终端的 I/O 0 引脚为高电平或者浮空状态，则串口波特率由用户预先设置，范围为 1 200～115 200 bit/s，具体设置方法参考 0xE0 指令。
- 上电时，如果终端的 I/O 0 引脚为低电平，则串口波特率固定在 921 600 bit/s。

1. 通信帧缓冲区（FIFO）

HMI 有一个 24 帧的通信帧缓冲区，通信帧缓冲区为 FIFO（先进先出存储器）结构，只要通信缓冲区不溢出，用户可以连续传送数据给 HMI。

HMI 有一个硬件引脚（用户接口中的“BUSY 引脚”）指示 FIFO 缓冲区的状态，正常时，BUSY 引脚为高电平（RS232 电平为负电压），当 FIFO 缓冲区只剩下一个帧缓冲区时，BUSY 引脚会立即变成低电平（RS232 电平为正电压）。

对于一般的应用，由于 HMI 的处理速度很快，用户用不着判断 BUSY 信号状态。但对于短时间需要传送多个数据帧的应用，比如一次需要高速刷新上百个屏幕参数，建议客户使用 BUSY 信号来控制串口发送，当 BUSY 信号为低电平时，就不要发送数据给 HMI。

如果用户使用 HMI 过程中，出现“丢帧”现象，即某些数据没有显示出来，可能就是缓冲区溢出了，这时需要用示波器检查 BUSY 信号是否跳变（使用示波器进行检测时要特别注意电源接地处理情况），如果有跳变，则需要减慢发送速度，或者增加硬件检测 BUSY 信号判忙处理。

2. 字节传送顺序

HMI 的所有指令或者数据都是十六进制（HEX）格式；对于字型（2 字节）数据，总是采用高字节先传送（MSB）方式。

比如，x 坐标为 100，其 HEX 格式数据为 0x0088，传送给 HMI 时，传送顺序为 0x00 0x88。

3. 传送方向

在 HMI 上，数据传送方向如图 2-8 所示，它按照下面的规则定义：

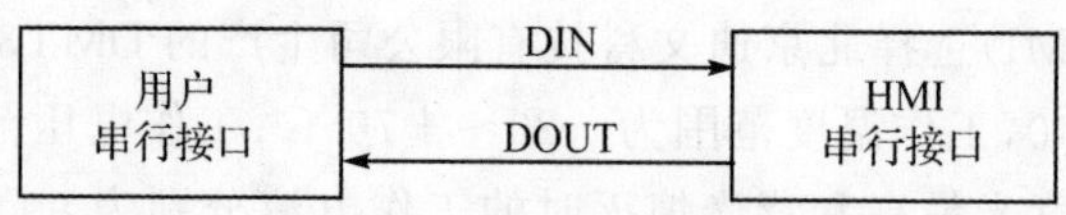

图 2-8　数据传送方向

- 下行(Tx)用户发送数据给 HMI，数据从 HMI 用户接口的“DIN 引脚”输入。
- 上行(Rx)HMI 发送数据给用户，数据从 HMI 用户接口的“DOUT 引脚”输出。

4. 串口电平的转换

图 2-9 是 3.3～5 V 电平的 TTL 串口转换电路，其中 SS14 可用其他压降小于 0.3 V 的肖特基二极管代替。

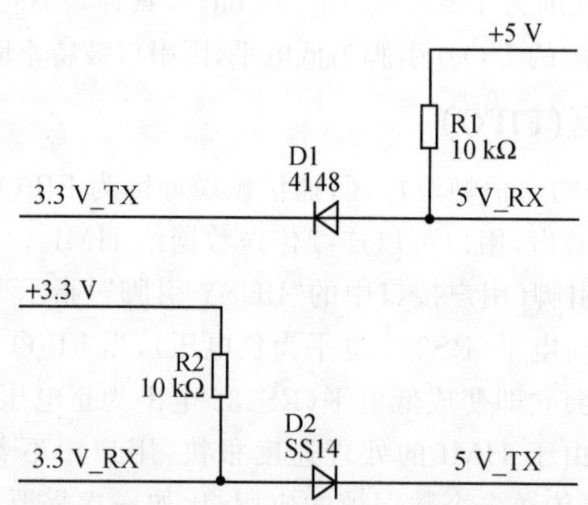

图 2-9　3.3～5 V 电平的 TTL 串口转换电路

图 2-10 是 3.3 V 或 5 V 电平的 TTL 串口到 RS232 电平串口的转换电路。

对于 PLC 等设备，或者信号需要远传时，往往需要使用抗干扰能力更好的 RS485 差分信号传输，常见的无源 RS232/RS485 转换器和 HMI 的连接电路如图 2-11所示。

有关 DMT32240S035_01WT 的更多内容请登录 http://www.dwin.com.cn，查询北京迪文科技有限公司的《智能显示终端开发指南》以及相关资料。

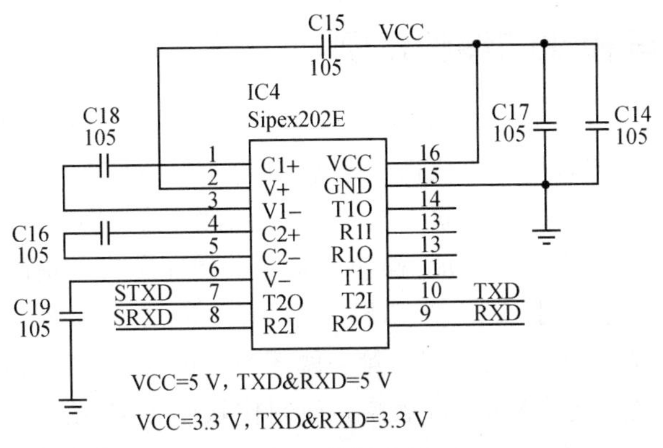

图2-10　TTL串口到RS232电平串口的转换电路

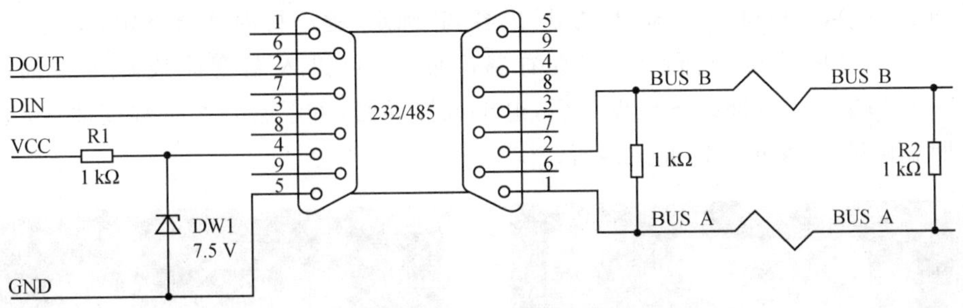

图2-11　无源RS232/RS485转换器和HMI的连接电路

2.3.2　LPC2148最小系统开发板与触摸屏模块的连接

LPC2148最小系统开发板与触摸屏模块通过UART接口进行连接，连接电路图如图2-12所示。

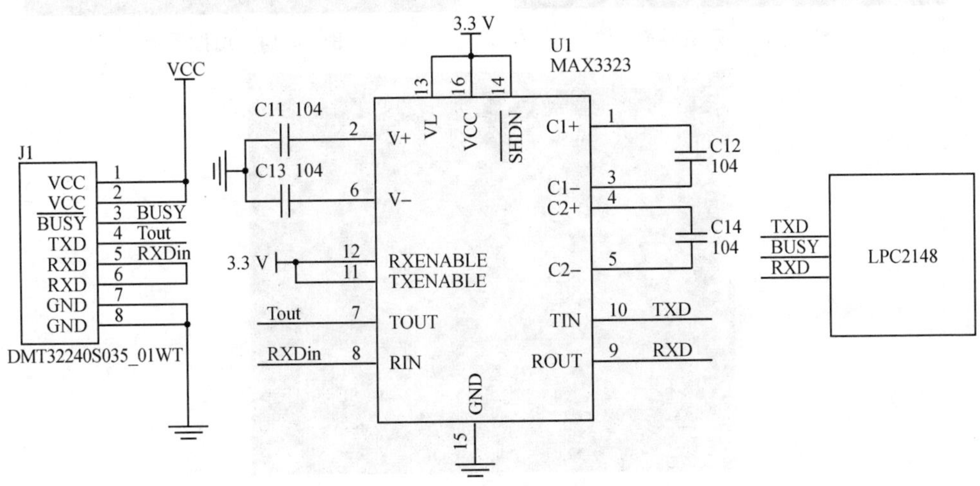

图2-12　LPC2148最小系统开发板与触摸屏模块连接电路

2.3.3　触摸屏模块的编程示例

该示例程序演示了如何使用LPC2148最小系统开发板控制DMT32240S035_01WT触摸屏模块。该触摸屏模块接上电源，连上串口即可工作，可以使用迪文公司提供的DWIN_DA_V2.exe软件进行与PC的通信作为调试。该模组使用一组指令集与微控制器通信来实现各种功能，操作简单。示例程序将模块的触控界面与非触控界面综合在一起，方便用户进一步拓展。

示例说明：事先设计好两幅图片，第一幅图为开机界面，第二幅图为温控系统界面，再将两幅图片经迪文HMI串口触控界面制作软件SysDef.exe进行添加按钮处理并配置到DMT32240S035_01WT内核中。开机后，进入触控界面，触摸屏的屏幕显示第一幅开机画面(见图2-13)，单击"点击进入"自动进入第二张温控界面画面(见图2-14)，用户可单击"输入键盘"内容和"确定"键进行温度值设定，单击"设定温度值"可返回开机画面，单击"查看温度曲线"可进入温度曲线显示界面(见图2-15)；进入温度曲线显示界面后，退出触控界面，可观察到2条温度曲线，单击"返回温控设定"又可返回温控界面。

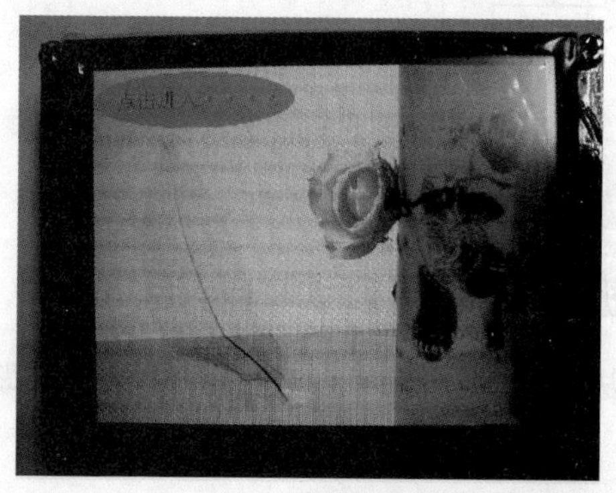

图2-13　开机界面

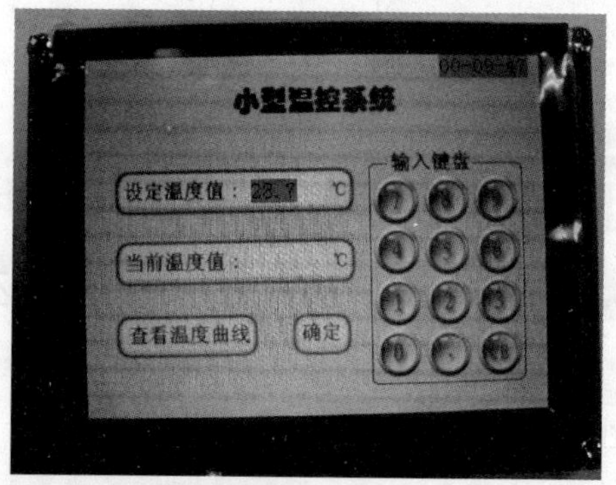

图2-14　温控界面

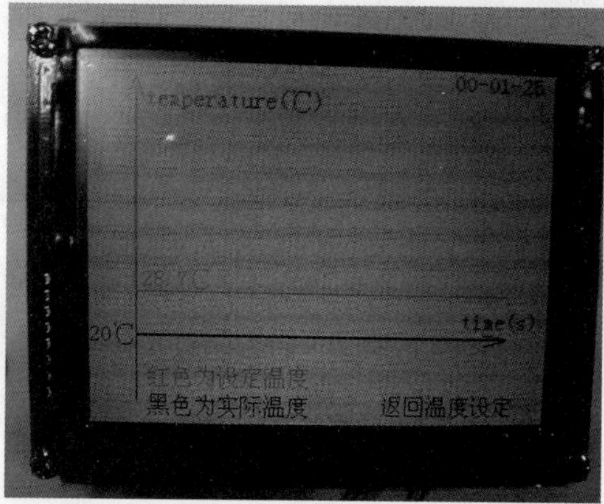

图2-15　温度曲线显示界面

1. 操作步骤

① 使用 ADS1.2 进行编程，将程序下载到 LPC2148 最小系统开发板上。

② 设计好图片，使用 SysDef.exe 配置到液晶屏中。

③ 将触摸屏模块通信接口与 LPC2148 最小系统开发板串口接口相连接。

④ 给触摸屏模块和 LPC2148 最小系统开发板上电，单击触摸屏上的相关区域，进行操作。

2. 程序流程图

触摸屏模块的编程示例程序流程图如图 2-16 所示。

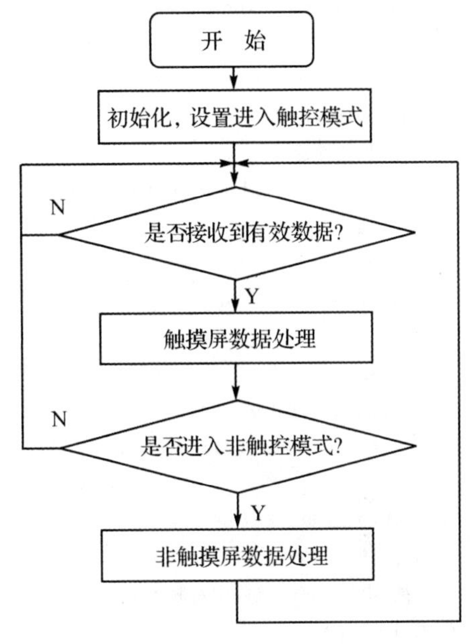

图 2-16　触摸屏模块的编程示例程序流程图

3. 示例程序

启动 ADS1.2，使用 ARM Executable Image for lpc2148 工程模板创建一个工程。使用过程中，需要 Startup.s 修改系统模式，堆栈设置为 0x5f，开启中断，详细代码如程序清单 2.4 所示。

程序清单 2.4　触摸屏的示例程序

```
#include "config.h"
#include <string.h>
#define  MAX_Uart1buf_Size   200                          // 定义串口数据长度
fp32 Temperature_Buf[3600] = {0};                         // 温度存储数组
fp64   Order_Temperature = 28.76;                         // 设定温度的存储变量
uint8    rcv_buf1[MAX_Uart1buf_Size] = {0},num1 = 0;    // 定义用于存放串口接收数据的数组
uint8    rcv_flag1 = 0;                                   // 定义用于表示串口接收到数据的标志变量
/*******************************************************************
*  名称:mDelaymS()
*  功能:延时 ms  毫秒
*  入口参数:ms  延时参数,值越大,延时越久
*  出口参数:无
*******************************************************************/
void mDelaymS( uint32 ms )
{
    uint32    i;
```

```
    while ( ms -- ) for ( i = 25000; i != 0; i -- );
}
/****************************************************************
 * 名称:DelayNS()
 * 功能:长软件延时
 * 入口参数:dly  延时参数,值越大,延时越久
 * 出口参数:无
****************************************************************/
void  DelayNS(uint32  dly)
{
    uint32   i;
    for(; dly>0; dly --)
        for(i=0; i<5000; i++);
}
/****************************************************************
 * 名称:IRQ_UART1()
 * 功能:串口 UART1 接收中断服务函数
 * 入口参数:无
 * 出口参数:无
****************************************************************/
void    __irq IRQ_UART1(void)
{
    uint8 i;
    switch(U1IIR&0x0f)                                // U0IIR 为中断标志,只看低四位
    {
        case 0x04:
            for(i=0; i<8; i++)                        // 触发点为 8
                rcv_buf1[num1++] = U1RBR;
            rcv_flag1 = 1;
            break;
        case 0x0c:
            while((U1LSR&0x01) == 1)                  // U0LSR 线状态寄存器
            {
                rcv_buf1[num1++] = U1RBR;
            }
            rcv_flag1 = 1;
            break;                                   // 发生 THRE 中断,读 U0IIR 或写 U0THR 中断复位
        default:                                     // 除以上四种情况
            break;
    }
   if(num1 >= MAX_Uart1buf_Size) num1 = 0;           // 数据接收个数限制
   VICVectAddr = 0x00;                               // 中断处理结束
}
```

```
/*************************************************************
*  名称:UART1_Ini()
*  功能:初始化串口 0
*  入口参数:bps    UART 通信波特率
*  出口参数:无
*************************************************************/
void UART1_Ini(uint32 bps)
{
    uint16 Fdiv;
    uint32 tmp;
  PINSEL0 = (PINSEL0 & 0xFFFCFFFF)|0x50000;    // 设置 I/O 为 UATR1
    if(bps == 115200)
    {
        U1FDR    = 1 | 12<<4;                   // LPC2148 串口除数校准波特率
    }
    tmp = bps;
    U1LCR = 0x83;                               // DLAB = 1,允许访问除数锁存寄存器
    Fdiv = (Fpclk/16)/tmp;                      // (UART_BPS)
    U1DLM = Fdiv/256;                           // 根据波特率设定 U0DLM 和 U0DLL 的值
    U1DLL =  Fdiv % 256;
    U1LCR = 0x03;                               // 禁止访问除数锁存寄存器
    U1FCR  =  0x81;                             // 使能 FIFO,并设置触发点为 8 字节
    U1IER  =  0x01;                             // 允许 RBR 中断,即接收中断
      /* 设置中断允许 */
    VICIntSelect  =  0x00000000;                // 设置所有通道为 IRQ 中断
    VICVectCntl0  =  0x27;                      // UART0 中断通道分配到 IRQ slot 0
    VICVectAddr0  =  (int)IRQ_UART1;            // 设置 UART0 向量地址
    VICIntEnable  =  1<<7;                      // 使能 UART0 中断
}
/*************************************************************
*  名称:UART1_SendByte
*  功能:向串口发送字节数据,并等待发送完毕
*  入口参数:data  要发送的数据
*  出口参数:无
*************************************************************/
void UART1_SendByte(uint8 data)
{
    U1THR = data;                               // 将要发送的数据传给寄存器
    while( (U1LSR&0x40 == 0) ) ;                // 等待数据发送完毕
    DelayNS(1);
}
```

```
/*************************************************************************
* 名称:UART1_SendBuf
* 功能:向串口发送数组数据,并等待发送完毕
* 入口参数:data  要发送的数组
* 出口参数:无
*************************************************************************/
void UART1_SendBuf(uint8 *str,uint16 no)          // 串口0发送字符串
{
    uint16 i;
    for(i = 0;i<no;i++)
     UART1_SendByte(str[i]);                       // 发送数据
}
/*************************************************************************
* 名称:LCD_ASK
* 功能:握手函数,发送一串数字,
*      会接收到 AA 00 4F 4B 5F 56 34 2E 32 0B 07 00 CC 33 C3 3C
* 入口参数:无
* 出口参数:无
*************************************************************************/
void LCD_ASK(void)
{
    uint8 tx_buf[10] = {0xAA,0x00,0xCC,0x33,0xC3,0x3C};
    UART1_SendBuf(tx_buf,6);
}
/*************************************************************************
* 名称:set_mode
* 功能:设置触摸屏工作模式
* 入口参数:所需触摸屏和键盘处理模式
* 出口参数:无
*************************************************************************/
void set_mode(uint8 mode)
{
    uint8 tx_buf[13] = {0xaa,0xe0,0x55,0xaa,0x5a,0xa5,0x0b,0x07,0x00,0xcc,0x33,0xC3,0x3C};
    tx_buf[8] = mode;
    UART1_SendBuf(tx_buf,13);
}
/*************************************************************************
* 名称:turn_light_max
* 功能:设置屏幕背光灯为最亮
* 入口参数:无
* 出口参数:无
*************************************************************************/
void turn_light_max()
```

```
{
    uint8 tx_buf[6] = {0xaa,0x5f,0xcc,0x33,0xc3,0x3c};
    UART1_SendBuf(tx_buf,6);
}
/***************************************************************
* 名称:set_colour
* 功能:设置屏幕颜色
* 入口参数:fore 前景色,2 字节  最大值 0xffff
*          black 背景色,2 字节  最大值 0xffff
* 出口参数:无
***************************************************************/
void set_colour(uint16 fore,uint16 black)
{
    uint8 tx_buf[10] = {0xaa,0x40,0xfc,0x1f,0x00,0x1f,0xcc,0x33,0xc3,0x3c};

    tx_buf[2] = fore/0xff;
    tx_buf[3] = fore % 0xff;

    tx_buf[4] = black/0xff;
    tx_buf[5] = black % 0xff;
    UART1_SendBuf(tx_buf,10);
}
/***************************************************************
* 名称:Disp_words
* 功能:显示文字
* 入口参数:word_style 0×53 用于显示 8×8 点阵 ASCII 字符串
*                     0x54 用于显示 16×16 点阵的扩展码汉字字符
*                     0x55 用于显示 32×32 点阵的内码汉字字符串
*                     0x6E 用于显示 12×12 点阵的扩展码汉字字符串
*                     0x6F 用于显示 24×24 点阵的内码汉字字符串
*          begin_x    首字符 x 坐标;
*          begin_y    首字符 y 坐标
* 出口参数:无
***************************************************************/
void Disp_words(uint8 word_style,uint16 begin_x,uint16 begin_y,unsigned char *words)
{
    uint8 tx_buf[500] = {0xaa};
    uint16 i = 0,length = 0;
    while(words[length])
        length ++ ;                                    //统计字符串长度
    for(i = 0;i<length;i ++ )
    {
```

```
        tx_buf[i + 6] = words[i];
    }
    tx_buf[1] = word_style;
    tx_buf[2] = begin_x/0xff;
    tx_buf[3] = begin_x % 0xff;
    tx_buf[4] = begin_y/0xff;
    tx_buf[5] = begin_y % 0xff;
    tx_buf[length + 6] = 0xcc;
    tx_buf[length + 7] = 0x33;
    tx_buf[length + 8] = 0xc3;
    tx_buf[length + 9] = 0x3c;
   UART1_SendBuf(tx_buf,length + 10);
}
/********************************************************************
 * 名称:LCD_Clear
 * 功能:清屏
 * 入口参数:无
 * 出口参数:无
********************************************************************/
void LCD_Clear(void)
{
    uint8 tx_buf[15] = {0xAA,0x52,0xCC,0x33,0xC3,0x3C};
    UART1_SendBuf(tx_buf,6);
}
/********************************************************************
 * 名称:LCD_curve
 * 功能:将 2 点之间用直线连接起来
 * 入口参数:old_x  前一点的 x 坐标;   old_y  前一点的 x 坐标;
 *          now_x  后一点的 x 坐标;    now_y  后一点的 y 坐标
 * 出口参数:无
********************************************************************/
void LCD_curve(uint16 old_x,uint16 old_y,uint16 now_x,uint16 now_y)
{
    uint8   tx_buf[14] = {0xaa,0x56,0x00,0x00,0x00,0x00,0x00,0x00,0x00,
                          0x00,0xcc0x33,0xC3,0x3C};
    tx_buf[2] = old_x/0xff;
    tx_buf[3] = old_x % 0xff;
    tx_buf[4] = old_y/0xff;                                //前一个点 y 坐标
    tx_buf[5] = old_y % 0xff;
    tx_buf[6] = now_x/0xff;
```

```
    tx_buf[7] = now_x % 0xff;
    tx_buf[8] = now_y/0xff;
    tx_buf[9] = now_y % 0xff;                          //后一个点 y 坐标
    UART1_SendBuf(tx_buf,14);
    mDelaymS(10);

}
/*************************************************************
* 名称:Disp_XY_Temperature
* 功能:在温度曲线上显示设定的温度
* 入口参数:y 坐标
* 出口参数:无
*************************************************************/
void Disp_XY_Temperature(uint8 y)
{
    uint8 DangQian[50] = "00.0℃";
    uint32 a;
    a = Order_Temperature * 100;
    DangQian[0] = a/1000 % 10 +'0';
    DangQian[1] = a/100 % 10 +'0';
    DangQian[2] = '.';
    DangQian[3] = a/10 % 10 +'0';
    Disp_words(0x54,36,y,DangQian);
}
/*************************************************************
* 名称:main()
* 功能:触摸屏示例程序主函数
*************************************************************/
int main(void)
{
    uint16 zuobiao_x = 0,y_old = 0,y_new = 0;
    uint8  buf[50] = "20℃";
    uint8  hongse[50] = {0xBa,0xec,0xC9,0xAB,0xCE,0xAA,0xC9,0xE8,0xB6,
                       0xA8,0xCE,0xC2,0xB6,0xC8,0};             // 红色为设定温度
    uint8  heise[50] = {0xBA,0xDA,0xC9,0xAB,0xCE,0xAA,0xCA,0xB5,0xBC,
                      0xCA,0xCE,0xC2,0xB6,0xC8,0};              // 黑色为实际温度

    uint8  words[50] = {0xB7,0xB5,0xBB,0xD8,0xCE,0xC2,0xB6,0xC8,0xC9,
                      0xE8,0xB6,0xA8,0};                        // 返回温度设定

    uint8  words_temperature[50] = {0x74,0x65,0x6D,0x70,0x65,0x72,0x61,0x74,0x75,
```

```
                                    0x72,0x65,0x28,0xA1,0xE6,0x29,0};// 显示 temperature(℃)
uint8   words_time[50] = {0x74,0x69,0x6D,0x65,0x28,0x73,0x29,0}; // 显示 time(s)
uint32 i,uint_a,uint_b;
uint16 tem_x = 115,dot_num = 0,dot = 0,temp = 0;
uint16 tem_integer = 0,tem_decimal = 0;
uint8   commen_flag = 0;
uint16 quxian[500] = {0};
unsigned char tep_words[5] = {0};                          // 用于存储按键
UART1_Ini(115200);
set_mode(0x30);                                            // 设置进入触控模式
mDelaymS(20);
while(1)
{
    if(rcv_flag1)
    {
        mDelaymS(5);
        rcv_flag1 = 0;
        num1 = 0;
        if((rcv_buf1[0] == 0xaa)&(rcv_buf1[1] == 0x78))
        {
            if((rcv_buf1[3]<10)&(tem_x<160))               // 按下 0～9 中的键
            {
                tep_words[0] = rcv_buf1[3] + 0x30;         // 显示数字
                Disp_words(0x54,tem_x,0x53,tep_words);
                tem_x = tem_x + 8;
                if(dot_num == 0)                           // 无小数位被按下
                tem_integer = tem_integer * 10 + rcv_buf1[3];
                else
                {
                    tem_decimal = tem_decimal * 10 + rcv_buf1[3];
                }
            }
            if((rcv_buf1[3] == 0x22)&(tem_x<160))          // 若按下的为"."
            {
                dot_num = 1;                               // 有小数点被按下时的标志位
                tep_words[0] = 0x2e;
                Disp_words(0x54,tem_x,0x53,tep_words);
                tem_x = tem_x + 8;
            }
            if((rcv_buf1[3] == 0x11)&(tem_x>115))          // 若按下清除键
            {
```

```
            tem_x = tem_x - 8;
            tep_words[0] = 0x20;
            Disp_words(0x54,tem_x,0x53,tep_words);
            if(dot_num == 1)                      // 若有小数被点按下
            {
                tem_decimal = tem_decimal/10;     // 变为个位
                if(tem_decimal == 0)              // 小数位减为 0
                {
                    dot_num = 0;
                    dot = 1;
                }
            }
            else
            {
                if(dot == 1)
                {
                    dot = 0;
                }
                else tem_integer = tem_integer/10;
            }
        }
        if(rcv_buf1[3] == 0x66)               // 若按下设定温度值,则返回开机界面
        {
            tem_x = 115;
            tem_integer = 0;
            tem_decimal = 0;
        }
        if(rcv_buf1[3] == 0x55)               // 进入查看温度变化曲线画面
        {
          LCD_Clear();
          set_mode(0x50);                     // 设置退出模式。开始进行温度曲线显示
          set_colour(0x00,0xfc08);            // 设黑色为前景色,橘色为背景色,画坐标
          LCD_Clear();
          commen_flag = 1;                    // 进入普通模式
        }
    }
  if((rcv_buf1[0] == 0xAA) & (rcv_buf1[1] == 0x73))
  {                 set_mode(0x30);           // 设置进入触控模式
                    LCD_Clear();
                    commen_flag = 0;
    }
 }
if(commen_flag == 1)
```

```
{
  if(commen_flag == 1)
{
  LCD_curve(0x0010,179,0x0120,180);        // x 坐标
  DelayNS(100);

  LCD_curve(0x0120,179,0x0110,176);
  DelayNS(5);

  LCD_curve(0x0120,180,0x0110,184);
  DelayNS(100);

  Disp_words(0x54,0x0100,160,words_time);
  DelayNS(10);

  LCD_curve(0x0020,0x0010,0x0020,0xe0); // y 坐标

  DelayNS(5);
  LCD_curve(0x0020,0x0009,0x001a,0x18);

  DelayNS(5);
  LCD_curve(0x0020,0x0009,0x0025,0x18);

  Disp_words(0x54,0x0027,0x10,words_temperature);
  DelayNS(10);

  Disp_words(0x54,0x00ca,0x00db,words); // 显示返回温度设定
  DelayNS(100);

  Disp_words(0x54,40,220,heise);
  Disp_words(0x54,0,172,buf);
  set_colour(0xf810,0xfc08);               // 设红色为前景色,橘色为背景色,画坐标
  Disp_words(0x54,40,200,hongse);
 uint_a = (Order_Temperature * 10 - 200)/3;
  uint_b = 240 - uint_a - 61;
  Disp_XY_Temperature(uint_b - 16);        // 显示设定温度值
  LCD_curve(32,uint_b,288,uint_b);         // 显示橘红色目标温度直线
  DelayNS(100);
  set_colour(0x00,0xfc08);                 // 黑色为前景色,橘色为背景色
  for(i = 0;i<= 288;i++)
  {
      uint_a = (Temperature_Buf[i * uint_b] * 10 - 200)/3;     // 抽样显示温度
      quxian[i] = 240 - uint_a - 61;
  }
  i = 0;
  for(zuobiao_x = 32;zuobiao_x<288;zuobiao_x++)
  {
    y_old = quxian[i++] + 61;
    y_new = quxian[i] + 61;
```

```
            LCD_curve(zuobiao_x,y_old,zuobiao_x+1,y_new);
          }
        commen_flag = 0;
      }
    }
  }
return (0);
}
/*************************************************************
**                        结束                             **
*************************************************************/
```

第 3 章

ADC 和 DAC 电路

3.1 ADC 电路的设计与制作

3.1.1 LPC214x 的 ADC 简介

1. LPC214x 的 ADC 特性

LPC214x 芯片内部带有 10 位逐次逼近式 ADC，在 LPC2141/2 中有一个，在 LPC2144 /6/8 中有两个；ADC0 和 ADC1 具有 6 或 8 个输入引脚(复用)；测量范围从 0～V_{REF}(通常为 3 V；不能够超过 V_{DDA} 电压电平)，10 位转换时间≥2.44 μs；可以选择一个或多个输入的突发转换模式，可选择由输入跳变或定时器匹配信号触发转换，可选择转换器的全局起始命令(仅对于 LPC2144/6/8)。

ADC 的基本时钟由 VPB 时钟提供。每个转换器包含一个可编程分频器，可将时钟调整至逐步逼近转换所需的 4.5 MHz(最大)。完全满足精度要求的转换需要 11 个这样的时钟。

2. 与 ADC 有关的引脚

与 ADC 有关的引脚和功能如表 3－1 所列。

表 3－1 与 ADC 有关的引脚和功能

引脚符号	类 型	功 能
AD0.7:6, AD0.4:1&AD1.7:0 (LPC2144/6/8)	输入	ADC 单元模拟电压输入。 **注意**：当使用 ADC 时，模拟输入引脚的信号电压在任何时候都不能大于 V_{DDA}，否则，读出的 A/D 值无效。如果在应用中未使用 A/D 转换器，则 A/D 输入引脚用作可承受 5 V 电压的数字 I/O 口。 **警告**：当 ADC 引脚指定为 5 V 电压时，ADC 模块中的模拟复用功能不可用。3.3 V(V_{DDA})＋10％以上的电压不应该加到选择作为 ADC 输入的任何引脚，否则 ADC 读操作将不正确。例如，若 AD0.0 和 AD0.1 作为 ADC0 输入，AD0.0＝4.5 V 而 AD0.1＝2.5 V，AD0.0 上过量的电压会造成 AD0.1 的错误读操作，尽管 AD0.1 输入电压在正确的范围以内

续表 3－1

引脚符号	类　型	功　能
V_{REF}	参考电压	参考电压输入，为 ADC 提供参考电压
V_{DDA}，V_{SSA}	电源	模拟电源和地。它们分别与标称为 V_{DD} 和 V_{SS} 的电压相同，但为了降低噪声和干扰，两者之间应当隔离，隔离电路如图 3－1 所示

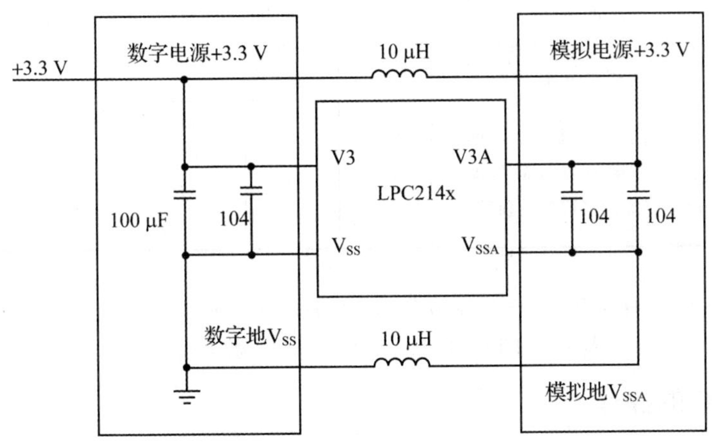

图 3－1　D/A 电路的电源和“地”的隔离

3. 与 ADC 有关的寄存器

与 ADC 有关的寄存器和功能如表 3－2 所列。

表 3－2　与 ADC 有关的寄存器和功能

符　号	功　能	访　问	复位值	AD0 地址 & 名称	AD1 地址 & 名称
ADCR	A/D 控制寄存器。A/D 转换开始前，必须写入 ADCR 寄存器来选择工作模式	R/W	0x0000 0001	0xE003 4000 AD0CR	0xE006 0000 AD1CR
ADGDR	A/D 全局数据寄存器。该寄存器包含 ADC 的 DONE 位和当前 A/D 转换的结果	R/W	NA	0xE003 4004 AD0GDR	0xE006 0004 AD1GDR
ADSTAT	A/D 状态寄存器。该寄存器包括所有 A/D 通道的 DONE 和 OVERRUN 标志，以及 A/D 中断标志	RO	0x0000 0000	0xE003 4030 AD0STAT	0xE006 0030 AD1STAT

续表 3-2

符　号	功　能	访　问	复位值	AD0 地址 & 名称	AD1 地址 & 名称
ADGSR	A/D 全局启动寄存器。可写该地址(在 A/D0 地址范围内)来同时启动 A/D 转换器的转换	WO	0x00	0xE003 4008 ADGSR	
ADINTEN	A/D 中断使能寄存器。该寄存器含有使能位,这些使能位允许每个 A/D 通道的 DONE 标志是否产生 A/D 中断	R/W	0x0000 0100	0xE003 400C AD0INTEN	0xE006 000C AD1INTEN
ADDR0~7	A/D 通道 0 ~通道 7 数据寄存器。该寄存器包括在通道 0~7 完成的当前的转换结果	RO	NA	0xE003 4010~0xE003 402C AD0DR0~AD0DR7	0xE006 0010~0xE006 002C AD1DR0~AD1DR7

有关 ADC 寄存器的更多内容请登录 http://www.zlgmcu.com 查询《PHILIPS 单片 16/32 位微控制器——LPC2141/42/44/46/48 数据手册》。

4. ADC 的操作

(1) 硬件触发转换

如果 ADCR 的 BURST 位为 0,且 START 字段的值包含在 010~111 之内,当所选引脚或定时器匹配信号发生跳变时,ADC 启动一次转换。也可选择在 4 个匹配信号中任何一个的指定边沿转换,或者在 2 个捕获/匹配引脚中任何一个的指定边沿转换。将所选端口的引脚状态或所选的匹配信号与 ADCR 位 27 相异或,所得的结果用作边沿检测逻辑。

(2) 中　断

当 DONE 位为 1 时,中断请求声明到向量中断控制器(VIC)。软件可通过 VIC 中 ADC 的中断使能位来控制是否产生中断。DONE 在 ADDR 读出时无效。

(3) 精度和数字接收器

对于 ADC 的输入引脚,只要在该引脚上选择了数字功能,其内部电路就会断开 ADC 硬件和相关引脚的连接。

当 ADC 用来测量 AIN 脚的电压时,相应的引脚必须选择 AIN 功能。通过禁能引脚的数字接收器来选择 AIN 功能可以提高转换精度。

3.1.2　LPC214x 的 ADC 编程示例

1. 单路 A/D 转换程序

以 ADC0.1 通道为例,给出单通道 A/D 的转换程序,转换结果通过串口输出,送

PC，EasyARM显示。具体程序如程序清单3.1所示。

程序清单3.1　单路A/D转换示例程序

```
/************************************************************************
* 文件名:ADC01.C
* 功能:使用ADC模块的通道0进行电压的测量,然后将转换结果从串口输出,上位机
*       使用EasyARM全仿真的DOS字符窗口观察测得的电压
* 说明:测量电压值可调;
*       通信格式  8位数据位,1位停止位,无奇偶校验,波特率为115 200
************************************************************************/
#include  "config.h"
#include  "stdio.h"
/************************************************************************
* 名称:DelayNS()
* 功能:长软件延时
* 入口参数:dly          延时参数,值越大,延时越久
* 出口参数:无
************************************************************************/
void  DelayNS(uint32  dly)
{
   uint32   i;

   for(; dly>0; dly--)
      for(i=0; i<5000; i++);
}
#define  UART_BPS    115200              /* 定义通信波特率 */
/************************************************************************
* 名称:UART0_Ini()
* 功能:初始化串口0,设置为8位数据位,1位停止位,无奇偶校验
* 入口参数:无
* 出口参数:无
************************************************************************/
void  UART0_Ini(void)
{
uint16   Fdiv;
   PINSEL0 = (PINSEL0&0xfffffff0) | 0x00000005;          // 设置P0.0、P0.1连接到UATR0
 if(UART_BPS == 115200)
   {
      U0FDR = 1 | 12<<4;                                   // LPC2148串口除数校准波特率
   }
   U0LCR = 0x83;                                           // DLAB=1,可设置波特率
   Fdiv = (Fpclk / 16) / UART_BPS;                         // 设置波特率
   U0DLM = Fdiv / 256;
   U0DLL = Fdiv % 256;
   U0LCR = 0x03;
}
/************************************************************************
```

```
 *  名称:UART0_SendByte()
 *  功能:向串口发送字节数据,并等待发送完毕
 *  入口参数:data  要发送的数据
 *  出口参数:无
****************************************************************************/
void  UART0_SendByte(uint8 data)
{
U0THR = data;                                              // 发送数据
   while( (U0LSR&0x40) == 0 );                            // 等待数据发送完毕
}
/****************************************************************************
 *  名称:PC_DispChar()
 *  功能:向PC发送显示字符
 *  入口参数:x      显示位置的纵坐标,0~79
 *           y      显示位置的横坐标,0~24
 *           chr    显示的字符,不能为0xff
 *           color  显示的状态包括前景色、背景色、闪耀位。它与DOS的字符显示状态一样,
 *                  即0~3位:前景色,4~6位:背景色,7位:闪耀位
 *  出口参数:无
****************************************************************************/
void  PC_DispChar(uint8 x, uint8 y, uint8 chr, uint8 color)
{
   UART0_SendByte(0xff);                                  // 发送起始字节
   UART0_SendByte(x);                                     // 发送字符显示坐标(x,y)
   UART0_SendByte(y);
   UART0_SendByte(chr);                                   // 发送显示字符
   UART0_SendByte(color);
}
/****************************************************************************
 *  名称:ISendStr()
 *  功能:向PC发送字串,以便显示
 *  入口参数:x     显示位置的纵坐标,0~79
 *           y     显示位置的横坐标,0~24
 *           color 显示的状态包括前景色、背景色、闪耀位。它与DOS的字符显示状态一样,
 *                 即0~3位:前景色,4~6位:背景色,7位:闪耀位
 *           str   要发送的字符串,字串以'\0'结束
 *  出口参数:无
****************************************************************************/
void  ISendStr(uint8 x, uint8 y, uint8 color, char * str)
{
   while(1)
   {  if( * str == '\0') break;                           // 若为'\0',则退出
      PC_DispChar(x++, y, * str++, color);                // 发送显示数据
      if(x>=80)
      {  x = 0;
         y++;
      }
```

```
    }
}
/****************************************************************
*  名称:main()
*  功能:进行通道0、1电压ADC转换,并把结果转换成电压值,然后发送到串口
*  说明:在CONFIG.H文件中包含stdio.h
****************************************************************/
int  main(void)
{
     uint32   ADC_Data;
     char     str[20];

     PINSEL1 = PINSEL1 & 0xFcFFFFFF | 0x01000000;        // 设置P0.28连接到AD0.1
     UART0_Ini();                                         // 初始化UART0
   /* 进行ADC模块设置,其中x<<n表示第n位设置为x(若x超过一位,则向高位顺延) */
   ADCR = (1 << 0)                       |               // SEL = 1,选择通道0
           ((Fpclk / 1000000 - 1) << 8)  |               // 转换时钟为1 MHz
           (0 << 16)                     |               // BURST = 0,软件控制转换操作
           (0 << 17)                     |               // CLKS = 0,使用11clock转换
           (1 << 21)                     |               // PDN = 1,非掉电模式
           (0 << 22)                     |               // TEST1:0 = 00,正常工作模式
           (1 << 24)                     |               // START = 1,直接启动ADC转换
           (0 << 27);                                    // EDGE = 0,DelayNS(10);

   ADC_Data = ADDR;                                      // 读取ADC结果,并清除DONE标志位

  while(1)
   {
      ADCR = (ADCR&0xFFFFFF00)|(1<<0)|(1 << 24);         // 切换通道并进行第一次转换
      while( (ADDR&0x80000000) == 0 );                   // 等待转换结束
      ADCR = ADCR | (1 << 24);                           // 再次启动转换
      while( (ADDR&0x80000000) == 0 );
      ADC_Data = ADDR;                                   // 读取ADC结果
      ADC_Data = (ADC_Data>>6) & 0x3FF;
      ADC_Data = ADC_Data * 2048;
      ADC_Data = ADC_Data / 1024;

      sprintf(str, "%4dmV at VIN1", ADC_Data);
      ISendStr(60, 23, 0x30, str);                       // 通过EasyARM显示转换结果

      DelayNS(800);
   }
   return(0);
}
```

2. 多路A/D转换程序

为说明多路A/D转换,此处采用AD0.1、AD0.7、AD1.3三路A/D转换,结果显示如单通道转换一样通过串口送PC,EasyARM显示。送上位机显示的串口程序与单路转化一样,故下面只给出ADC初始化、开启通道转换函数和主函数部分,详

细代码如程序清单 3.2 所示。

程序清单 3.2 多路 A/D 转换示例程序

```
/*****************************************************************
*  函数名:ADC_Init()
*  功能:ADC 引脚初始化,清除所用 ADC 通道的 DONE 标志位
*  说明:调用
*****************************************************************/
void ADC_Init(void)
{
    uint32     ADC_Data;
    PINSEL1 = (PINSEL1 & 0xFcFFFFFF )| 0x01000000;     // 设置 P0.28 连接到 AD0.1
    PINSEL0 = (PINSEL0 & 0xFcFFF3FF )| 0x03000c00;     // 设置 P0.5 连接到 AD0.7,设置 P0.12
                                                        // 连接到 AD1.3
/*************************清零 AD0DR7 中标志位*****************************/
   ADCR =    ( 1 ≪ 7 )                     |           // 选择 AD0.7
          ((Fpclk / 1000000 - 1) ≪ 8)      |           // 转换时钟为 1 MHz
          (0 ≪ 16)                         |           // BURST = 0,软件控制转换操作
          (0 ≪ 17)                         |           // CLKS = 0,使用 11clock 转换
          (1 ≪ 21)                         |           // PDN = 1,正常工作模式
          (0 ≪ 22)                         |           // TEST1:0 = 00,正常工作模式
          (1 ≪ 24)                         |           // START = 1,直接启动 ADC 转换
          (0 ≪ 27);                                    // CAP/MAT 引脚下降沿触发 ADC 转
   DelayNS(10);
   ADC_Data = ADDR;                                    // 读取 ADC 结果,并清除 DONE 标志位
/*************************清零 AD1GDR 中标志位*****************************/
   AD1CR = ( 1 ≪ 3 )                       |           // 选择 AD1.3
          ((Fpclk / 1000000 - 1) ≪ 8)      |           //转换时钟为 1 MHz
          (0 ≪ 16)                         |           // BURST = 0,软件控制转换操作
          (0 ≪ 17)                         |           // CLKS = 0,使用 11 clock 转换
          (1 ≪ 21)                         |           // PDN = 1,正常工作模式
          (0 ≪ 22)                         |           // TEST1:0 = 00,正常工作模式
          (1 ≪ 24)                         |           // START = 1,直接启动 ADC 转换
          (0 ≪ 27);                                    // CAP/MAT 引脚下降沿触发 ADC 转换
   DelayNS(10);
   ADC_Data   =   AD1GDR;                              // 读取 ADC 结果,并清除 DONE 标志位
/*************************清零 AD0DR1 中标志位*****************************/
   ADCR = ( 1 ≪ 1 )                        |           // SEL.1 = 1,选择 AD0.1
          ((Fpclk / 1000000 - 1) ≪ 8)      |           // 转换时钟为 1 MHz
          (0 ≪ 16)                         |           // BURST = 0,软件控制转换操作
          (0 ≪ 17)                         |           // CLKS = 0,使用 11 clock 转换
          (1 ≪ 21)                         |           // PDN = 1 正常工作模式
          (0 ≪ 22)                         |           // TEST1:0 = 00,正常工作模式
          (1 ≪ 24)                         |           // START = 1,直接启动 ADC 转换
          (0 ≪ 27);                                    // CAP/MAT 引脚下降沿触发 ADC 转换
DelayNS(10);
ADC_Data = ADDR;                                       // 读取 ADC 结果,并清除 DONE 标志位
```

```
}
/**************************************************************
* 函数名:   AD0_1_Date()
* 功  能:   通道切换,进行 AD0.1 转换
* 入口参数: 无
* 出口参数: AD0.1 通道转换结果
* 说  明:   调用
**************************************************************/
uint32 AD0_1_Date(void)
{
    uint32   AD0_1Date;

    ADCR = (ADCR&0xFFFFFF00)| (1<<1) | (1 << 24);          // 切换通道并进行第一次转换
    while( (ADDR&0x80000000) == 0 );                       // 等待转换结束
    ADCR = ADCR | (1 << 24);                               // 再次启动转换
    while( (ADDR&0x80000000) == 0 );
    AD0_1Date = ADDR;                                      // 读取 ADC 结果

    AD0_1Date = (AD0_1Date>>6) & 0x3FF;
    AD0_1Date = AD0_1Date * 2048;                          // 基准电压为 2 048 mV
    AD0_1Date = AD0_1Date / 1024;

    return AD0_1Date;
}
/**************************************************************
* 函数名:   AD0_7_Date()
* 功  能:   通道切换,进行 AD0.7 转换结果
* 入口参数: 无
* 出口参数: AD0.7 通道转换结果
* 说  明:   调用
**************************************************************/
uint32 AD0_7_Date(void)
{
    uint32   AD0_7Date;

    ADCR = (ADCR & 0xFFFFFF00)| (1<<7) | (1 << 24);        // 切换通道并进行第一次转换
    while( (ADDR & 0x80000000) == 0 );                     // 等待转换结束
    ADCR = ADCR | (1 << 24);                               // 再次启动转换
    while( (ADDR & 0x80000000) == 0 );
    AD0_7Date = ADDR;                                      // 读取 ADC 结果
    AD0_7Date = (AD0_7Date>>6) & 0x3FF;
    AD0_7Date = AD0_7Date * 2048;                          // 基准电压为 2 048 mV
    AD0_7Date = AD0_7Date / 1024;

    return   AD0_7Date;
}
/**************************************************************
* 函数名:   AD1_3_Date()
* 功  能:   通道转换,进行 AD1.3 转换
* 入口参数: 无
```

```
* 出口参数：AD1.3 通道转换结果
* 说  明：  调用
***********************************************************************/
uint32 AD1_3_Date(void)
{
    uint32  AD1_3Date;
    AD1CR = (AD1CR & 0xFFFFFF00)|(1<<3)|(1 << 24);        // 切换通道并进行第一次转换
    while( (AD1GDR & 0x80000000) == 0 );                  // 等待转换结束
    AD1CR = AD1CR | (1 << 24);                            // 再次启动转换
    while( (AD1GDR & 0x80000000) == 0 );
    AD1_3Date = AD1GDR;                                   // 读取 ADC 结果
    AD1_3Date = (AD1_3Date>>6) & 0x3FF;
    AD1_3Date = AD1_3Date * 2048;                         // 基准电压为 2 048 mV
    AD1_3Date = AD1_3Date / 1024;
    return  AD1_3Date;
}
/***********************************************************************
* 名称:main()
* 功能:进行通道 0、1 电压 ADC 转换,并把结果转换成电压值,然后发送到串口
* 说明:在 CONFIG.H 文件中包含 stdio.h
***********************************************************************/
int  main(void)
{
    uint32  ADC_Data;
    char    str[20];
    UART0_Ini();                                          // 初始化 UART0
    ADC_Init();                                           // 初始化 ADC
    while(1)
    {
        ADC_Data = AD0_1_Date();                          // 读取 AD0.1 转换值
        sprintf(str, "%4dmV at VIN1",ADC_Data);
        ISendStr(20, 23, 0x30, str);                      // 送 PC,EasyARM 显示转换结果
        DelayNS(800);
        ADC_Data = AD0_7_Date();                          // 读取 AD0.7 转换值
        sprintf(str, "%4dmV at VIN1",ADC_Data);
        ISendStr(40, 23, 0x30, str);                      // 送 PC,EasyARM 显示转换结果
        DelayNS(800);
        ADC_Data = AD1_3_Date();                          // 读取 AD1.3 转换值
        sprintf(str, "%4dmV at VIN1",ADC_Data);
        ISendStr(60, 23, 0x30, str);                      // 送 PC,EasyARM 显示转换结果
        DelayNS(800);
    }
    return(0);
}
```

3. 硬件触发转换——P0.16

A/D转换同样可以通过外部引脚触发,下面给出使用P0.16硬件触发转换的程序清单,同样只给出初始化和主函数部分,详细代码如程序清单3.3所示。

程序清单3.3 P0.16硬件触发A/D转换示例程序

```
/**********************************************************************
* 名称:ADC_init()
* 功能:初始化ADC,选择在通道AD0.1完成A/D转化,转化开启由P0.16引脚上电平变化决定
* 入口参数:无
* 出口参数:无
**********************************************************************/
void ADC_init(void)
{
       uint32  ADC_Data;
       PINSEL1 = PINSEL1 & 0xFcFFFFFF | 0x01000000;    //设置P0.28连接到AD0.1
   /* 进行ADC模块设置,其中x<<n表示第n位设置为x(若x超过一位,则向高位顺延) */
       ADCR = ( 1 << 1 )                        |      // SEL.1 = 1,选择AD0.1
           ((Fpclk / 1000000 - 1) << 8)         |      // 转换时钟为1 MHz
           (0 << 16)                            |      // BURST = 0,软件控制转换操作
           (0 << 17)                            |      // CLKS = 0,使用11 clock转换
           (1 << 21)                            |      // PDN = 1正常工作模式
           (0 << 22)                            |      // TEST1:0 = 00,正常工作模式
           (2 << 24)                            |      // START = 2,P0.16电平变化触发
           (0 << 27);                                  // CAP/MAT引脚下降沿触发ADC转换
  DelayNS(10);
  ADC_Data = ADDR;                                     // 读取ADC结果,并清除DONE标志位
}
/**********************************************************************
* 名称:main()
* 功能:进行A/D转化,转换结果送上位机显示
**********************************************************************/
int  main(void)
{
     uint32  ADC_Data;
     char    str[20];
     ADC_init();
     UART0_Ini();
     while(1)
     {
        // 切换通道并等待P0.16上有下降沿出现时进行第一次转换
        ADCR = (ADCR&0xFFFFFF00)| (1 << 1) | (2 << 24);
        while( (ADDR&0x80000000) == 0 );               // 等待转换结束
        ADC_Data = ADDR;                               // 读取ADC结果
        ADC_Data = (ADC_Data>>6) & 0x3FF;
        ADC_Data = ADC_Data * 2048;
```

```
        ADC_Data = ADC_Data / 1024;                         // 电压计算,单位 mV
        sprintf(str, " % 4dmV at VIN1",ADC_Data);
        ISendStr(20, 23, 0x30, str);                        // 送 PC,EasyARM 显示转换结果
        DelayNS(800);
    }
    return    (0);
}
```

4. 硬件触发转换——MAT1.0

A/D 转换的硬件触发模式除了外部引脚触发外,还能够通过定时器翻转功能触发,这样可以实现无需 CPU 干预就能精确定时触发 ADC 转化,并且无需占用定时器中断。下面给出使用定时器 1 匹配通道 0(MAT1.0)产生下降沿来启动 ADC 的程序清单,同样只给出初始化和主函数部分。定时器 1 每隔 1 s 触发 MAT1.0 翻转,故每隔 2 s 产生一个下降沿,即每隔 2 s 开启一次 A/D 转换,详细代码如程序清单 3.4所示。

程序清单 3.4　MAT1.0 硬件触发 A/D 转换示例程序

```
/**************************************************************************
 * 名称:ADC_init()
 * 功能:初始化 ADC 在通道 AD0.1 完成 A/D 转化,转化开启 MAT1.0 引脚产生下降沿开启
 * 入口参数:无
 * 出口参数:无
**************************************************************************/
void ADC_init(void)
{
   uint32   ADC_Data;
   PINSEL1 = PINSEL1 & 0xFcFFFFFF | 0x01000000;      // 设置 P0.28 连接到 AD0.1
   /* 进行 ADC 模块设置,其中 x≪n 表示第 n 位设置为 x(若 x 超过一位,则向高位顺延) */
   ADCR = ( 1 ≪ 1 )                    |              // SEL.1 = 1,选择 AD0.1
          ((Fpclk / 1000000 - 1) ≪ 8)  |              // CLKDIV = Fpclk / 1 000 000 - 1
           (0 ≪ 16)                    |              // BURST = 0,软件控制转换操作
           (0 ≪ 17)                    |              // CLKS = 0,使用 11 clock 转换
           (1 ≪ 21)                    |              // PDN = 1,正常工作模式
           (0 ≪ 22)                    |              // TEST1:0 = 00,正常工作模式
           (6 ≪ 24)                    |              // 当下降沿出现在 MAT1.0 时启动转换
           (0 ≪ 27);                                  // CAP/MAT 引脚下降沿触发 ADC 转换
   DelayNS(10);
  ADC_Data = ADDR;
}
/**************************************************************************
 * 名称:Time1Init()
 * 功能:初始化定时器 1,设定每隔 1 s 使 MAT1.0 翻转
 * 入口参数:无
 * 出口参数:无
```

```
*****************************************************************/
void Time1Init(void)

{
    T1PR   = 99;                       // 设定定时器分频为 100 分频,得 11.059 2 MHz/100
    T1MCR  = 0x02;                     // 设置匹配通道 0 匹配时复位 T1TC
    T1MR0  = 120000;                   // 设置预分频寄存器值,定时 1 s
    T1EMR  = 3<<4;                     // T1MR0 匹配后 MAT1.0 输出翻转
    T1TCR  = 0x03;                     // 设置定时器计数器和预分频计数器启动并复位
    T1TCR  = 0x01;                     // 开启定时器工作
}
/*****************************************************************
*  名称:main()
*  功能:进行 A/D 转化,转换结果送上位机显示
*****************************************************************/
int main(void)
{
     uint32   ADC_Data;
     char     str[20];
     ADC_init();
     Time1Init();
     UART0_Ini();

     while(1)
     {
        // 切换通道并等待 MAT1.0 有下降沿出现时进行第一次转换
        ADCR = (ADCR&0xFFFFFF00)| (1 << 1) | (6 << 24);
        while( (ADDR&0x80000000) == 0 );  // 等待转换结束
        ADC_Data = ADDR;

        // 读取 ADC 结果
        ADC_Data = (ADC_Data>>6) & 0x3FF;
        ADC_Data = ADC_Data * 2048;
        ADC_Data = ADC_Data / 1024;
        // 送上位机显示
        sprintf(str, "%4dmV at VIN1",ADC_Data);
        ISendStr(20, 23, 0x30, str);       // 送 PC,EasyARM 显示转换结果
     }
     return    (0);
}
```

3.2 DAC 电路设计与制作

3.2.1 LPC214x 的 DAC 简介

LPC2142/4/6/8 器件带有 DAC,它是一个 10 位的数/模转换器,采用电阻串联结构,具有输出缓冲,掉电模式,可选择速率与功率。

1. 与 DAC 有关的引脚

与 DAC 有关的引脚和功能如表 3-3 所列。

表 3-3 与 DAC 有关的引脚和功能

引脚符号	类 型	功 能
AOUT	输出	模拟输出。当 DACR 写入一个新值选择好设定时间后,该引脚上的电压(相对 V_{SSA})为 $V_{ALUE}/1\,024 \times V_{REF}$
V_{REF}	参考电压	参考电压。该引脚为 DAC 提供一个参考电压
V_{DDA},V_{SSA}	电源	模拟电源和地。它们分别与标称为 V_{DD} 和 V_{SSD} 的电压相同,但为了降低噪声和干扰,两者应当隔离,隔离电路如图 3-1 所示

PINSEL1 寄存器(引脚功能选择寄存器 1——PINSEL1,0xE002C004)的位 19:18控制着 DAC 的使能和引脚 P0.25/AD0.4/AOUT 的状态。当这两位的值为 10 时,DAC 接通且有效。

当 AOUT 引脚电容负载不超过 100 pF 时,BIAS 位描述中指定的设定时间才有效。大于该值(100 pF)的负载阻抗值将使设定时间比规定的时间更长。

2. DAC 寄存器

DAC 寄存器 DACR 位描述如表 3-4 所列,该寄存器是一个读/写寄存器,它包含用于模拟转换的数字值和一个用来调节转换性能和功率两者之间关系的位。

表 3-4 DAC 寄存器 DACR 位描述(地址 0xE006 C000)

位	符 号	值	描 述	复位值
5:0 31:17	—		保留,用户软件不要向其写入 1。从保留位读出的值未被定义	NA
15:6	VALUE	0~1 023	当该字段写入一个新值选择好设定时间后,AOUT 引脚上的电压(相对 V_{SSA})为 $V_{ALUE}/1\,024 \times V_{REF}$	0
16	BIAS	0 1	0:DAC 的设定时间最大为 1 μs,最大电流为 700 μA; 1:DAC 的设定时间为 2.5 μs,最大电流为 350 μA	0

3.2.2 LPC214x 的 DAC 编程示例

下面给出通过 D/A 转换输出正弦波的程序示例,如程序清单 3.5 所示,设置基准电压为2.048 V,P0.25 为 DAC 的输出。

程序清单 3.5 D/A 转换示例程序

```
#include "config.h"
/****************************************************************************
```

```
*  名称:main()
*  功能:控制 LPC2148 的 10 位 D/A 输出正弦波
*************************************************************************/
int main(void)
{
    uint32 i = 0,j;
    uint16   sines[1000] = {                                    // 定义正弦波数组
    512, 524, 537, 549, 562, 574, 587, 599, 611, 624,
    636, 648, 660, 672, 684, 696, 707, 719, 730, 742,
    753, 764, 775, 785, 796, 806, 816, 826, 836, 846,
    855, 865, 874, 882, 891, 899, 907, 915, 923, 930,
    937, 944, 951, 957, 963, 969, 974, 980, 985, 989,
    994, 998, 1001, 1005, 1008, 1011, 1014, 1016, 1018,1020,
    1021, 1022, 1023, 1023, 1023, 1023, 1023, 1022, 1021,1020,
    1018, 1016, 1014, 1011, 1008, 1005, 1001, 998, 994, 989,
    985, 980, 974, 969, 963, 957, 951, 944, 937, 930,
    923, 915, 907, 899, 891, 882, 874, 865, 855, 846,
    836, 826, 816, 806, 796, 785, 775, 764, 753, 742,
    730, 719, 707, 696, 684, 672, 660, 648, 636, 624,
    611, 599, 587, 574, 562, 549, 537, 524, 511, 499,
    486, 474, 461, 449, 436, 424, 412, 399, 387, 375,
    363, 351, 339, 327, 316, 304, 293, 281, 270, 259,
    248, 238, 227, 217, 207, 197, 187, 177, 168, 158,
    149, 141, 132, 124, 116, 108, 100, 93, 86, 79,
    72, 66, 60, 54, 49, 43, 38, 34, 29, 25,
    22, 18, 15, 12, 9, 7, 5, 3, 2, 1,
    0, 0, 0, 0, 0, 1, 2, 3, 5, 7,
    9, 12, 15, 18, 22, 25, 29, 34, 38, 43,
    49, 54, 60, 66, 72, 79, 86, 93, 100, 108,
    116, 124, 132, 141, 149, 158, 168, 177, 187, 197,
    207, 217, 227, 238, 248, 259, 270, 281, 293, 304,
    316, 327, 339, 351, 363, 375, 387, 399, 412, 424,
    436, 449, 461, 474, 486, 499,};
   PINSEL1 = PINSEL1 & 0xfff3ffff | 0x00080000;                  // 设置 P0.25 为 DAC 的输出引脚
   while(1)
   {
       for(i = 0;i<256;i++)
       {
           V1V = sines[i];
           DACR = V1V << 6;
           for(j = 0;j<0x7FFFF;j++)  ;                           // 等待 D/A 转换完
       }
   }

}
```

第4章

电机控制

4.1 LPC214x的定时器/计数器和脉宽调制器

4.1.1 定时器/计数器(定时器0和定时器1)

1. LPC214x定时器/计数器的基本特性

LPC214x定时器/计数器的基本特性如下。

① 带可编程32位预分频器的32位定时器/计数器。

② 计数器或定时器操作。

③ 每个定时器具有多达4路32位的捕获通道。当输入信号跳变时可取得定时器的瞬时值,也可选择使捕获事件产生中断。

④ 4个32位匹配寄存器:

- 匹配时定时器继续工作,可选择产生中断;
- 匹配时停止定时器,可选择产生中断;
- 匹配时复位定时器,可选择产生中断。

⑤ 多达4个对应于匹配寄存器的外部输出,特性如下:

- 匹配时设置为低电平;
- 匹配时设置为高电平;
- 匹配时翻转;
- 匹配时无动作。

LPC214x的定时器/计数器对外设时钟(PCLK)或外部提供的时钟周期进行计数,可选择产生中断或根据4个匹配寄存器的设定,在到达指定的定时值时执行其他动作。它还包括4个捕获输入,用于在输入信号发生跳变时捕获定时器值,并可选择产生中断。

LPC214x可用于对内部事件进行计数的间隔定时器,通过捕获输入实现脉宽调制,或者作为自由运行的定时器使用。定时器/计数器0和定时器/计数器1除了外设基地址以外,其他都相同。

2. 与定时器/计数器有关的引脚

与定时器/计数器有关的引脚如表 4-1 所列。

表 4-1 与定时器/计数器有关的引脚

引脚符号	类 别	功 能
CAP0.3.0 CAP1.3.0	输入	捕获信号。捕获引脚的跳变可配置为将定时器值装入一个捕获寄存器,并可选择产生一个中断。可选择多个引脚用作捕获功能。当有多个引脚被选择用作一个TIMER0/1 通道的捕获输入时,使用编号最小的引脚。例如,30 引脚(P0.6)和 46 引脚(P0.16)选择用作 CAP0.2 时,TIMER0 只选择 30 引脚执行 CAP0.2 功能。 所有捕获信号及其可选择的引脚如下所示: CAP0.0(3 引脚):P0.2,P0.22 和 P0.30; CAP0.1(2 引脚):P0.4 和 P0.27; CAP0.2(3 引脚):P0.6,P0.16 和 P0.28; CAP0.3(1 引脚):P0.29; CAP1.0(1 引脚):P0.10; CAP1.1(1 引脚):P0.11; CAP1.2(2 引脚):P0.17 和 P0.19; CAP1.3(2 引脚):P0.18 和 P0.21。 计数器/定时器可选择一个捕获信号作为时钟源来代替 PCLK 分频时钟。详见计数控制寄存器(CTCR,TIMER0:T0CTCR,0xE000 4070 和 TIMER1:T1TCR,0xE000 8070)
MAT0.3.0 MAT1.3.0	输出	外部匹配输出 0/1。当匹配寄存器 0/1(MR3:0)等于定时器/计数器(TC)时,该输出可翻转,变为低电平、变为高电平或不变。外部匹配寄存器(EMR)控制该输出的功能。可选择多个引脚并行用作匹配输出功能。例如,同时选择 2 个引脚并行提供 MAT1.3 功能。 所有匹配信号及其可选择的引脚如下所示: MAT0.0(2 引脚):P0.3 和 P0.22; MAT0.1(2 引脚):P0.5 和 P0.27; MAT0.2(2 引脚):P0.16 和 P0.28; MAT0.3(1 引脚):P0.29; MAT1.0(1 引脚):P0.12; MAT1.1(1 引脚):P0.13; MAT1.2(2 引脚):P0.17 和 P0.19; MAT1.3(2 引脚):P0.18 和 P0.20

3. 与定时器/计数器有关的寄存器

与定时器/计数器有关的寄存器如表 4-2 所列。

表 4-2 与定时器/计数器有关的寄存器

符 号	功 能	访 问	复位值	定时器/计数器 0 地址 & 名称	定时器/计数器 1 地址 & 名称
IR	中断寄存器。可以写 IR 来清除中断。可读取 IR 来识别哪个中断源被挂起	R/W	0	0xE000 4000 T0IR	0xE000 8000 T1IR
TCR	定时器控制寄存器。TCR 用于控制定时器/计数器功能。定时器/计数器可通过 TCR 禁止或复位	R/W	0	0xE000 4004 T0TCR	0xE000 8004 T1TCR
TC	定时器/计数器。32 位 TC 每经过 PR+1 个 PCLK 周期就加 1。TC 通过 TCR 进行控制	R/W	0	0xE000 4008 T0TC	0xE000 8008 T1TC
PR	预分频寄存器。32 位 TC 每经过 PR+1 个 PCLK 周期就加 1	R/W	0	0xE000 400C T0PR	0xE000 800C T1PR
PC	预分频计数器。32 位 PC 是加 1 到 PR 内的存储值的计数器。当到达 PR 中保存的值时,TC 加 1 且 PC 被清除。通过总线接口可观察和控制 PC	R/W	0	0xE000 4010 T0PC	0xE000 8010 T1PC
MCR	匹配控制寄存器。MCR 用于控制在匹配时是否产生中断或复位 TC	R/W	0	0xE000 4014 T0MCR	0xE000 8014 T1MCR
MR0～MR3	匹配寄存器 0～3。MR0～MR3 可通过 MCR 设定为在匹配时复位 TC,停止 TC 和 PC 和/或产生中断	R/W	0	0xE000 4018～0xE000 4024 T0MR0～T0MR3	0xE000 8018～0xE000 8024 T1MR0～T1MR3
CCR	捕获控制寄存器。CCR 控制用于装载捕获寄存器的捕获输入边沿以及在发生捕获时是否产生中断	R/W	0	0xE000 4028 T0CCR	0xE000 8028 T1CCR
CR0～CR3	捕获寄存器 0～3。当在 CAPn.0 上产生捕获事件时,CR0～CR3 装载 TC 的值	RO	0	0xE000 402C～0xE000 4038 T0CR0～ T0CR3	0xE000 802C～0xE000 8038 T1CR0～T1CR3

续表 4-2

符　号	功　能	访　问	复位值	定时器/计数器 0 地址 & 名称	定时器/计数器 1 地址 & 名称
EMR	外部匹配寄存器。EMR 控制外部匹配引脚 MATn.0～3（分别为 MAT0.0～3 和 MAT1.0～3）	R/W	0	0xE000 403C T0EMR	0xE000 803C T1EMR
CTCR	计数控制寄存器。CTCR 用来选择定时器或计数器模式，在计数器模式下选择计数的信号和边沿	R/W	0	0xE000 4070 T0CTCR	0xE000 8070 T1CTCR

有关定时器/计数器寄存器的更多内容请登录 http://www.zlgmcu.com 查询《PHILIPS 单片 16/32 位微控制器——LPC2141/42/44/46/48 数据手册》。

4.1.2　脉宽调制器

LPC2141/2/4/6/8 的脉宽调制器（PWM）建立在 4.1.1 小节的标准定时器/计数器 0/1 之上，应用可在 PWM 和匹配功能当中进行选择。

1. 脉宽调制器的基本特性

LPC2141/2/4/6/8 脉宽调制器的基本特性如下。

- 7 个匹配寄存器，可实现 6 个单边沿控制或 3 个双边沿控制的 PWM 输出，或这两种类型的混合输出。匹配寄存器也允许：
  - ◆ 连续操作，可选择在匹配时产生中断；
  - ◆ 匹配时停止定时器，可选择产生中断；
  - ◆ 匹配时复位定时器，可选择产生中断。
- 每个匹配寄存器对应一个外部输出，并具有下列特性：
  - ◆ 匹配时设置为低电平；
  - ◆ 匹配时设置为高电平；
  - ◆ 匹配时翻转；
  - ◆ 匹配时无动作。
- 支持单边沿控制和/或双边沿控制的 PWM 输出。单边沿控制 PWM 输出，在每个周期开始时总是为高电平，除非输出保持恒定低电平；双边沿控制 PWM 输出，可在一个周期内的任何位置产生边沿。这样可同时产生正、负脉冲。

- 脉冲周期和宽度可以是任何的定时器计数值。这样可实现灵活的分辨率和重复速率的设定。所有PWM输出都以相同的重复率发生。
- 双边沿控制的PWM输出可编程为正脉冲或负脉冲。
- 匹配寄存器更新与脉冲输出同步,防止产生错误的脉冲。软件必须在新的匹配值生效之前将它们释放。
- 如果不使能PWM模式,可作为一个标准定时器。
- 带可编程32位预分频器的32位定时器/计数器。
- 当输入信号跳变时4个32位捕获通道可取得定时器的瞬时值,捕获事件也可选择产生中断。

2. 功能描述

PWM基于标准的定时器模块并具有其所有特性。不过LPC2141/2/4/6/8只将其PWM功能输出到引脚。定时器对外设时钟(PCLK)进行计数,可选择产生中断或基于7个匹配寄存器,在到达指定的定时值时执行其他动作。它还包括4个捕获输入,用于在输入信号发生跳变时捕获定时器值,并可选择在事件发生时产生中断。PWM功能是一个附加特性,建立在匹配寄存器事件基础之上。

独立控制上升和下降沿位置的能力使PWM可以应用于更多的领域。例如,多相位电机控制通常需要3个非重叠的PWM输出,而这3个输出的脉宽和位置需要独立进行控制。

两个匹配寄存器可用于提供单边沿控制的PWM输出。一个匹配寄存器(PWMMR0)通过匹配时重新设置计数值来控制PWM周期率。另一个匹配寄存器控制PWM边沿的位置。每个额外的单边沿控制PWM输出只需要一个匹配寄存器,因为所有PWM输出的重复速率是相同的。多个单边沿控制PWM输出在每个PWM周期的开始,当PWMMR0发生匹配时,都有一个上升沿。

3个匹配寄存器可用于提供一个双边沿控制PWM输出。也就是说,PWMMR0匹配寄存器控制PWM周期速率,其他匹配寄存器控制两个PWM边沿位置。每个额外的双边沿控制PWM输出只需要两个匹配寄存器,因为所有PWM输出的重复速率是相同的。

使用双边沿控制PWM输出时,指定的匹配寄存器控制输出的上升沿和下降沿。这样就产生了正PWM脉冲(当上升沿先于下降沿时)和负PWM脉冲(当下降沿先于上升沿时)。

PWM值与波形输出之间的关系示例如图4-1所示。下面所示的波形是单个PWM周期,它演示了在下列条件下的PWM输出。

- 定时器配置为PWM模式。
- 匹配寄存器0配置为在发生匹配事件时复位定时器/计数器。
- 控制位PWMSEL2和PWMSEL4置位。

● 匹配寄存器的值如下：
MR0＝100（PWM 速率）；
MR1＝41，MR2＝78(PWM2 输出）；
MR3＝53，MR4＝27(PWM4 输出）；
MR5＝65(PWM5 输出）。

PWM 输出逻辑允许通过 PWMSELn 位选择单边沿或者双边沿控制的 PWM 输出。

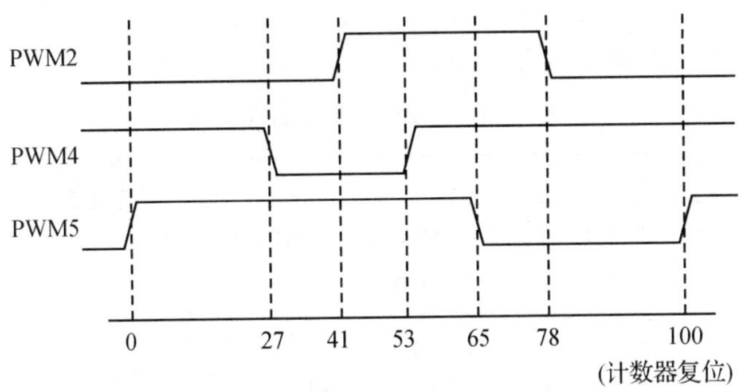

图 4－1　PWM 值与波形输出之间的关系示例

(1) 单边沿控制的 PWM 输出规则

单边沿控制的 PWM 输出规则如下：

● 所有单边沿控制的 PWM 输出在 PWM 周期开始时都为高电平，除非它们的匹配值等于 0。
● 每个 PWM 输出在到达其匹配值时都会变为低电平。如果没有发生匹配(即匹配值大于 PWM 速率)，则 PWM 输出将一直保持高电平。

(2) 双边沿控制的 PWM 输出规则

当一个新的周期将要开始时，使用以下规则来决定下一个 PMW 输出的值：

● 在一个 PWM 周期结束时(与下一个 PWM 周期的开始重合的时间点)，使用下一个 PWM 周期的匹配值。
● 等于 0 或与当前 PWM 速率(与匹配通道 0 的值相同)的匹配值等效。例如，在 PWM 周期开始时的下降沿请求与 PWM 周期结束时的下降沿请求等效。
● 当匹配值正在改变时，如果其中有一个"旧"匹配值等于 PWM 速率，并且新的匹配值不等于 0 或 PWM 速率，旧的匹配值不等于 0，那么旧的匹配值将再次被使用。
● 如果同时请求 PWM 输出置位和清零，清零优先。当置位和清零匹配值相同时，或者置位或清零值等于 0 并且其他值等于 PWM 速率时，即同时请求 PWM 输出置位和清零时，清零优先。

● 如果匹配值超出范围(大于 PMW 速率值),将不会发生匹配事件,匹配通道对输出不起作用。也就是说 PWM 输出将一直保持一种状态,可以为低电平、高电平或是“无变化”输出。

3. 与 PWM 有关的引脚

与 PWM 有关的引脚有 PWM1~PWM6,作为 PWM 通道 1~6 的输出。

4. 与 PWM 有关的寄存器

如前所述,LPC214x 的 PWM 基于标准的定时器模块并具有其所有特性,不过 PWM 功能增加了一些新的寄存器和寄存器位,如表 4-3 所列。

表 4-3 与 PWM 有关的寄存器

符 号	功 能	访 问	复位值	地 址
PWMIR	PWM 中断寄存器。可以写 PWMIR 来清除中断,也可读取 PWMIR 来识别哪个中断源被挂起	R/W	0	0xE001 4000
PWMTCR	PWM 定时器控制寄存器。TCR 用于控制定时器/计数器功能。定时器/计数器可通过 TCR 禁止或复位	R/W	0	0xE001 4004
PWMTC	PWM 定时器/计数器。32 位 TC 每经过 PR+1 个 PCLK 周期就加 1。TC 通过 TCR 进行控制	R/W	0	0xE001 4008
PWMPR	PWM 预分频寄存器。TC 每经过 PR+1 个 PCLK 周期就加 1	R/W	0	0xE001 400C
PWMPC	PWM 预分频计数器。每当 32 位 PC 的值增加到等于 PR 中保存的值时,TC 加 1。可通过总线接口观察和控制 PC	R/W	0	0xE001 4010
PWMMCR	PWM 匹配控制寄存器。MCR 用于控制在匹配时是否产生中断或复位 TC	R/W	0	0xE001 4014
PWMMR0	PWM 匹配寄存器 0。MR0 可通过 MCR 设定为在匹配时复位 TC,停止 TC 和 PC 和/或产生中断。此外,MR0 和 TC 的匹配将置位所有单边沿模式的 PWM 输出,并置位双边沿模式下的 PWM1 输出	R/W	0	0xE001 4018
PWMMR1~PWMMR5	PWM 匹配寄存器 1~5。MR1~MR5 可通过 MCR 设定为在匹配时复位 TC,停止 TC 和 PC 和/或产生中断。此外,MR1 和 TC 的匹配将清零单边沿模式或双边沿模式下的 PWM1~5,并置位双边沿模式下的 PWM2~6 输出	R/W	0	0xE001 401C~0xE001 4044

续表 4-3

符　号	功　能	访　问	复位值	地　址
PWMMR6	PWM 匹配寄存器 6。MR6 可通过 MCR 设定为在匹配时复位 TC,停止 TC 和 PC 和/或产生中断。此外,MR6 和 TC 的匹配将清零单边沿模式或双边沿模式下的 PWM6	R/W	0	0xE001 4048
PWMPCR	PWM 控制寄存器。使能 PWM 输出并选择 PWM 通道类型为单边沿或双边沿控制	R/W	0	0xE001 404C
PWMLER	PWM 锁存使能寄存器。使能使用新的 PWM 匹配值	R/W	0	0xE001 4050

有关定时器/计数器寄存器和 PWM 寄存器的更多内容请登录 http://www.zlgmcu.com 查询《PHILIPS 单片 16/32 位微控制器——LPC2141/42/44/46/48 数据手册》。

4.2　直流电机控制

4.2.1　直流电机电枢的调速原理与调速方式

1. 直流电机电枢的调速原理

根据电机学可知,直流电机转速 n 的表达式为

$$n=(U-IR)/(K\Phi) \tag{4.2.1}$$

式中,U 为电枢端电压;I 为电枢电流;R 为电枢电路总电阻;Φ 为每极磁通量;K 为电动机结构参数。

由式(4.2.1)可知,直流电动机的转速控制方法可分为两大类:对励磁磁通进行控制的励磁控制法,对电枢电压进行控制的电枢控制法。其中励磁控制法在低速时受磁极饱和的限制,在高速时受换向火花和换向器结构强度的限制,并且励磁线圈电感较大,动态响应较差,所以这种控制方法较少使用。现在,大多数应用场合都使用电枢电压控制法。目前多采用在保证励磁恒定不变的情况下,利用 PWM 来实现直流电动机的调速方法。

2. 直流电机电枢的调速方式

直流电机的控制方法比较简单,只需给电机的两根控制线加上适当的电压即可使电机转动起来,电压越高则电机转速越高。对于直流电机的速度调节,可以采用改变电压的方法,也可采用 PWM 调速方法。

在对直流电动机电枢电压的控制和驱动中,半导体功率器件在使用上可以分为

两种方式:线性放大驱动方式(改变电压)和开关驱动方式(PWM 调速)。线性放大驱动方式,输出波动小,线性好,对邻近电路干扰小,但存在效率低和散热等问题。开关驱动方式是使半导体功率器件工作在开关状态,通过脉宽调制(PWM)来控制电动机的电枢电压,从而实现电动机转速的控制。PWM 调速就是使加在直流电机两端的电压为方波形式,通过改变方波的占空比实现对电机转速的调节。

直流电动机 PWM 调速控制原理图和输入、输出电压波形如图 4-2 所示。在图 4-2(a)中,当开关管的驱动信号为高电平时,开关管导通,直流电动机电枢绕组两端有电压 U_S。t_1 秒后,驱动信号变为低电平,开关管截止,电动机电枢两端电压为 0。t_2 秒后,驱动信号重新变为高电平,开关管的动作重复前面的过程。对应输入电平的高低,直流电动机电枢绕组两端的电压波形如图 4-2(b)所示。

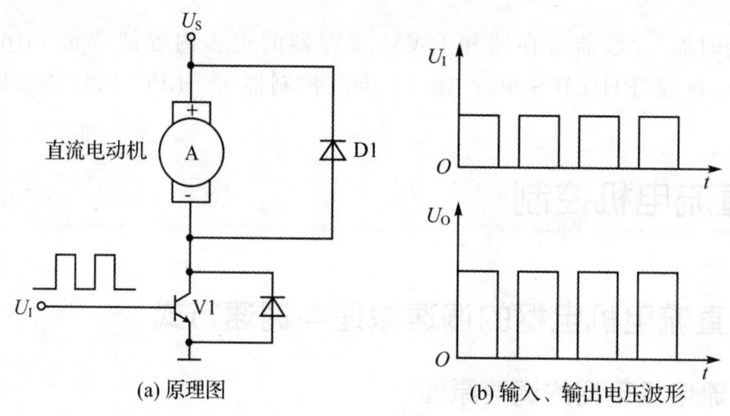

图 4-2 PWM 调速控制原理和电压波形图

电动机电枢绕组两端的电压平均值 U_O 为

$$U_O=(t_1 \times U_S+0)/(t_1+t_2)=(t_1 \times U_S)/T=DU_S \qquad (4.2.2)$$

式中,D 为占空比,$D=t_1/T$。

占空比 D 表示了在一个周期 T 里开关管导通的时间与周期的比值。D 的变化范围为 $0 \leqslant D \leqslant 1$。由式(4.2.2)可知,当电源电压 U_S 不变的情况下,电枢两端电压的平均值 U_O 取决于占空比 D 的大小,改变 D 值也就改变了电枢两端电压的平均值,从而达到控制电动机转速的目的,即实现 PWM 调速。

在 PWM 调速时,占空比 D 是一个重要的参数。改变占空比的方法有定宽调频法、调宽调频法和定频调宽法等。定频调宽法,同时改变 t_1 和 t_2,但周期 T(或频率)保持不变。

4.2.2 直流电机驱动电路设计

直流电机驱动电路的选择

直流电机驱动可以采用晶体管或者场效应管构成的 H 桥式驱动电路、专用直流电机驱动集成电路芯片(如 LM18200T 等),以及 L298 N 双全桥驱动器电路。

L298N 是 SGS 公司生产的双全桥 2 A 电机驱动芯片,采用 Multiwatt Vert (L298N)、Multiwatt Horiz(L298HN)、PowerSO209(L298P)封装,可以方便地用来驱动 2 个直流电机,或 1 个两相步进电机。

采用 L298N 构成的直流电机驱动电路如图 4-3 所示,电路中 D1～D4 采用快速恢复二极管(t_{rr}≤200 ns),R_S 为过流检测电阻(0.5 Ω),控制信号关系如表 4-4 所列。如果需要更大的电机驱动电流,L298N 可以采用并联输出形式,采用 L298N 构成的并联直流电机驱动电路如图 4-4 所示,图中通道 1 与通道 4 并联,通道 2 与通道 3 并联。

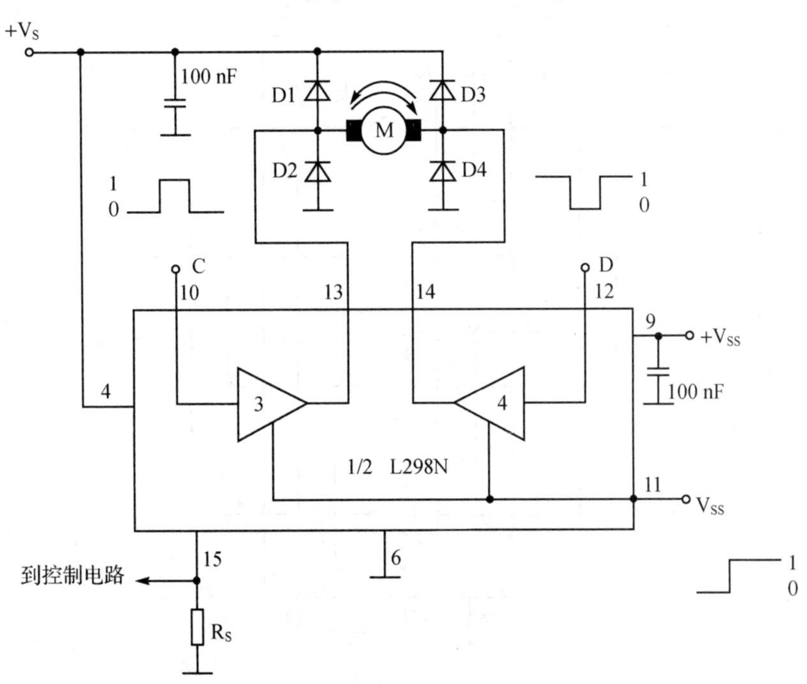

图 4-3　采用 L298N 构成的直流电机驱动电路

表 4-4　L298N 控制信号关系

输入控制信号		功　能
Ven=高电平	C=高电平;D=低电平	右转
	C=高电平;D=高电平	左转
	C=D	电机快速停止
Ven=低电平	C=x;D=C	电机自由停止

一个采用L298N双全桥驱动器构成的双直流电机驱动电路原理图和印制板图如图4-5所示。驱动电路输出端口通过J1的A、B、C、D分别接LPC2148的4个输出端,作为电机的驱动控制信号,通过改变A、B的输出电平和C、D的输出电平,可分别控制左右两个电机;微控制器提供的控制信号通过光耦TLP521-6送入L298N,用来隔离微控制器电路与电机驱动电路。当改变A、B、C、D的PWM的占空比时,可控制电机的转速。L298N控制电机正反转状态如表4-5所列。8个快速恢复二极管采用1N5822,用来泄放电机绕组电流,保护L298N芯片。

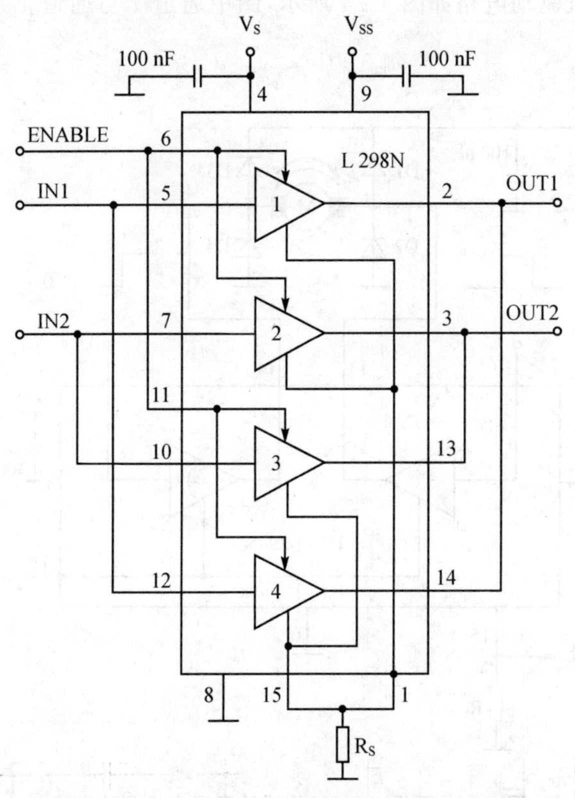

图4-4 采用L298N构成的并联直流电机驱动电路

表4-5 L298N控制电机正反转状态表

端口名称	A	B	左电机状态	C	D	右电机状态
电平状态	高电平	低电平	正转	高电平	低电平	正转
	低电平	高电平	反转	低电平	高电平	反转
	低电平	低电平	停止	低电平	低电平	停止

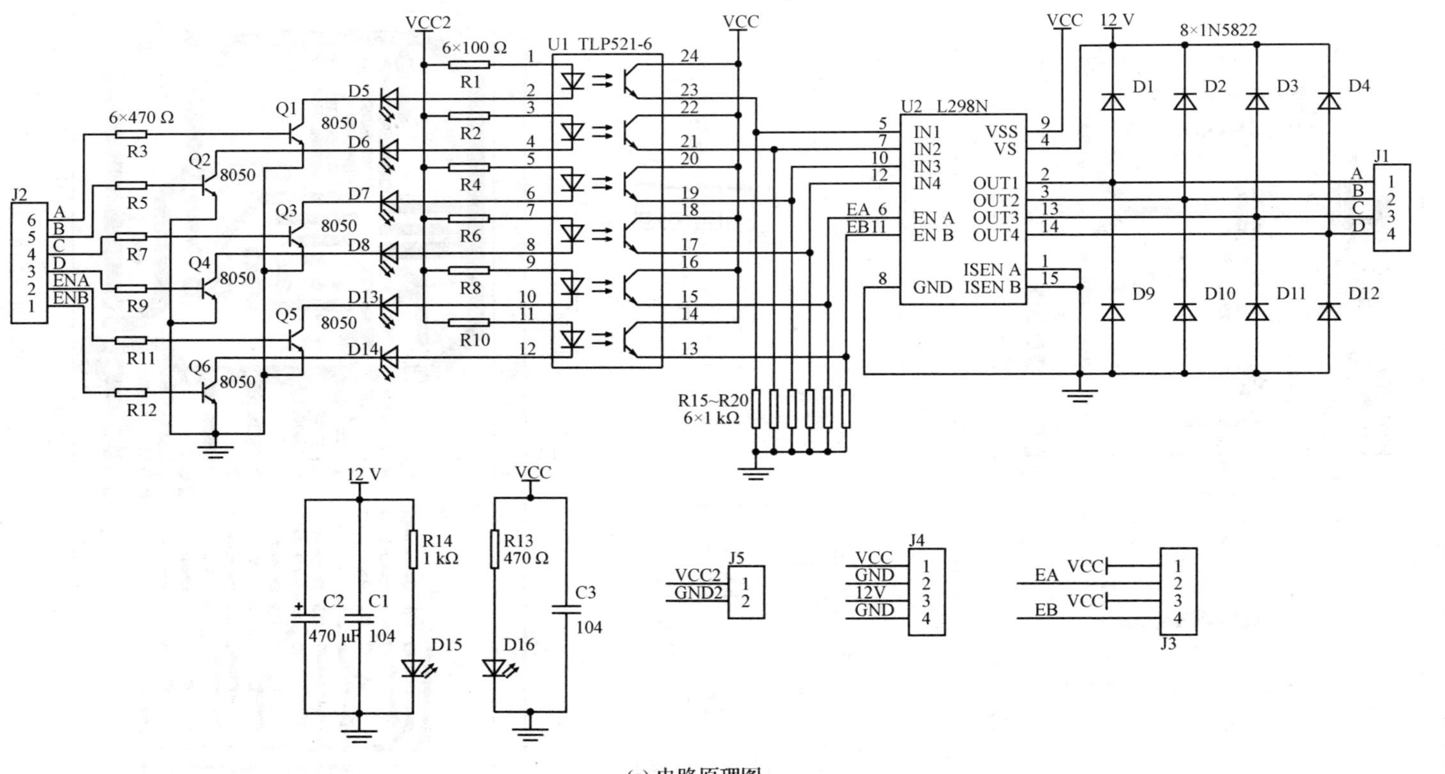

(a) 电路原理图

图 4-5 采用L298N构成的双直流电机驱动电路

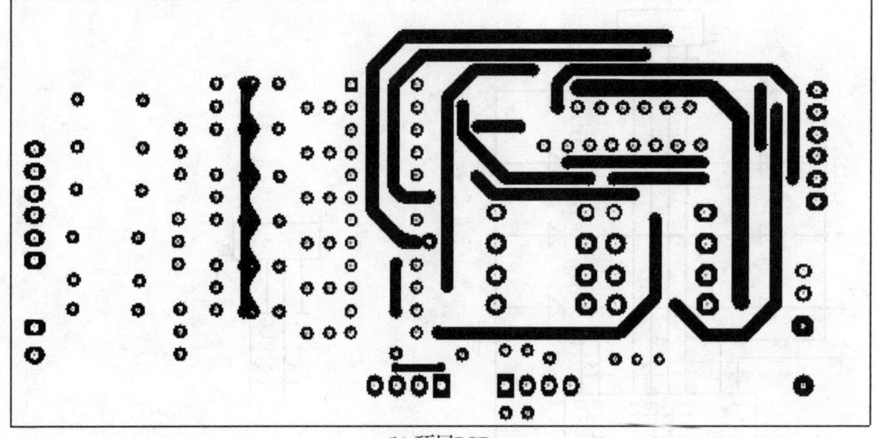

(b) 顶层PCB

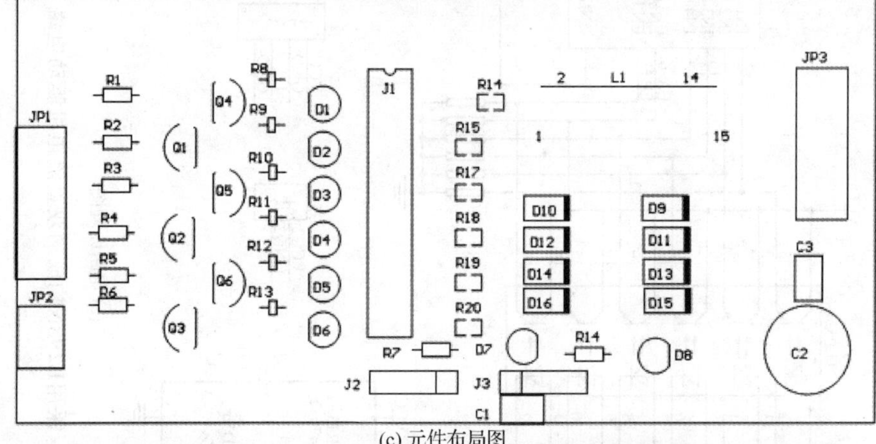

(c) 元件布局图

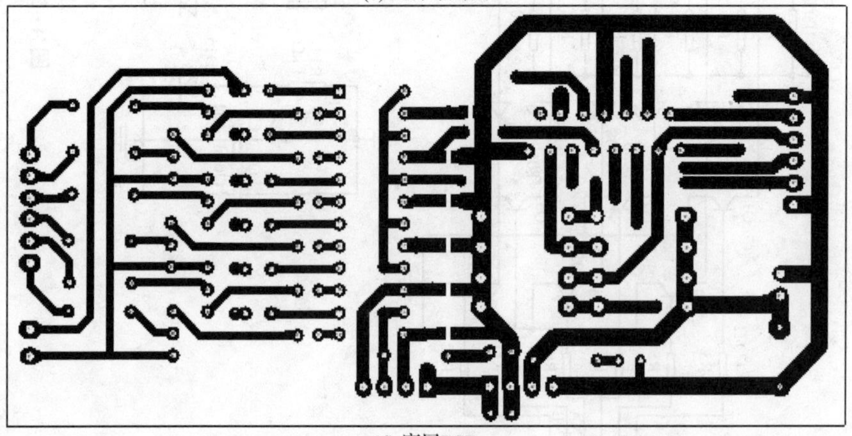

(d) 底层PCB

图 4－5　采用 L298N 构成的双直流电机驱动电路(续)

4.2.3　直流电机与 LPC214x 的连接

直流电机与 LPC214x 的连接示意图如图 4－6 所示，LPC241x 通过光耦隔离的 L298N 驱动电路控制直流电机旋转。

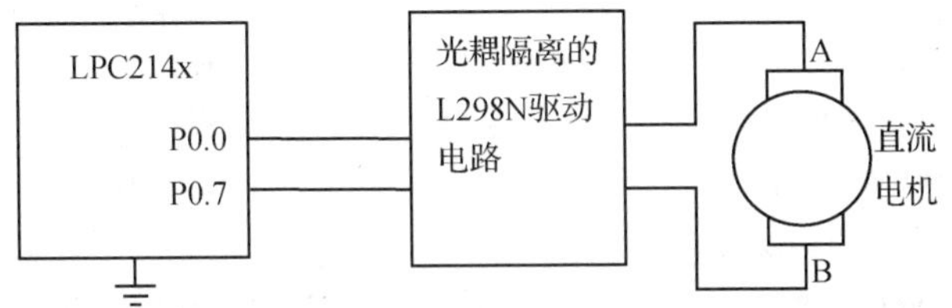

图 4－6　直流电机与 LPC214x 连接示意图

4.2.4　直流电机控制编程示例

直流电机控制编程示例如程序清单 4.1 所示。

程序清单 4.1　直流电机示例程序

```
/***************************************************************
* 文件名:DC_MOTO.C
* 功能:  直流电机控制软件包,采用 PWM 控制 2 个 I/O 口,控制 1 个直流电机是否运转及运转速度
* 说明:  MOTO_A    P0.0 电控制机 A 端口
         MOTO_B    P0.7 电控制机 B 端口
***************************************************************/
#include  "config.h"
/***************************************************************
* 名称:DelayNS()
* 功能:长软件延时
* 入口参数:dly  延时参数,值越大,延时越久
* 出口参数:无
***************************************************************/
void  DelayNS(uint32  dly)
{
    uint32  i;
    for(; dly>0; dly--)
        for(i=0; i<5000; i++);
}
/***************************************************************
* 函数名称:PWM_Init()
* 函数功能:初始化 PWM,并开启向量中断
* 入口参数:freq  初始化 PWM 频率
          Duty  占空比
* 出口参数:无
***************************************************************/
void PWM_Init(uint32 freq,uint32 Duty)
```

```
{
    PINSEL0 = (PINSEL0 & 0xffff3fff) | 0x00008000;          // 将 PWM2 连接到 P0.7 引脚
    PINSEL0 = (PINSEL0 & 0xfffffffc) | 0x00000002;          // 将 PWM1 连接到 P0.0 引脚
  /* 初始化 PWM2 */
    PWMPCR |= (1<<10);                                      // 允许 PWM2 输出,单边 PWM
    PWMPR = 0x00;                                           // 不分频,计数频率为 Fpclk
    PWMMCR = 0x02;                                          // 设置 PWMMR0 匹配时复位 PWMTC
    PWMMR0 = Fpclk/freq;                                    // 设置 PWM 频率
    PWMMR2 = PWMMR0 * Duty/100;                             // 设置 PWM 占空比
    PWMLER |= (0x01 | 1<<2);                                // PWMMR0、PWMMR6 锁存
    PWMTCR = 0x09;                                          // 启动定时器,PWM 使能
  /* 初始化 PWM1 */
    PWMPCR |= (1<<9);                                       // 允许 PWM1 输出,单边 PWM
    PWMPR = 0x00;                                           // 不分频,计数频率为 Fpclk
    PWMMCR = 0x02;                                          // 设置 PWMMR0 匹配时复位 PWMTC
    PWMMR0 = Fpclk/freq;                                    // 设置 PWM 频率
    PWMMR1 = PWMMR0 * Duty/100;                             // 设置 PWM 占空比
    PWMLER |= (0x01 | 1<<1);                                // PWMMR0、PWMMR6 锁存
    PWMTCR = 0x09;                                          // 启动定时器,PWM 使能
}
/******************************************************************************
 * 名称:MOTO_A_PWM()
 * 功能:电机控制端口 A 配置 PWM
 * 入口参数:Duty  PWM 占空比
 * 出口参数:无
******************************************************************************/
void MOTO_A_PWM(uint32 Duty)
{
    PWMMR1 = PWMMR0 * Duty/100;                             // 设置 PWM 占空比
    PWMLER |= (0x01 | 1<<1);                                // PWMMR0、PWMMR6 锁存
    PWMTCR = 0x09;                                          // 启动定时器,PWM 使能
}
/******************************************************************************
 * 名称:MOTO_B_PWM()
 * 功能:电机控制端口 B 配置 PWM
 * 入口参数:Duty  PWM 占空比
 * 出口参数:无
******************************************************************************/
void MOTO_B_PWM(uint32 Duty)
{
    PWMMR2 = PWMMR0 * Duty/100;                             // 设置 PWM 占空比
    PWMLER |= (0x01 | 1<<2);                                // PWMMR0、PWMMR6 锁存
    PWMTCR = 0x09;                                          // 启动定时器,PWM 使能
}
```

```
/****************************************************************
* 名称:MOTO_Init()
* 功能:电机控制端口初始化
* 入口参数:无
* 出口参数:无
****************************************************************/
void MOTO_Init(void)
{
    PWM_Init(200,0);
}
/****************************************************************
* 名称:STOP_R()
* 功能:停止电机
* 入口参数:无
* 出口参数:无
****************************************************************/
void STOP(void)
{
    MOTO_A_PWM(0);                              // 设置 A 端口占空比为 0
    MOTO_B_PWM(0);                              // 设置 B 端口占空比为 0
}
/****************************************************************
* 名称:MOTO_R_forward()
* 功能:右电机正转
* 入口参数:Duty  占空比
* 出口参数:无
****************************************************************/
void MOTO_forward(uint32 Duty)
{
    MOTO_A_PWM(Duty);
    MOTO_B_PWM(0);                              // 设置 B 端口占空比为 0
}
/****************************************************************
* 名称:MOTO_R_back()
* 功能:右电机反转
* 入口参数:无
* 出口参数:无
****************************************************************/
void MOTO_back(uint32 Duty)
{
    MOTO_A_PWM(0);                              // 设置 A 端口占空比为 0
    MOTO_B_PWM(Duty);
}
/****************************************************************
* 名称:main()
* 功能:控制直流电机正转、反转、停止示例程序主函数
****************************************************************/
```

```
int main(void)
{
    PWM_Init(200,0);
    while(1)
    {
        MOTO_forward(50);                       // 控制直流电机占空比为50%,正转
        DelayNS(5000);
        MOTO_back(50);                          // 控制直流电机占空比为50%,反转
        DelayNS(5000);
        STOP();                                 // 控制直流电机停止
        DelayNS(5000);
    }
    return 0;
}
```

4.3 步进电机控制

4.3.1 步进电机的工作原理及方式简介

步进电机是一种将电脉冲转化为角位移的执行机构。当步进电机接收到一个脉冲信号,它就按设定的方向转动一个固定的角度(称为"步距角")。可以通过控制脉冲个数来控制角位移量,从而达到准确定位的目的;同时还可以通过控制脉冲频率实现步进电机的调速。

步进电机的转子为多极分布,定子上嵌有多相星形连接的控制绕组,由专门电源输入电脉冲信号,每输入一个脉冲信号,步进电机的转子就旋转一步。由于输入的是脉冲信号,输出的角位移是断续的,所以又称为脉冲电机。步进电机的种类很多,按结构可分为反应式、永磁式和混合式步进电机三种;按相数分则可分为单相、两相和多相三种。

如果给处于错齿状态的相通电,则转子在电磁力的作用下,将向磁导率最大(或磁阻最小)的位置转动,即趋于对齿的状态转动。

为使步进电机获得更大的转动力矩和更高的精度,电机内部加入了减速装置,通过变速齿轮组将电机的输出转速减小至1/64。这样,步进电机每接收到一个脉冲信号,输出转动的角度(即步距角)可以达到15°的1/64。

以35BYJ26型永磁步进电机为例,其工作方式为双极性两相四拍。电机共引出4根控制线,如表4-6所列。

表4-6 步进电机控制线

控制线颜色	蓝	黄	粉	橙
控制线名称	1A	1B	2A	2B

表 4－6 中，1A 与 1B 是电机内部一组线圈的两个抽头，2A 与 2B 是另一组线圈的两个抽头。只需以一定的顺序控制两组线圈中的电流方向即可使步进电机按指定方向转动。例如，图 4－7 所示的通电时序可使电机逆时针转动。

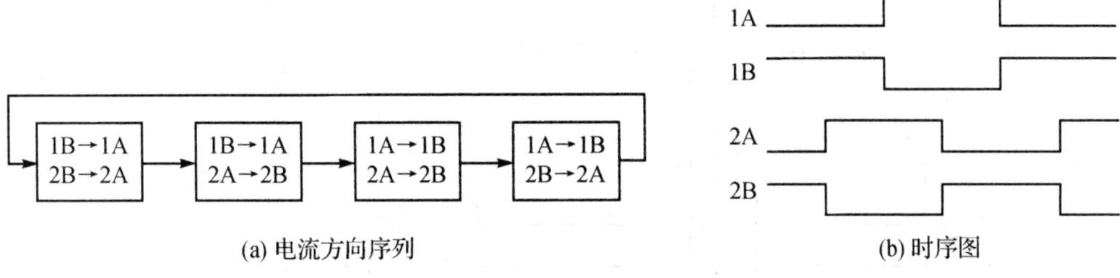

(a) 电流方向序列　　(b) 时序图

图 4－7　使步进电机逆时针转动的控制时序

可以在步进电机的轴上加装刻度盘和指针，以便于观察步进电机的转动情况。

4.3.2　基于“L297＋L298N”的步进电机驱动与控制电路

一个采用“L297＋L298N”构成的步进电机驱动电路原理图如图 4－8 所示，PCB 图如图 4－9 所示。L298N 可以驱动直流电机和两个二相电机，也可以驱动一个四相电机，可直接通过电源来调节输出电压。最大输入电流 DC 2 A，最高输入电压为 DC 50 V，最大输出功率 25 W。L297 译码器能将控制器的控制信号译成所需的相序，并将产生的四相 A、B、C、D 或控制信号 $\overline{\text{INH1}}$ 和 $\overline{\text{INH2}}$ 输入到 L298N 进行步进电机驱动控制。

1. 隔离电路

图 4－8 采用的是光耦隔离微控制器电路与电机驱动电路。一般步进电机转速不超过 1 000 r/s，查询资料可得光耦 TLP521 的响应时间最慢为 10 μs，计算公式为

$$f = \frac{1\ \text{s}}{1\times 10^{-5}\ \text{s}} = 1\times 10^{5} \tag{4.3.1}$$

$$v = \frac{f}{360°/1.8°} = 5\times 10^{4}\ \text{r/s} > 1\,000\ \text{r/s} \tag{4.3.2}$$

选用 TLP521 光耦可达到设计要求，如需提高电机转速，则只需将光耦更换为高速光耦。图 4－8 中 R7～R11 为限流电阻，阻值为 1 kΩ。U1 的 2、4、6、8、10 引脚分别接控制器的 5 个 I/O 口，U1 的 20、18、16、14、12 引脚与 L297 相接；当控制器给这 5 个 I/O 口输出低电平时，每路光耦内部发光二极管导通发光，使光敏三极管导通，从而使 U1 的 19、17、15、13、11 引脚输出低电平。

2. 驱动电路

步进电机驱动电路采用“L297＋L298N”构成。

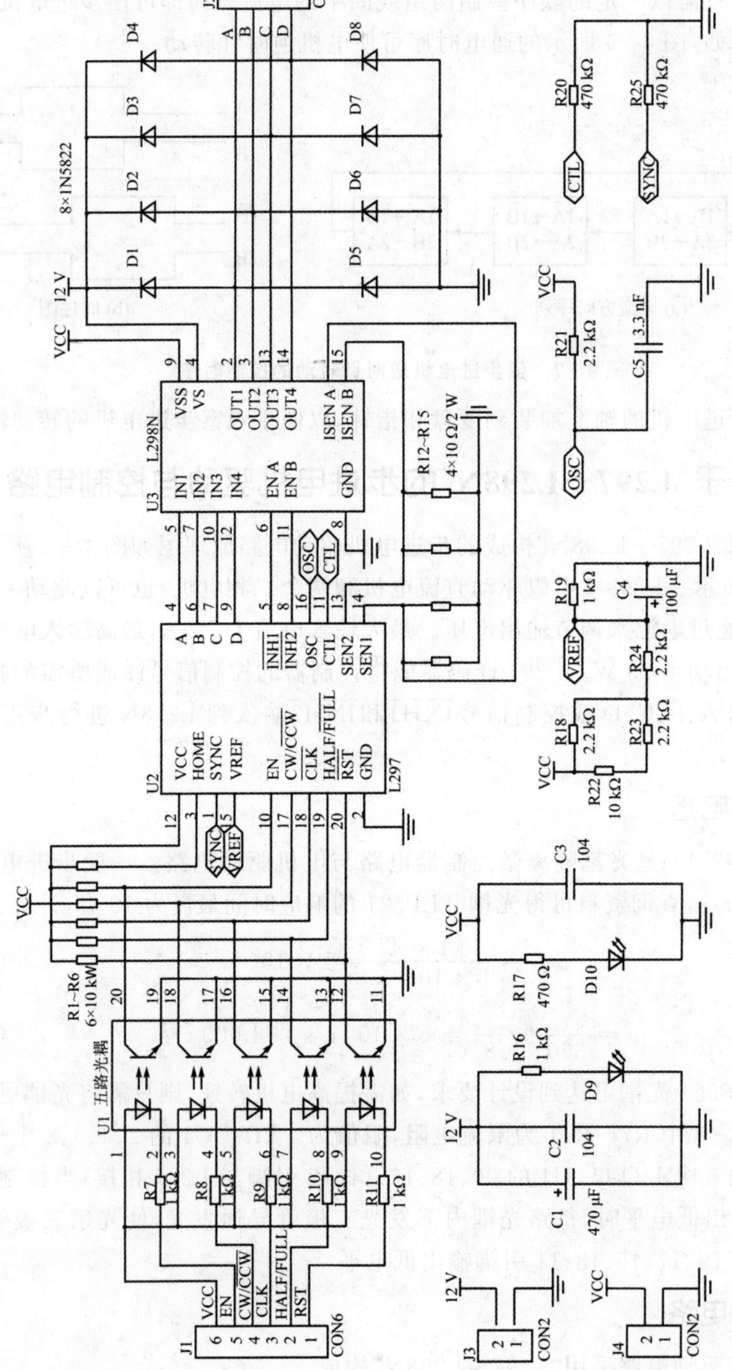

图 4-8　采用“L297+L298N”构成的步进电机驱动电路

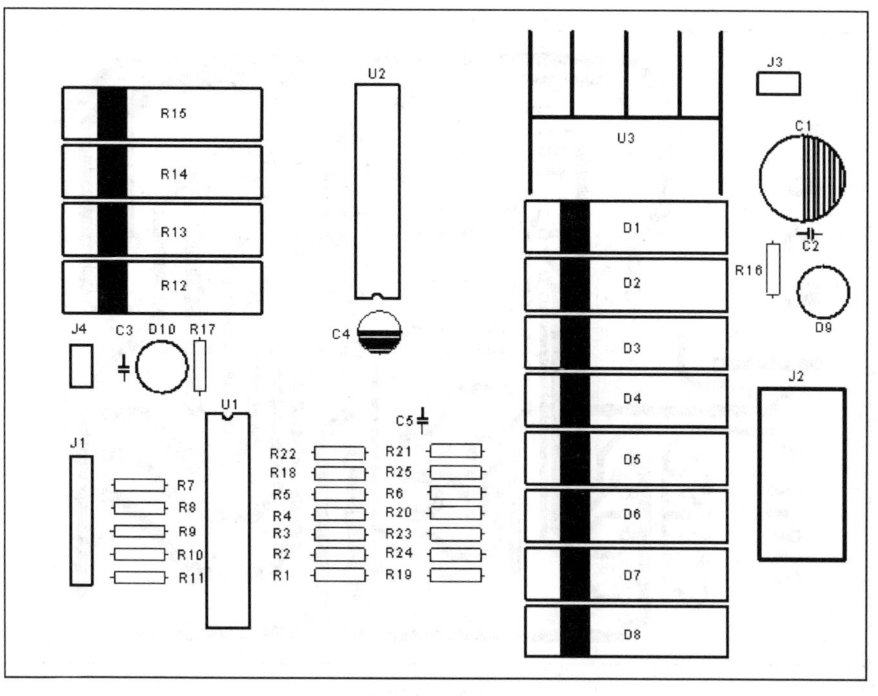

(a) 元器件布局图

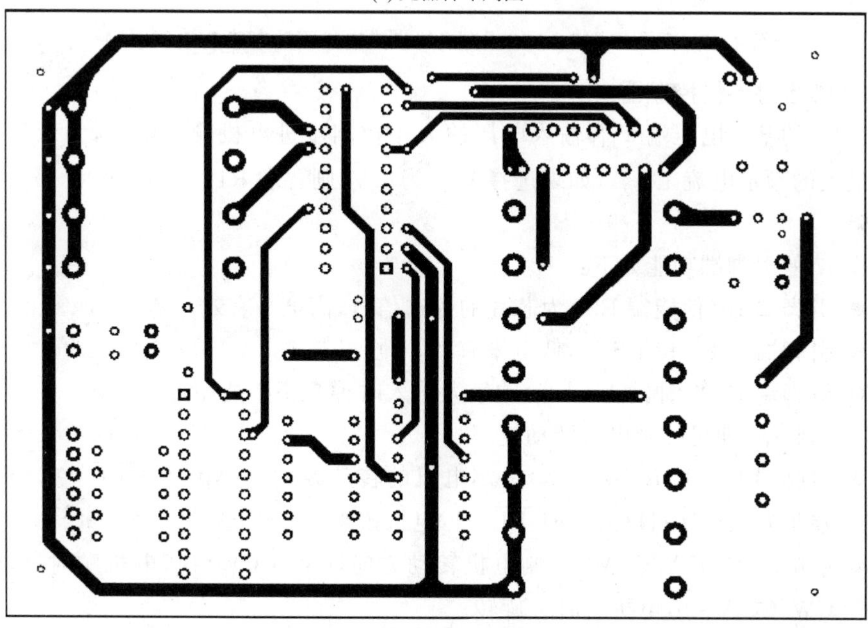

(b) 顶层PCB图

图 4-9 采用“L297+L298”构成的步进电机驱动电路 PCB 图

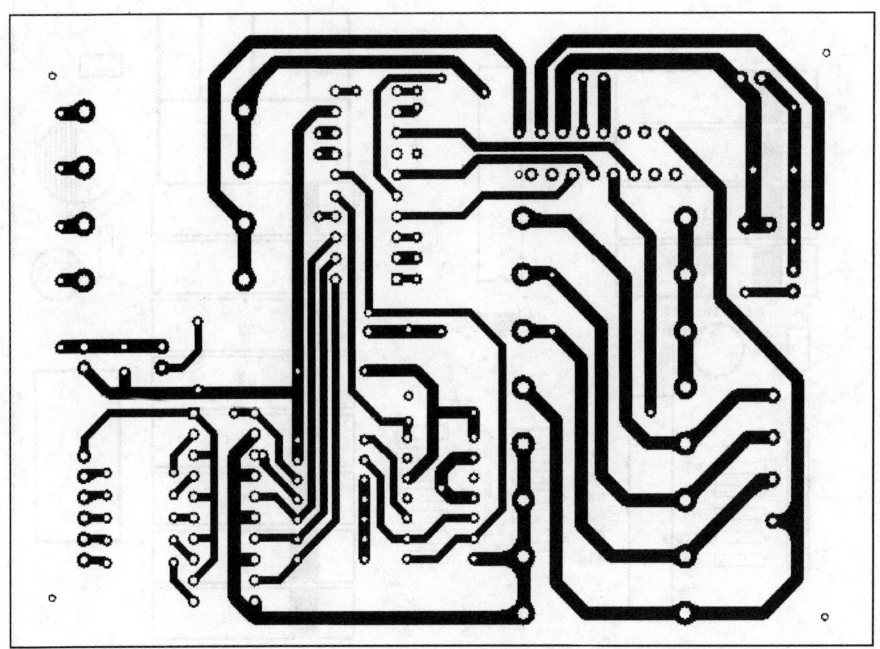

(c) 底层PCB图

图 4-9　采用“L297+L298”构成的步进电机驱动电路 PCB 图(续)

L297 步进电机控制器

L297 的基准电压端 V_{REF} 输入电压的大小控制步进电机输入电流，为保证步进电机最大的额定电流 1.5 A，如果选择 V_{REF} 为 1 V，则要求 R12～R15 选用 10 Ω/2 W 的电阻。

L297 的引脚端功能如下：

- 引脚端 10(使能端 EN)为芯片的片选信号，高电平有效。
- 引脚端 20(复位 RST)，低电平有效。
- 引脚端 18(时钟输入 CLK)的最大输入时钟频率不能超过 5 kHz。控制时钟的频率，即可控制电机转动速率。
- 引脚端 19(HALF/$\overline{\text{FULL}}$)决定电机的转动方式，HALF/$\overline{\text{FULL}}$=0，电机按整步方式运转；HALF/$\overline{\text{FULL}}$=1，电机按半步方式运转。
- 引脚端 17(CW/$\overline{\text{CCW}}$)控制电机转动方向，CW/$\overline{\text{CCW}}$=1，电机顺时针旋转；CW/$\overline{\text{CCW}}$=0，电机逆时针旋转。
- 当给 L297 的复位引脚$\overline{\text{RST}}$输入低电平时，L297 复位。

电机控制具体状态如表 4-7 所列。

控制信号 CW/$\overline{\text{CCW}}$经 L297 处理后，产生的四相 A、B、C、D 或$\overline{\text{INH1}}$和$\overline{\text{INH2}}$输入到 L298N 双全桥驱动器进行功率放大，经 L298N 功率放大后的四相控制信号输

出到步进电机,控制步进电机运动。D1～D8 采用快速恢复二极管 1N5822,用来泄放绕组电流。需要注意的是,L297 和 L298N 的引脚端不能接反或接错,不同接线模式,决定不同的分频模数。

表 4-7　控制电机状态表

端口名称	光耦输入端	L297 端口	电机响应状态
$\overline{\text{RST}}$	低电平	低电平	停止
	高电平	高电平	启动
使能端 EN	低电平	高电平	输入控制信号可以旋转
	高电平	低电平	停止旋转并保持
HALF/$\overline{\text{FULL}}$	低电平	高电平	半步工作模式
	高电平	低电平	全步工作模式
CW/$\overline{\text{CCW}}$	低电平	高电平	顺时针旋转
	高电平	低电平	逆时针旋转

4.3.3　基于"L297＋L298N"的步进电机控制编程示例

一个采用 LPC2148 微控制器控制步进电机驱动模块的程序设计如程序清单 4.2所示。其程序流程如图 4-10 所示。

程序清单 4.2　基于"L297＋L298N"的步进电机控制示例程序

```
/******************************************************************
*  文件名：L297 + L298N.c
*  功能:包含电机初始化及各种电机控制的子函数
*  说明:需添加 L297 + L298.h 文件
******************************************************************/
#include   "config.h"
#define    EN     1<<13              // L297 控制电机使能端 P0.13
#define    CCW    1<<15              // 正转或反转控制端 P0.15
#define    H_F    1<<17              // 全速或半速选择 P0.17
#define    RST    1<<19              // L297 复位端 P0.19
#define    MTIO   EN|RST|H_F|CCW
uint32     PWM_Num = 0;
uint8      PWM_irq_flag = 0;
/******************************************************************
*  函数名称:Dianji_init()
*  函数功能:初始化电机 I/O 口及电机状态
*  入口参数:无
*  出口参数:无
******************************************************************/
void Dianji_init(void)
```

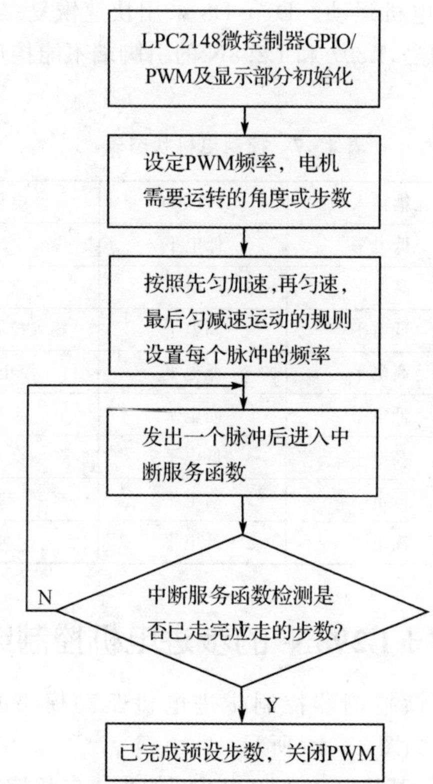

图 4-10 程序流程图

```
{
    IO0DIR |= MTIO;
    IO0SET = EN;
    IO0SET = RST;                      // 开启复位
    IO0CLR = CCW;                      // 设置为顺时针旋转
    IO0SET = H_F;                      // 设置为全速
    IO0CLR = RST;                      // 关闭复位
}
/**************************************************************
* 函数名称:Z_Z()
* 函数功能:使电机正转
* 入口参数:无
* 出口参数:无
**************************************************************/
void Z_Z(void)
{
    IO0SET = CCW;                      // 设置为顺时针旋转
    STAR();
}
/**************************************************************
```

```
 *  函数名称:F_Z()
 *  函数功能:使电机反转
 *  入口参数:无
 *  出口参数:无
*******************************************************************/
void F_Z(void)
{
   IO0CLR = CCW;                    // 设置为逆时针旋转
   STAR();
}
/*******************************************************************
 *  函数名称:Half_Speed()
 *  函数功能:使步进电机半速运转
 *  入口参数:无
 *  出口参数:无
*******************************************************************/
void Half_Speed(void)
{
   IO0CLR = H_F;
}
/*******************************************************************
 *  函数名称:Full_Speed()
 *  函数功能:使步进电机全速运转
 *  入口参数:无
 *  出口参数:无
*******************************************************************/
void Full_Speed(void)
{
    IO0SET = H_F;
}
/*******************************************************************
 *  函数名称:STOP()
 *  函数功能:使电机停止转动
 *  入口参数:无
 *  出口参数:无
*******************************************************************/
void STOP(void)
{
    IO0SET = EN;                    // 禁止 L297 工作
}
/*******************************************************************
 *  函数名称:STAR()
 *  函数功能:使电机开始转动
 *  入口参数:无
 *  出口参数:无
*******************************************************************/
void STAR(void)
```

```
{
    IO0CLR = EN;                    // 使能 L297
}
/*********************************************************************
* 函数名称:IRQ_PWM()
* 函数功能:步进步数检测,若应走的步数 PWM_Num 没走完,则清零 PWMTC 计数器,使电机继续下一步;
          若应走的步数已走完,则不清零,电机停转
* 入口参数:无
* 出口参数:无
*********************************************************************/
void    __irq IRQ_PWM(void)
{
   PWM_irq_flag = 1;
   if(PWM_Num>0)
   {
     PWM_Num--;                     // 步数减 1
     PWMTCR = 0x09;                 // 重新开启 PWM 定时器
     PWMTC = 0;                     // 清零 PWM 计数器
     PWMIR = 0x01 | 1<<10;          // 清除 PWMMR0 中断标志
     VICVectAddr = 0;               // 向量中断结束
}
/*********************************************************************
* 函数名称:PWM_IRQ_Ini()
* 函数功能:初始化 PWM6 连接到 P0.9 引脚,并开启向量中断
* 入口参数:freq 初始化 PWM 频率
* 出口参数:无
*********************************************************************/
void PWM_IRQ_Ini(uint32 freq)
{
    PINSEL0 = (PINSEL0 & 0xfff3ffff) | 0x00080000;
    PWMPCR |= 0x4000;               // 允许 PWM6 输出,单边 PWM
    PWMPR   = 0x00;                 // 不分频,计数频率为 Fpclk
    PWMMCR = 0x02;                  // 设置 PWMMR0 匹配时复位 PWMTC
    PWMMR0 = Fpclk/freq;            // PWM 频率
    PWMMR6 = PWMMR0/2;              // 设置 PWM 占空比 1/2
    PWMLER |= 0x41;                 // PWMMR0、PWMMR6 锁存
    PWMTCR = 0x09;                  // 启动定时器,PWM 使能
  /* 打开 PWM 中断(设置向量控制器,即使用向量 IRQ) */
   VICIntSelect = 0x00000000;       // 设置所有中断分配为 IRQ 中断
   VICVectCntl1 = 0x20 | 8;         // 分配 EINT1 中断到向量中断 0
   VICVectAddr1 = (int)IRQ_PWM;     // 设置中断服务程序地址
   VICIntEnable = 1<<8;             // 使能 EINT1 中断
}
/*********************************************************************
* 函数名称:PWM_X_Y()
```

```
*  函数功能:PWM 以一定频率发出 step 个脉冲,使电机转动 step 步
*  入口参数:freq   PWM 频率
             step   转动的步数,每步 1.8° 由 PWM 中断服务函数检测应走的步数是否走完
*  出口参数:无
*  注意事项:此函数法为中断方式控制电机,即电机运转过程中不需要等待,电机运转
*            过程中微控制器可继续做其他事情
******************************************************************************/
void PWM_X_Y(uint32 freq,uint32 step)
{
    PWM_Num = step - 1;              // 将要走的步数传给 PWM 向量中断服务函数
    PWMTC = 0;
    PWMPR   = 0x00;                  // 不分频
    PWMMR0 = Fpclk/freq;             // 设置频率
    PWMMR6 = PWMMR0/2;               // PWM 占空比为 50%
    PWMLER |= 0x41;                  // 锁存 PWMMR0,PWMMR6
    PWMPCR = 0x4000;                 // 允许 PWM6 输出,单边 PWM
    PWMMCR = 0x02 | 0x04 | 0x01;     // 匹配时复位、停止、中断 PWMMR0
    PWMTCR = 0x09;                   // 计数器使能,PWM 使能
    /*  打开 PWM 中断(设置向量控制器,即使用向量 IRQ) */
   VICIntSelect = 0x00000000;        // 设置所有中断分配为 IRQ 中断
   VICVectCntl1 = 0x20 | 8;          // 分配 EINT1 中断到向量中断 0
   VICVectAddr1 = (int)IRQ_PWM;      // 设置中断服务程序地址
   VICIntEnable = 1<<8;              // 使能 EINT1 中断
}
/*****************************************************************************
* 函数功能:步进 Y 步,先匀加速,再匀速,最后匀减速,可达到稳定启动,不失步稳定停止,不多步
* 入口参数:x   需要步进的步数,大于 100 步,若小于 100 步,则电机不转
* 出口参数:无
* 注    意:调用此函数时,程序会长时间等待,单片机无法处理其他事件,直到电机运转完毕,除非有
*            中断程序打断电机运转,若电机运转过程不能被中断,需要调用此函数前关闭所有中断
******************************************************************************/
uint8 PWM_X(uint32 x)
{
    uint16 a;
    uint32 num = 0,intnum = 0;
    if(x<=100) return 0;             // 若步数很小,无法匀变速运动,则直接退出
    /* 匀加速 50 步 */
    for(a=4;a<200;a+=4)
    {
        PWM_X_Y(a,1);
        while(PWM_irq_flag == 0)    // 等待走完 50 步
        {
            if(PWM_irq_flag == 1)
            PWM_irq_flag = 0;
        }
```

```
        PWM_irq_flag = 0;
    }
    if(x>100)
    {
        /* 匀速 x-100 步 */
        num = x-100;
        intnum = num;
        PWM_X_Y(200,num);
            while(intnum > 0)
            {
            if(PWM_irq_flag == 1)
            {PWM_irq_flag = 0;
            intnum -= 1;
            }
        }
        PWM_irq_flag = 0;
    }
        /* 匀减速 50 步 */
        for(a=200;a>4;a-=4)
    {
        PWM_X_Y(a,1);
        while(PWM_irq_flag == 0)  // 等待走完 50 步
        {
            if(PWM_irq_flag == 1)
            PWM_irq_flag = 0;
        }
        PWM_irq_flag = 0;
    }
    return 0;
}
/*****************************************************************************
* 函数名:PWM_X_Wait()
* 函数功能:步进 x 步,电机匀速旋转
* 入口参数:x  需要步进的步数
* 出口参数:无
* 注    意:调用此函数时,程序会长时间等待,单片机无法处理其他事件,直到电机运转完毕,除非有
*          中断程序打断电机运转,若电机运转过程不能被中断,需要调用此函数前关闭所有中断
*****************************************************************************/
uint8 PWM_X_Wait(uint32 fp,uint32 x)
{
    uint32 intnum = 0;
    /*匀速 x 步*/
    intnum = x;
    PWM_X_Y(fp,x);
    /*等待走完*/
        while(intnum > 0)
```

```
    {
        if(PWM_irq_flag == 1)
        {PWM_irq_flag = 0;
        intnum -= 1;
        }
    }
    PWM_irq_flag = 0;
     return 0;
}
/*************************** END ************************************/
```

4.3.4 基于 TA8435H 的步进电机驱动与控制电路

1. TA8435H 简介

TA8435H 是东芝公司推出的一款单片正弦细分二相步进电机驱动专用芯片，可以驱动二相步进电机，且电路简单，工作可靠。工作电压为 10～40 V，平均输出电流可达 1.5 A，峰值可达 2.5 A，具有整步、半步、1/4 细分、1/8 细分运行方式可供选择，采用脉宽调制式斩波驱动方式，具有正/反转控制功能，带有复位和使能引脚，可选择使用单时钟输入或双时钟输入。TA8435H 采用 ZIP25 封装形式，其引脚功能如表 4-8 所列。

表 4-8　TA8435H 引脚功能

引脚号	引脚名称	引脚功能
1	SG	信号地
2	$\overline{\text{RST}}$	复位端，低电平有效
3	$\overline{\text{EN}}$	使能端，低电平有效，高电平时，各相输出被强制关闭
4	OSC	该引脚外接电容的典型值可决定芯片内部驱动级的斩波频率
5	CW/$\overline{\text{CCW}}$	正、反转控制引脚
6，7	CLK2，CLK1	时钟输入端，可选择单时钟输入或双时钟输入，最大时钟输入为 5 kHz
8，9	IM1，IM2	选择激励方式，00—整步；10—半步；01—1/4 细分；11—1/8 细分
10	Ref	VNF 输入控制，接高电平时 VNF 为 0.8 V，低电平时为 0.5 V
11	MO	输出监视，用于监视输出电流峰值位置
13	VCC	逻辑电路供电引脚，一般为 5 V
15，24	VMB、VMA	MAB 相和 A 相负载地
16，19	B，$\overline{\text{B}}$	B 相输出引脚
17，22	PG-B，PG-A	B 相和 A 相负载地

续表 4-8

引脚号	引脚名称	引脚功能
18,21	NFB,NFA	B相和A相电流检测端,由该引脚外接电阻和REF-IN引脚控制输出电流为:I=VNF/RNF
20,23	A,$\overline{\text{A}}$	A相输出引脚

2. TA8435H 应用电路

使用TA8435H构成的驱动二相步进电机的驱动电路原理图如图4-11所示,驱动电路的PCB图如图4-12所示。

(1) 隔离电路

TA8435H的4路信号,包括使能信号、正反转控制信号、时钟输入信号和复位信号,它们都需要通过光耦隔离后才与微控制器的4个I/O口相连,R9～R12为限流电阻器,阻值为1 kΩ,对微控制器起保护作用。U2的2、4、6、8引脚端分别连接微控制器的4个I/O口,16、14、12、10引脚端与TA8435H对应引脚相连;当微控制器给这4个I/O口输出低电平时,TLP521-4内部4个发光二极管导通发光,使光敏三极管导通,从而使光耦的19、17、15、13、11引脚输出低电平。因此,通过光耦便将微控制器与驱动电路相隔离,起到了保护微控制器的作用。

(2) 驱动电路

TA8435H具有控制电机以整步、半步、1/4细分、1/8细分方式运动的功能,由TA8435H的第8、9引脚IM1、IM2决定,在本电路设计中,采用硬件设计来选择激励方式,IM1和IM2引脚分别接2 kΩ上拉电阻,同时通过JP1引出,可使用跳线帽将IM1和IM2接地,具体激励方式选择如表4-9所列。

表4-9 激励方式选择

IM1电平值	IM2电平值	激励方式
低电平	低电平	整步
高电平	低电平	半步
低电平	高电平	1/4细分
高电平	高电平	1/8细分

同时,TA8435H还具有控制电机正转或反转的功能,且能通过时钟输入信号控制电机的运转速度,同时,TA8435H还有一个使能控制端和一个复位端,分别用于使能和复位TA8435H工作。这4路控制信号都通过光耦隔离后分别与控制器的4个I/O口相连,因此控制器对这4个I/O输出的电平决定了电机的运动状态,其中时钟输入信号CLK的频率不能超过5 kHz,其他3个控制信号控制电机运动状态如表4-10所列。

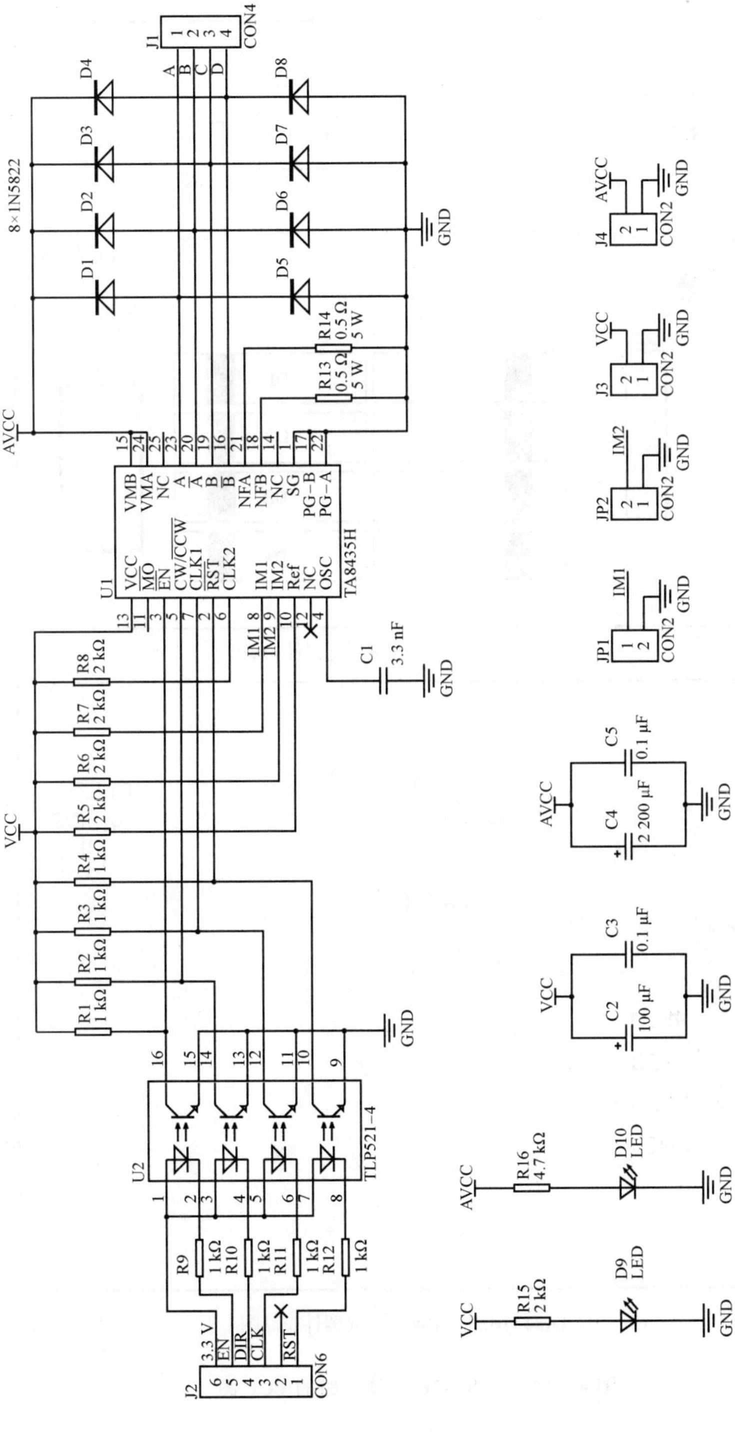

图 4-11　TA8435H驱动电路原理图

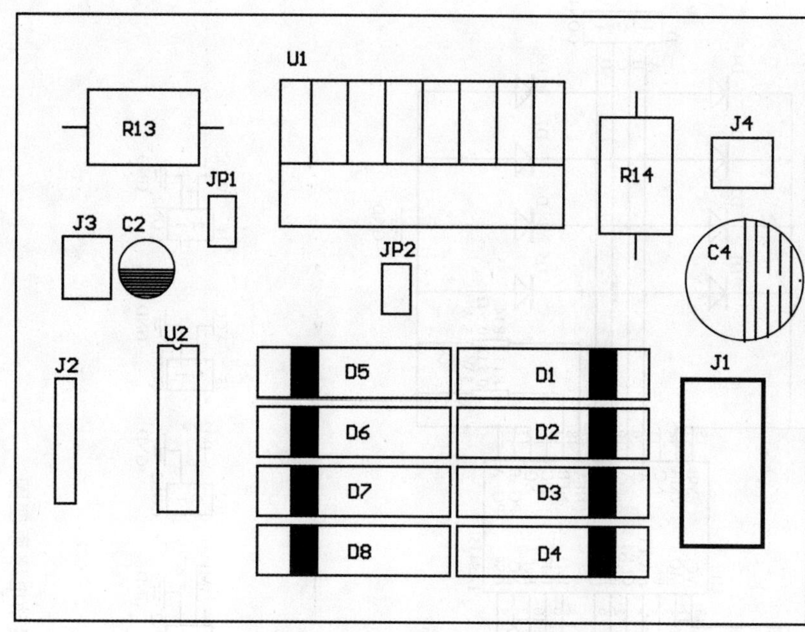

(a) TA8435H驱动电路顶层元器件布局图

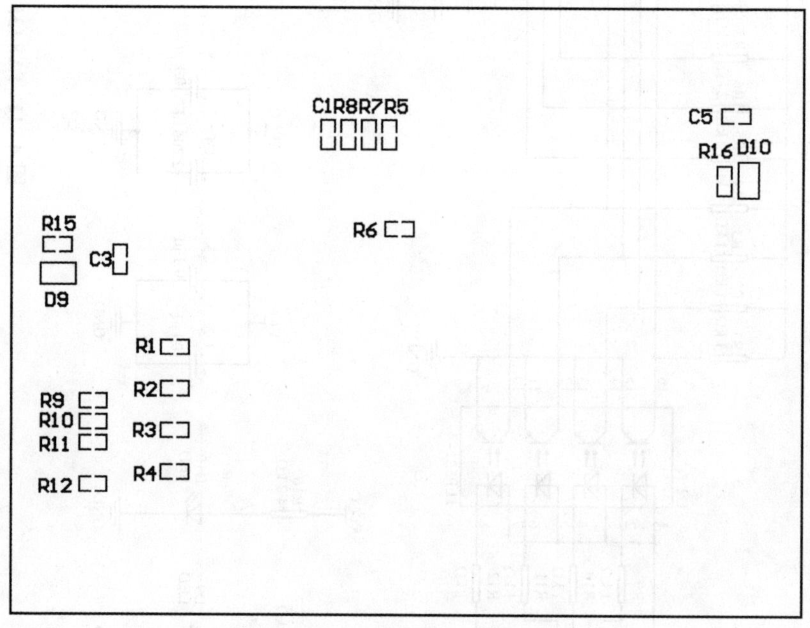

(b) TA8435H驱动电路底层元器件布局图

图 4-12 TA8435H 驱动电路的 PCB 图

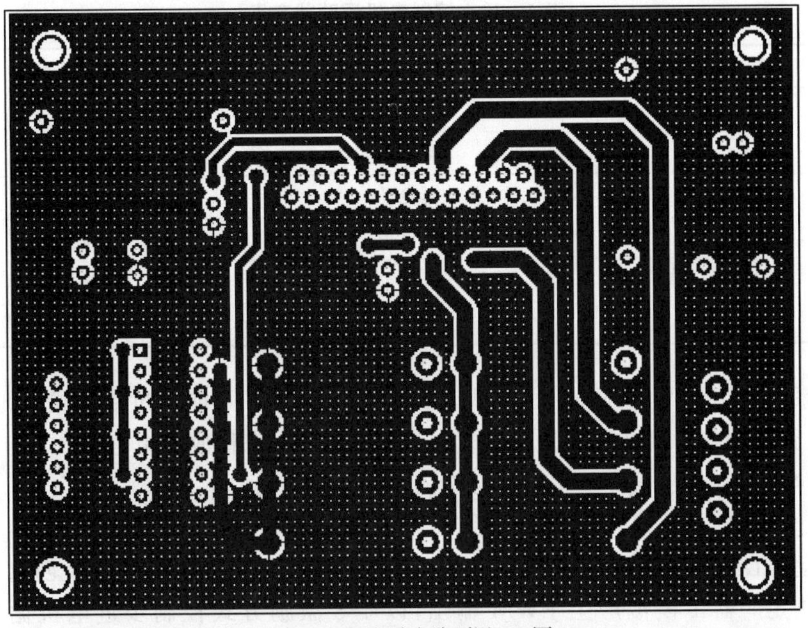

(c) TA8435H驱动电路顶层PCB图

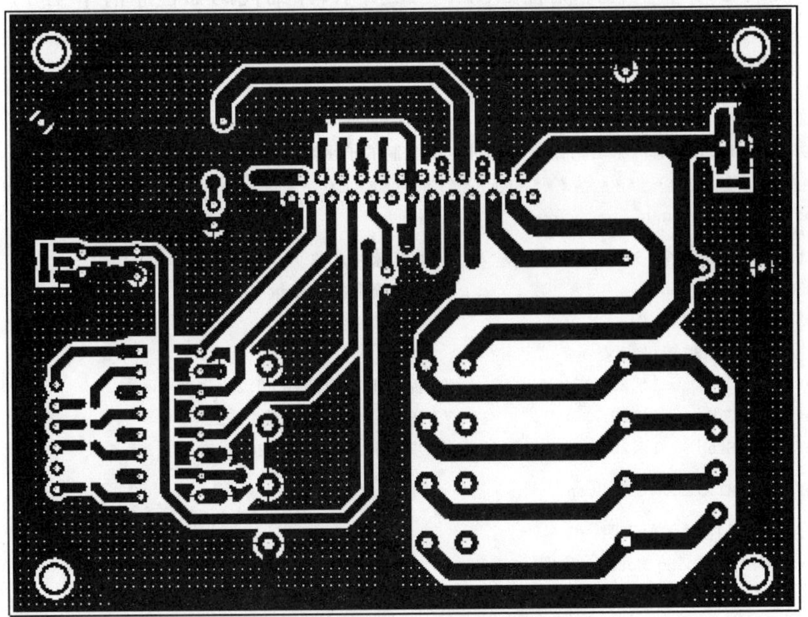

(d) TA8435H驱动电路底层PCB图

图 4－12　TA8435H 驱动电路的 PCB 图(续)

表 4-10 控制电机运动状态表

TA8435H 端口名称	相应控制器 I/O 口	电机响应状态
复位端 ($\overline{RST}$)	高电平	启动
	低电平	停止
使能端 (EN)	低电平	停止
	高电平	启动
正反转控制端 (CW/$\overline{CCW}$)	低电平	正转
	高电平	反转

本系统选用 1/8 细分的激励方式,因此 IM1 和 IM2 不能通过跳线帽与地短接,另外,电路中的 D1～D8 使用快速恢复二极管 1N5822 来释放绕组电流。

4.3.5 基于 TA8435H 的步进电机控制编程示例

启动 ADS1.2,使用 ARM Executable Image for lpc2148 工程模板创建一个工程。采用定时器 1 模拟产生 PWM 输出,以 1/8 激励方式控制步进电机循环正转 200 步再反转 200 步。程序流程如图 4-13 所示,详细代码如程序清单 4.3 所示。

程序清单 4.3 基于 TA8435H 的步进电机控制示例程序

```
/*****************************************************************
 * 文件名:Stepping_Motor.c
 * 功能:含 TA8435 电机驱动初始化及各种电机控制的子函数
*****************************************************************/
#include   "config.h"
volatile  uint32 PWM_Num = 0;               // PWM 脉冲计数
uint32 MOTO_Frequency;                      // PWM 频率控制变量
uint32   MOTO_PWM = 0;                      // 计数变量
/* 步进电机控制端口定义 */
#define   EN      1<<21                     // 控制电机使能端 P0.21
#define   CCW     1<<22                     // 正转或反转控制端 P0.22
#define   Step    1<<20                     // 步数脉冲引脚 P0.20
#define   MTIO  EN|CCW|Step
/* 步进电机控制函数声明 */
extern void Stepping_Motor_init(void);
extern void Z_Z(void);
extern void F_Z(void);
extern void STOP(void);
extern void STAR(void);
extern void PWM_X_Y(uint32 freq,uint32 step);
extern uint8 PWM_X_Wait(uint32 fp,uint32 x);
/*****************************************************************
 * 函数名称:Stepping_Motor_init()
```

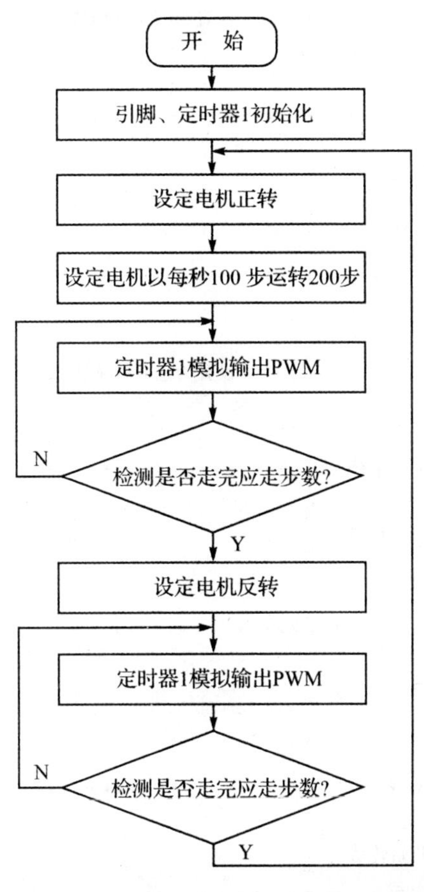

图 4-13 基于 TA8435H 的步进电机控制程序流程图

```
* 函数功能:初始化步进电机 I/O 口及电机状态
* 入口参数:无
* 出口参数:无
*****************************************************************/
void Stepping_Motor_init(void)
{
    IO0DIR |= MTIO;                     // 设置为输出
    IO0CLR = EN;
    IO0SET = CCW;                       // 设置为顺时针旋转
}
/*****************************************************************
* 函数名称:Z_Z()
* 函数功能:使步进电机正转
* 入口参数:无
* 出口参数:无
*****************************************************************/
void Z_Z(void)
{
```

```
    IO0CLR = CCW;                         // 设置为顺时针旋转
    STAR();
}
/*********************************************************************
 * 函数名称:F_Z()
 * 函数功能:使电机反转
 * 入口参数:无
 * 出口参数:无
*********************************************************************/
void F_Z(void)
{
   IO0SET = CCW;                          // 设置为逆时针旋转
   STAR();
}
/*********************************************************************
 * 函数名称:STOP()
 * 函数功能:使步进电机停止转动
 * 入口参数:无
 * 出口参数:无
*********************************************************************/
void STOP(void)
{
    IO0SET = EN;
}
/*********************************************************************
 * 函数名称:STAR()
 * 函数功能:使步进电机开始转动
 * 入口参数:无
 * 出口参数:无
*********************************************************************/
void STAR(void)
{
    IO0CLR = EN;                          // 使能
}
/*********************************************************************
 * 函数名称:PWM_X_Y()
 * 函数功能:向步进电机输出方波脉冲
 * 入口参数:freq      方波频率
 *          step      脉冲个数
 * 出口参数:无
*********************************************************************/
void PWM_X_Y(uint32 freq,uint32 step)
{
    PWM_Num = step;                       // 将要走的步数传给 PWM 向量中断服务函数
    MOTO_Frequency = freq;
}
/*********************************************************************
```

```
*  函数名称:PWM_X_Wait()
*  函数功能:向步进电机输出方波脉冲,并等待脉冲输出完毕
*  入口参数:freq      方波频率
*           step      脉冲个数
*  出口参数:无
******************************************************************/
void PWM_X_Wait(uint32 fp,uint32 x)
{
    /* 以每秒 fp 步的速度,匀速转动 x 步 */
    PWM_X_Y(fp,x);
    /* 等待走完 */
    while((PWM_Num > 0) );
}
/******************************************************************
 * 函数名称:IRQ_Time1()
 * 函数功能:定时器 1 中断服务程序,因 LPC2148 专用 PWM 资源不够,故采用定时器模拟 PWM
 * 入口参数:无
 * 出口参数:无
 * 说明:
******************************************************************/
void __irq IRQ_Time1(void)
{
    T1IR = 0x01;                    // 清除中断标志
    Time1_Flag = 1;
    if(PWM_Num>0)
    {
        MOTO_PWM++;
        if (MOTO_PWM == 10000/MOTO_Frequency/2)
        {
            IO0SET = Step;          // 拉低 I/O 口电平
        }
        if (MOTO_PWM == 10000/MOTO_Frequency)
        {
            MOTO_PWM = 0;
            PWM_Num--;
            IO0CLR = Step;          // 拉低 I/O 口电平
        }
    }
    VICVectAddr = 0;                // 通知 VIC 中断处理结束
}
/******************************************************************
 * 函数名称:Time1Init()
 * 函数功能:初始化定时器,定时时钟为 10 000 Hz
 * 入口参数:无
 * 出口参数:无
******************************************************************/
void Time1Init(void)
```

```
{
    T1PR = 11;                              // 设定定时器分频为 12 分频,得 1 MHz
    T1MCR = 0x03;                           // 设置匹配通道 0 匹配中断并复位 T1TC
    T1MR0 = 100;                            // 设置预分频寄存器值,得定时时钟 10 000 Hz
    T1TCR = 0x03;                           // 设置定时器/计数器和预分频计数器启动并复位
    T1TCR = 0x01;                           // 开启定时器工作
    /* 设置定时器 1 向量中断 IRQ */
    VICIntSelect = 0x00000000;              // 设置所有中断分配为 IRQ 中断
    VICVectCntl3 = 0x25;                    // 给定时器 1 中断分配向量中断
    VICVectAddr3 = (uint32)IRQ_Time1;       // 设置中断服务程序地址
    VICIntEnable = 1<<5;                    // 使能定时器 1 中断
}
/***********************************************************************
 * 名称:main()
 * 功能:LPC2148 使用 PWM 控制步进电机示例程序主函数
 ***********************************************************************/
int main(void)
{
    PINSEL0 = 0;
    PINSEL1 = 0;
    PINSEL2 &= ~(0x00000006);
    IO0DIR = 0;
    IO1DIR = 0;
    Stepping_Motor_init();
    Time1Init();                            // 定时器 1 初始化,用于模拟 PWM 控制步进电机
    while(1)
    {
    Z_Z();
    PWM_X_Wait(100,200);                    // 以每秒 100 步的速度正转 200 步,并等待转动完毕
    F_Z();
    PWM_X_Wait(100,200);                    // 以每秒 100 步的速度反转 200 步,并等待转动完毕
    }
}
```


4.4 舵机控制

4.4.1 舵机简介

1. 舵机的基本特性

舵机是一种位置伺服的驱动器。它接收一定的控制信号,输出一定的角度,适用于那些需要角度不断变化并可以保持的控制系统。在微机电系统和航模中,它是一个基本的输出执行机构。舵机是遥控模型控制动作的动力来源,不同类型的遥控模

型所需的舵机种类也随之不同。如何选择合乎需求而且经济的舵机，是一个费心的事情。

厂商所提供的舵机规格资料，都会包含外形尺寸(cm)、扭力(kg/cm)、速度(s/60°)、测试电压(V)及质量(g)等技术参数。

扭力的单位是 kg/cm，是指在摆臂长度 1 cm，能吊起几公斤重的物体。这就是指力臂的能力，一般摆臂长度愈长，则扭力愈小。速度的单位是 s/60°，是指舵机转动 60°所需要的时间。

电源电压会直接影响舵机的性能，例如 Futaba S－9001 在 4.8 V 时扭力为 3.9 kg/cm，速度为 0.22 s/60°；在 6.0 V 时扭力为 5.2 kg/cm，速度为 0.18 s/60°。速度快、扭力大的舵机，除了价格贵，还会伴随着耗电高的特点。因此在使用高级的舵机时，务必搭配高品质、高容量的镍镉电池，它能提供稳定且充裕的电流，使舵机发挥应有的性能。

2. 舵机的结构

舵机主要由外壳、电路板、无核心马达、齿轮与位置检测器所构成。其工作原理是由接收机发出信号给舵机，经由电路板上的 IC 判断转动方向，再驱动无核心马达开始转动，透过减速齿轮将动力传至摆臂，同时由位置检测器送回信号，判断是否已到达定位。

位置检测器其实就是可变电阻，当舵机转动时电阻值也会跟着改变，测量电阻值便可知转动的角度。

一般的伺服马达是将细铜线缠绕在转子上，当电流流过线圈时便会产生磁场，与转子外围的磁铁产生排斥作用，进而产生转动的作用力。依据物理学原理，物体的转动惯量与质量成正比，因此要转动质量愈大的物体，所需的作用力也愈大。舵机为求转速快、耗电小，于是将细铜线缠绕成极薄的中空圆柱体，形成一个质量极轻的五极中空转子，并将磁铁置于圆柱体内，这就是无核心马达。

为了适合不同的工作环境，有防水及防尘设计的舵机。并且因应用不同的负载需求，舵机的齿轮有塑胶及金属的区分。较高级的舵机会装置滚珠轴承，使得转动时能更轻快精准。滚珠轴承有一颗及两颗的区别，当然是两颗的比较好。

3. 舵机的控制

标准的舵机有电源线、地线、控制线 3 条连接导线。电源线和地线用于提供舵机内部的直流电机和控制线路所需的能源，电源电压通常为 4～6 V，一般取 5 V。**注意：**给舵机供电电源应能提供足够的功率。控制线的输入信号是一个宽度可调的周期性方波脉冲信号，方波脉冲信号的周期为 20 ms(即频率为 50 Hz)。当方波的脉冲宽度改变时，舵机转轴的角度发生改变，角度变化与脉冲宽度的变化成正比。以 180°(－90°～90°)伺服为例，某型舵机的输出轴转角与输入信号的脉冲宽度之间的关系如图 4－14 所示。

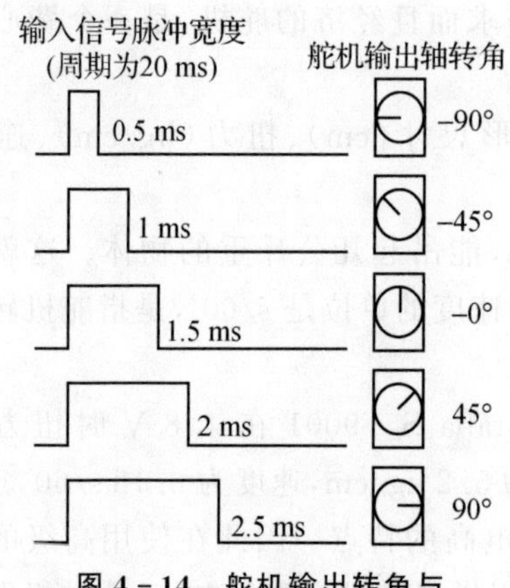

图4-14 舵机输出转角与输入信号脉冲宽度的关系

精确地控制舵机的输出转角,要比图4-14所示的复杂。很多舵机的位置(输出转角)等级可以达到1 024个,如果舵机的有效角度范围为180°,则其输出转角的控制精度可以达到180°/1 024,大约等于0.18°,输入脉冲的宽度精度要求为(20 000/1 024)μs,大约为20 μs。

利用微控制器产生精密的PWM脉冲波形,可以实现对舵机的精确控制。

4. 数码舵机

现在,几乎所有主流厂商都生产数码舵机。数码舵机比模拟舵机,在速度、保持力方面有较大的优势。数码舵机内置微控制器,在控制舵机的动作上,速度要比模拟舵机快6倍,几乎不存在“无反应区”。“无反应区”是指输出轴必须从当前位置旋转一定的角度后,才能引起控制器发出指令来修正它的位置的时间。数码舵机的信号刷新频率通常是300次/秒,而模拟舵机只有50~60次/秒,因此数码舵机可以更有力地保持正确的位置,这就是所谓的“保持力”。

数码舵机生产厂商与产品特点如表4-11所列。

表4-11 数码舵机生产厂商与产品特点

生产厂商	产品特点	网 址
Airtronics/Sanwa	Airtroncis提供少数几款数码舵机,舵机使用标准尺寸的外壳及无核心马达,94758Z舵机具有0.06 s转动速度,是市面上转动速度较快的舵机之一。94755Z使用塑料和金属的齿轮组合,其他款采用的是全金属的齿轮组合	http://www.airtronics.net
Futaba	Futaba提供10多款数码舵机,有塑料齿轮组合、金属齿轮组合、小尺寸、标准尺寸、大尺寸、薄的标准尺寸,以及低重心形式的各种舵机。其中,S9254是市面上转动速度较快的舵机之一	http://www.futaba-rc.com
Hitec	Hitec的产品具有价格优势,在10多款产品当中,大部分是标准尺寸的,也有少数迷你舵机,全部都使用金属齿。其中,HS-5996、HS-5997和HS-5998采用钛合金齿轮。HS-5955具有333 oz.-in(2.331 N·m)扭力,是最为强力的舵机之一	http://www.hitecrcd.com

续表 4-11

生产厂商	产品特点	网　址
JR Racing	JR Racing 的数码舵机全部采用标准尺寸、金属齿，以及高效率的无核心马达。其中，DS9000T 具有 282 oz. - in (1.974 N·m)的扭力	http://www.jrradios.com
KO Propo	KO Propo 提供不同款的舵机，标有“ICS”的是标准的数码舵机，标有“FET”的是使用 FET 管增强动力技术的舵机。KO Propo 还提供一系列相对便宜的全塑料齿轮舵机，有标准尺寸舵机，也有 1/12 用的迷你舵机，以及大舵机。所有 KO Propo 的舵机都使用无核心马达	http://www.kopropo.com

4.4.2 舵机与 LPC214x 的连接

舵机与 LPC214x 连接的示意图如图 4-15 所示，采用 LPC214x 的 PWM 控制舵机输出转角。

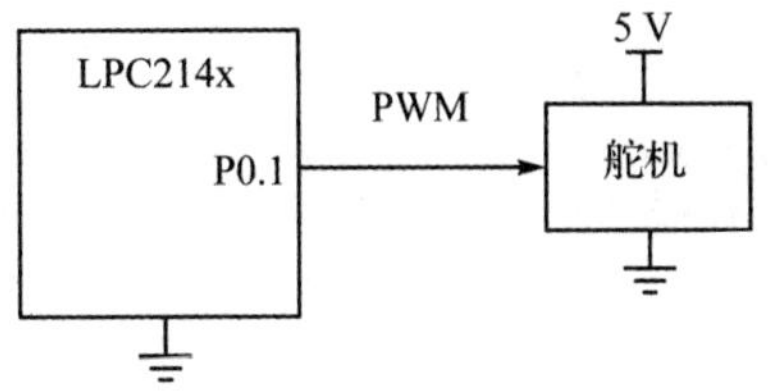

图 4-15　舵机与 LPC214x 连接示意图

4.4.3 舵机控制编程示例

舵机控制程序示例如程序清单 4.4 所示。

程序清单 4.4　舵机控制程序示例

```
/*****************************************************************
*  文件名:Rudder.C
*  功能:舵机控制软件包,采用 PWM 舵机到达不同位置
*  说明:P0.1 为舵机控制端口
*****************************************************************/
#include "config.h"
/*****************************************************************
*  函数名称:PWM_IRQ_Ini()
*  函数功能:初始化 PWM,并开启向量中断
*  入口参数:freq   初始化 PWM 频率
            Duty  占空比为 Duty/1 000
*  出口参数:无
```

```
**************************************************************************/
void PWM_Ini(uint32 freq,uint32 Duty)
{
    PINSEL0 = (PINSEL0 & 0xfffffff3) | 0x00000008;  // 将 PWM3 连接到 P0.1 引脚
    /* 初始化 PWM3 */
    PWMPCR |= (1<<11);                              // 允许 PWM3 输出,单边 PWM
    PWMPR    = 0x00;                                // 不分频,计数频率为 Fpclk
    PWMMCR  = 0x02;                                 // 设置 PWMMR0 匹配时复位 PWMTC
    PWMMR0  = Fpclk/freq;                           // 设置 PWM 频率
    PWMMR3  = PWMMR0 * Duty/1000;                   // 设置 PWM 占空比为 Duty/1 000
    PWMLER |= (0x01 | 1<<3);                        // PWMMR0、PWMMR6 锁存
    PWMTCR   = 0x09;                                // 启动定时器,PWM 使能
}
/**************************************************************************
 * 函数名称:Change_PWM()
 * 函数功能:改变 PWM 占空比
 * 入口参数:Duty  占空比为 Duty/1 000
 * 出口参数:无
**************************************************************************/
void Change_PWM(uint32 Duty)
{
    PWMMR3  = PWMMR0 * Duty/1000;                   // 设置 PWM 占空比为 Duty/1 000
    PWMLER |= (0x01 | 1<<3);                        // PWMMR0、PWMMR6 锁存
    PWMTCR  = 0x09;                                 // 启动定时器,PWM 使能
}
/**************************************************************************
 * 函数名称:Rudder_0()
 * 函数功能:控制舵机转到 0°位置
 * 入口参数:无
 * 出口参数:无
**************************************************************************/
void Rudder_0(void)
{
    Change_PWM(31);                                 // 设置占空比为 31/1 000,即 3.1%
}
/**************************************************************************
 * 函数名称:Rudder_90()
 * 函数功能:控制舵机转到 90°位置
 * 入口参数:无
 * 出口参数:无
**************************************************************************/
void Rudder_90(void)
{
    Change_PWM(75);                                 // 设置占空比为 75/1 000,即 7.5%
}
/**************************************************************************
```

```
* 函数名称:Rudder_180()
* 函数功能:控制舵机转到180°位置
* 入口参数:无
* 出口参数:无
******************************************************************/
void Rudder_180(void)
{
    Change_PWM(100);                        // 设置占空比为100/1 000,即10%
}
******************************************************************/
* 函数名称:main()
* 函数功能:LPC2148使用PWM控制舵机示例程序主函数
******************************************************************/
int main(void)
{
    PWM_Ini(50,0);
    Rudder_90();                            // 控制舵机到达90°位置
    while(1);
    return 0;
}
```

第 5 章

传感器电路

5.1 光电传感器及其应用

5.1.1 光电传感器选型

光电传感器品种繁多，有红外发光二极管（LED）、光电接收二极管、光电接收三极管、阻挡弱光的光电三极管、光电接收达林顿管、光电施密特接收管、反射式光电组件、光电施密特对射组件、对射式编码检测器和条形码传感器等。光电接收器有光电接收二极管、光电接收三极管、阻挡弱光的光电三极管、光电接收达林顿管、光电施密特接收管等。光电传感器的种类繁多，生产厂商也很多，以“光电传感器”为关键词，可以在相关网站查询到大量的资料。

1. 光电传感器的主要类型

光电传感器根据检测模式的不同可分为以下几种：

① 反射式光电传感器将发光器与光敏器件置于一体，发光器发射的光被检测物反射到光敏器件。

② 透射式光电传感器将发光器与光敏器件置于相对的两个位置，光束也是在两个相对的物体之间，当物体穿过发光器与光接收器件时，穿过发光器与光敏器件之间的被检测物体会阻断光束，并启动受光器。

③ 聚焦式光电传感器将发光器与光敏器件聚焦于特定距离，只有当被检测物体出现在聚焦点时，光敏器件才会接收到发光器发出的光束。

2. 集成的光电传感器

(1) 分　类

集成的光电传感器主要有反射式光电开关、会聚式光电开关、透射式光电开关、反射板式光电开关、光纤穿透式开关、光纤反射式开关等几种，一般采用前三种。

(2) 工作光源

集成的光电传感器采用的工作光源主要有可见红光（650 nm）、可见绿光（510 nm）和红外光（800～940 nm）。不同的光源在具体情况下各有长处。例如，在不考虑被测物体颜色的情况下，红外光有较宽的敏感范围，而可见红光或绿光特别适合于反差检测，光源的颜色必须根据被测物体的颜色来选择，红色物体与红色标记宜用绿光（互补色）进行检测。

(3) 外　形

按照外壳形状可分为螺纹圆柱形系列、圆柱形系列、方形系列、槽形贯穿形系列等不同的系列。

3. 安装各种不同类型的光电传感器需要注意的问题

① 对反射式光电传感器的安装，要根据不同的检测材料，确定适当的距离。具体的距离和具体的位置必须在现场调试。

② 对聚焦式光电传感器的安装，最主要的就是要确定聚焦点的位置，如果位置选择的不合适，就会使传感器失去作用。

③ 对透射式光电传感器的安装，一定要安装好遮光片。安装时，一是要选择好材料；二是要特别注意其安装的位置。

5.1.2　利用反射式光电传感器检测障碍物

1. 工作原理

可以利用反射式光电传感器进行障碍物检测。用于障碍物检测的反射式光电传感器也称为红外避障传感器。红外避障传感器具有一对红外信号发射与接收的二极管，发射管发射一定频率的红外信号，接收管接收这种频率的红外信号，当红外的检测方向遇到障碍物（反射面）时，红外信号反射回来被接收管接收，经过处理之后，通过数字传感器接口返回到微控制器，微控制器可利用红外波的返回信号来识别周围环境的变化。

红外是通过发射端发射红外信号，接收端接收由障碍物反射回来的红外信号，来判断是否有障碍物，如图 5－1 所示。

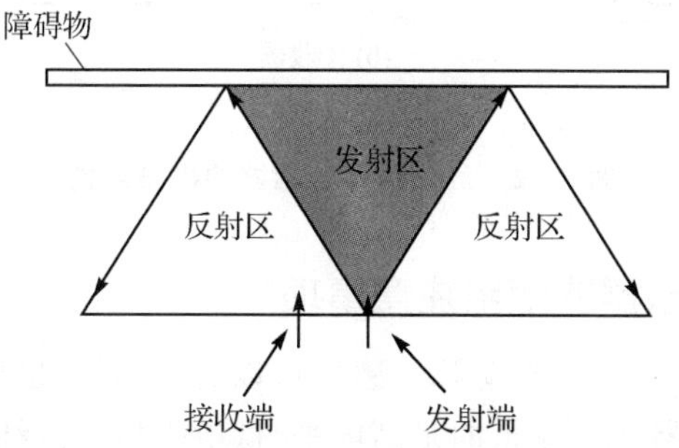

图 5－1　反射式光电传感器进行障碍物的检测示意图

如图 5－1 所示，红外处于工作状态时，不停地以 30°左右的散角向外发射红外信号，当红外接收管处于反射区内时，便能接收到障碍物（反射面）反射回来的红外信号,即认为此时有障碍,这种设计方法多用于前方有障碍的检测。

对于不同颜色的障碍物(反射面)其对红外信号的反射能力不同,导致传感器对不同颜色的障碍物的检测范围不同。另外,不同公司生产的产品其检测范围也不同,如某公司红外避障传感器的检测范围如表5-1所列。

表5-1 红外避障传感器的检测范围

反射面颜色	红外传感器最小检测范围/cm	红外传感器最大检测范围/cm
白色	1	40
黑色	1	25
红色	近似于0	45

2. 电路结构

红外避障传感器的电路结构如图5-2所示,其中调制的频率在几kHz～几十kHz之间,高的有上百kHz。通常在红外避障传感器的发射和接收端都会加装聚焦镜,使传感器具有更远的探测距离,避免受可见光干扰。

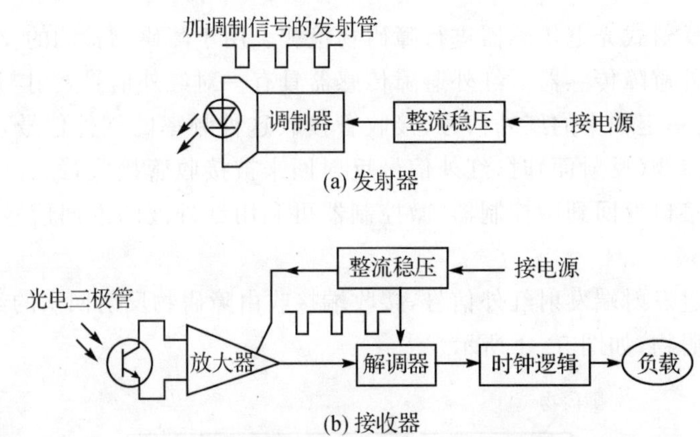

图5-2 红外避障传感器的电路结构

3. 红外避障传感器的使用注意事项

① 在电子设计竞赛中,红外避障传感器可以自己制作,也可以采用现成的产品。

② 红外避障传感器通常输出的是高电平/低电平信号。红外接收管只有在接收到一定强度的红外信号时才会有高电平/低电平的变化。障碍物(反射面)太小时,红外避障传感器会检测不到;障碍物(反射面)颜色为黑色或深色时,会被吸收大部分的红外信号,而只反射回一小部分,导致红外接收管接收到的红外信号强度不够,不足以产生有障碍物(反射面)的信号。

③ 红外在暖光源的照射下(如白炽灯、太阳光)检测受到很大影响,它会受到所有相近红外信号的干扰,白炽灯和太阳光中含有红外信号成分较多,对红外的影响也较大。红外相互之间也存在干扰,因而在使用时需要注意。

④ 红外采用的是发射、接收原理,不同反射面对红外信号的吸收与散射将影响其检测范围。根据测试红色的反射面效果最佳,白色其次,黑色最差;同时反射面的粗糙度和平整度也会影响检测的效果。

⑤ 另外需要注意的一点就是:千万不要把接收管外的黑色塑料皮割掉,如果没有这层黑色塑料皮,它将会一直受到发射管发射出的红外线的干扰,就会出现一直检测到障碍的情况。

5.1.3　利用反射式光电传感器检测黑线

1. 检测电路与工作原理

利用反射式光电传感器检测黑线(也可以检测黑白物体)的电路如图5-3所示。

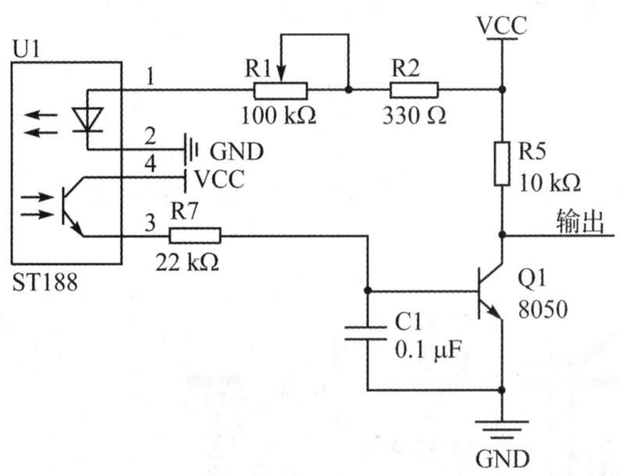

图5-3　反射式光电传感器检测黑白物体(检测黑线)的电路

检测电路使用型号为ST188的一体化反射型光电传感器(也可以选择其他型号的一体化反射型光电传感器),其发射器是一个砷化镓红外发光二极管,而接收器是一个高灵敏度、硅平面光电三极管,其内置可见光过滤器能减小离散光的影响,体积小,结构紧凑,能排除外界光线的影响,因而工作性能稳定。

由于黑色物体和白色物体的反射系数不同,调节反射式光电传感器与检测对象之间的距离,使光敏三极管就只能接受到白色物体反射回来的光束。而对于黑色物体由于其反射系数小,所反射回来的光束很弱,光敏三极管无法接收到反射光。利用反射光可以使光敏三极管实现导通和关断,从而实现对黑白物体的分辨。

检测电路的工作过程如下：当被测物体是黑色物体时，红外发光二极管发射出的光，被反射回来得很弱，光敏三极管无法导通，所以此时A点输出为高电平。当被测物体是白色物体时，红外发光二极管发射出的光，被反射回来得很强，光敏三极管导通，所以此时A点输出为低电平；微控制器检测A点输出的电平，即可以判断此时被检测物体是白色物体还是黑色物体。

应用光电检测电路时应注意以下几点：

① 注意发光器的光强度。发光器的光强度可以通过选择适当的型号，改变加在发光器的限流电阻，或者在发光器和光敏器件的外面加上聚光装置。

② 注意反射式光电传感器与检测对象之间的距离。不同物体表面对光线的反射能力不同，应仔细调节反射式光电传感器与检测对象之间的距离。

③ 注意工作环境条件。由于无法改变工作环境，必须考虑光电传感器的安装位置。

2. 车载黑线检测电路实例

车载黑线检测电路如图5-4所示，装在小车上，共使用5个ST188，用于黑线寻迹。

3. 检测程序

黑线检测程序用来完成黑线的检测与判断，黑线检测程序流程图如图5-5所示。

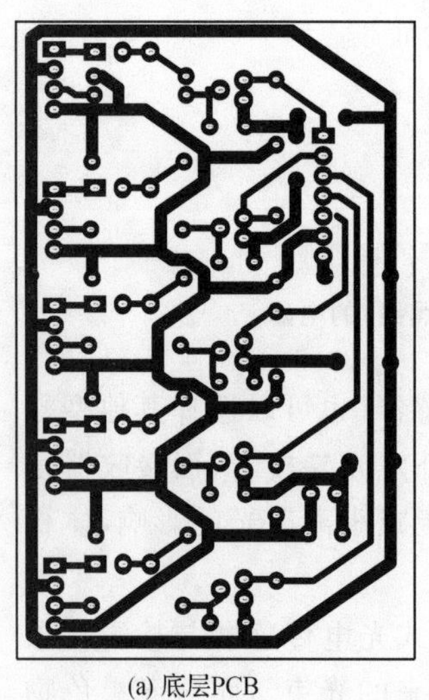

(a) 底层PCB

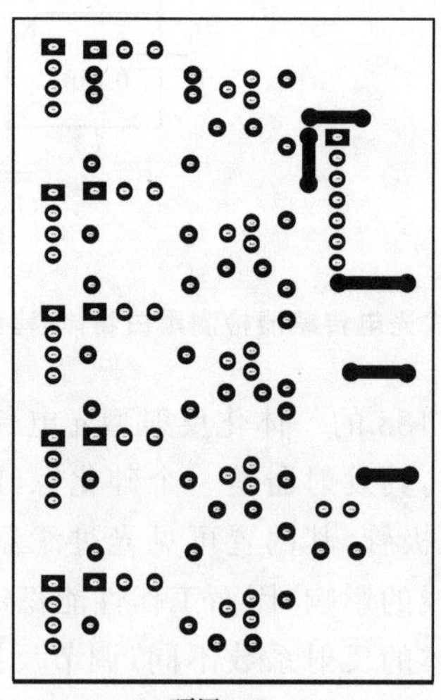

(b) 顶层PCB

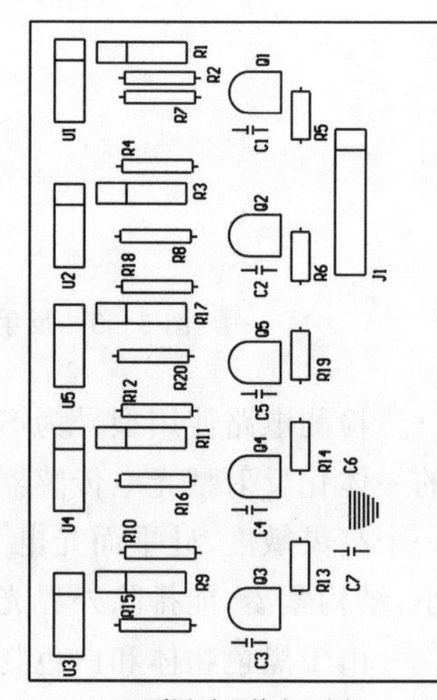

(c) 顶层元器件布局图

图5-4 车载黑线检测电路

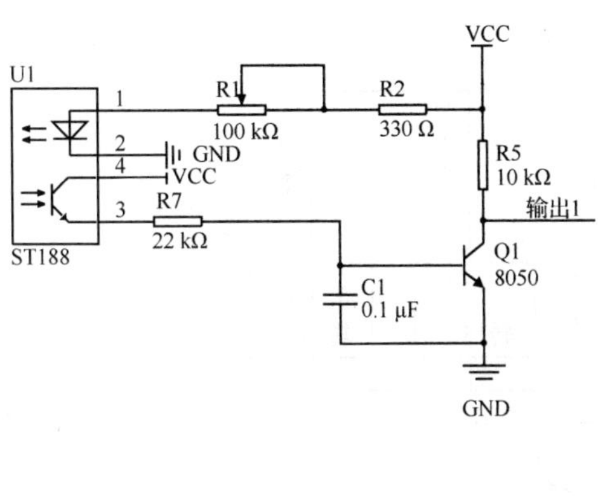

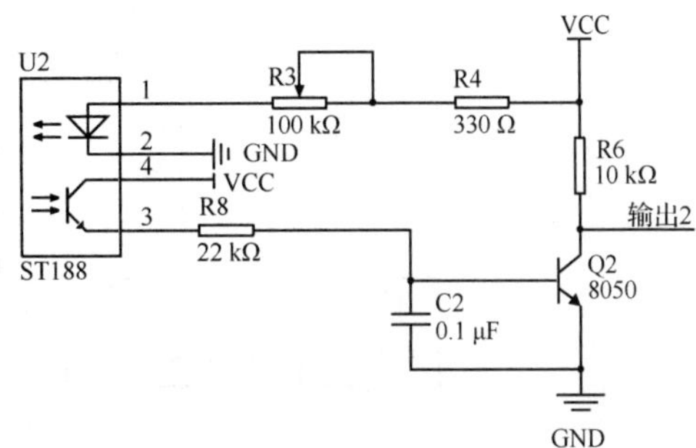

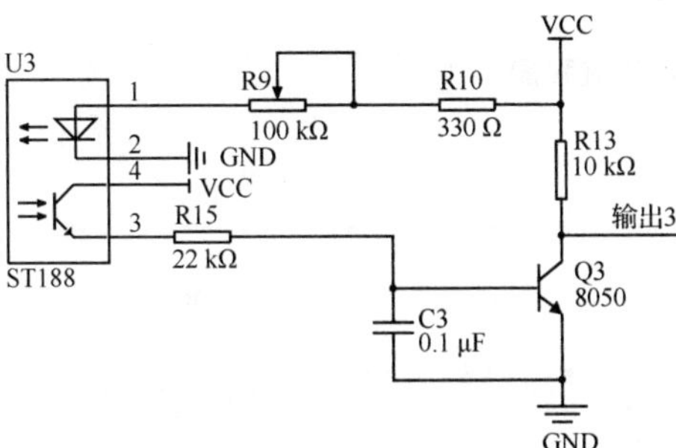

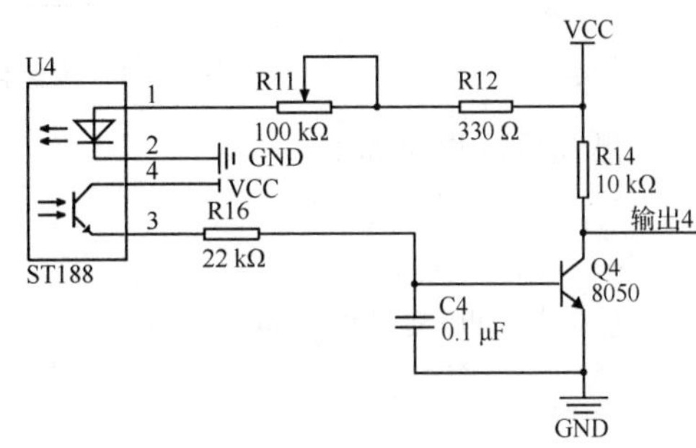

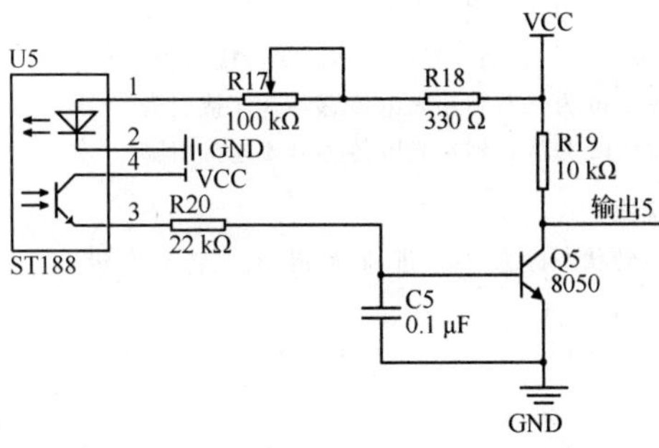

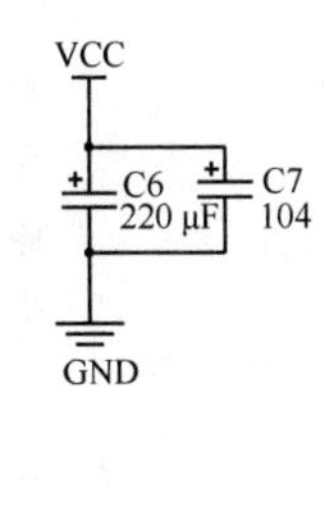

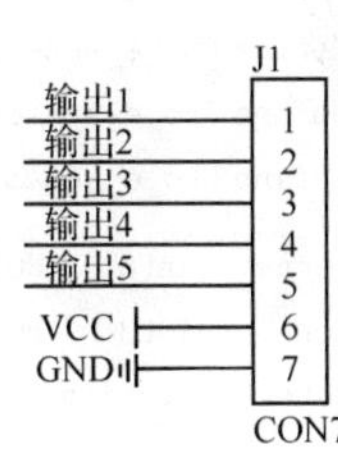

(d) 黑线检测电路

图 5-4　车载黑线检测电路(续)

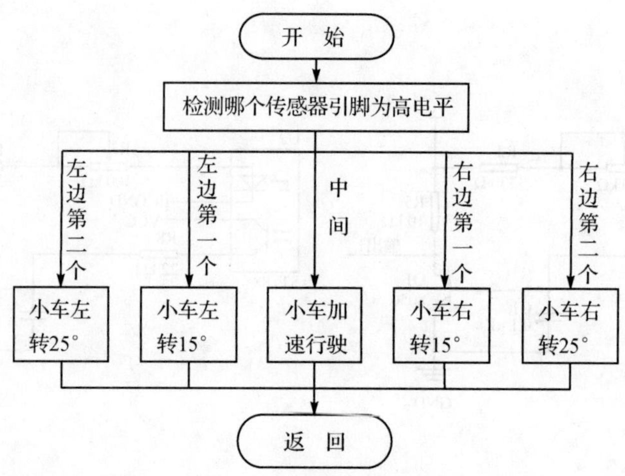

图 5-5 黑线检测程序流程图

黑线检测程序如下：

```
/**********************************************************************
*  文件名:HeiXian_check
*  功能:  用于检测黑线,控制小车前进
**********************************************************************/
#include   "config.h"
#define Pho_e_sensor1   1≪27               // 定义 P1.27 为从右到左光电传感器 1 连接引脚
#define Pho_e_sensor2   1≪20               // 定义 P0.20 为从右到左光电传感器 2 连接引脚
#define Pho_e_sensor3   1≪7                // 定义 P0.7 为从右到左光电传感器 3 连接引脚
#define Pho_e_sensor4   1≪30               // 定义 P0.30 为从右到左光电传感器 3 连接引脚
#define Pho_e_sensor5   1≪18               // 定义 P0.18 为从右到左光电传感器 4 连接引脚
extern   uint8   MOTO_STOP_Flag;           // 停机标志
extern   uint8   MOTO_Flag;                // 小车运转状态标志,1:前进,2:后退,3:左转 4:右转
extern   volatile uint8   MOTO_done_Flag;
/**********************************************************************
*  函数名称:HeiXian_check
*  函数功能:用于检测黑线,控制小车前进
*  入口参数:无
*  出口参数:无
**********************************************************************/
uint8 HeiXian_check(void)
{
    if((IO0PIN & Pho_e_sensor3) ! = 0 ) // 若中间检测到黑线,其他的不再检测
    {
        Speed_Grade_Change(1);           // 加快速度
        Car_Go_Forward_mm(100);
        return 0;
    }
```

```
    else
    {
        if((IO0PIN & Pho_e_sensor2) != 0 )
        {
            Speed_Grade_Change(2);              // 减慢速度

            MOTO_Turn_Right_Angle(15);          // 原地向右旋转
            while(MOTO_done_Flag == 0);         // 等待电机旋转完毕
        }
        else if((IO0PIN & Pho_e_sensor4) != 0 )
        {
            MOTO_Turn_Left_Angle(15);           // 左转
            while(MOTO_done_Flag == 0);         // 等待电机旋转完毕
        }
        else if((IO0PIN & Pho_e_sensor5) != 0 )
        {
            MOTO_Turn_Left_Angle(25);           // 左转
            while(MOTO_done_Flag == 0);         // 等待电机旋转完毕
        }
        else if((IO1PIN & Pho_e_sensor1) != 0 )
        {
            Speed_Grade_Change(2);              // 减慢速度等级

            MOTO_Turn_Right_Angle(25);          // 原地向右旋转
            while(MOTO_done_Flag == 0);         // 等待电机旋转完毕
        }
        else
        {
            Car_Go_Forward_mm(50);
        }
    }
    return 0;
}
```

5.1.4　利用光电传感器检测光源

1. 光源检测电路方案 1

(1) 电路结构与工作原理

一个光源检测电路如图 5-6 所示。

由光敏二极管 D2(也可以采用光敏三极管)对光源进行检测,当光敏二极管接收到光源发出的光时,VT1 和 VT2 导通,A 点为低电平,VT3 不能导通,B 点为高电平,此时微控制器接收到的电平为高电平;当光敏二极管未接收到光源时,VT1 和 VT2 不导通,A 点为高电平,VT3 导通,B 点输出低电平,此时微控制器接收到的电平为低电平。故微控制器通过检测输入端的电平值便可判断是否检测到了光源。

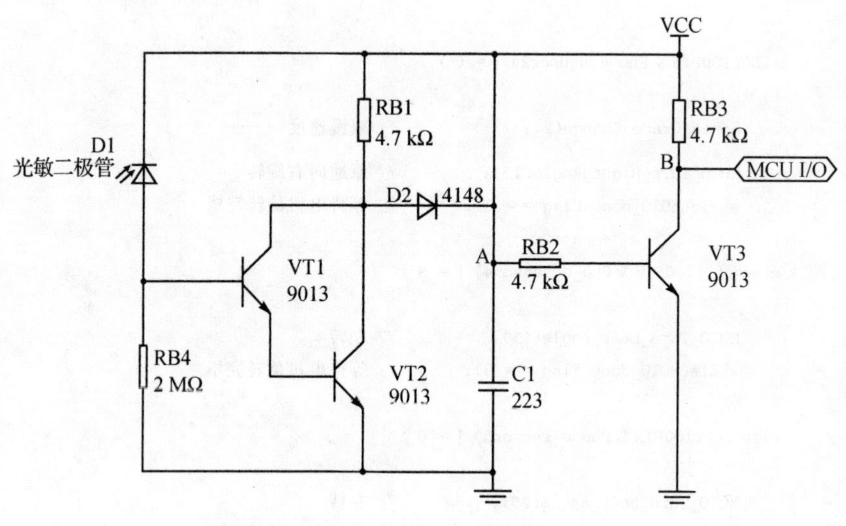

图 5-6 光源检测电路 1

(2) 应用实例

光源检测电路用来判断光源的位置。例如在 2003 年 E 题“简易智能电动车”中要求小车在光源的引导下,通过障碍区进入停车区并到达车库。光源探测电路的功能主要是引导小车朝光源行驶,使小车具有追踪光源的功能。由于光源会发出光线和热量,可以采用光敏二极管(或者光敏三极管)传感器和热释电传感器实现追踪光源的功能。光敏二极管传感的检测电路简单,因此在设计中采用了光敏传感器。

由于光源采用的是白炽灯,光线是射散的,为了便于小车能够在偏离光源一定角度的情况下仍能检测到光线,使用了三个互成一定角度的传感器组,这样增加了小车的检测范围。同时为提高传感器的方向性,在光敏二极管感应平面的前端固定一根 2 cm 长的塑料筒,光源检测实物示意图如图 5-7 所示。

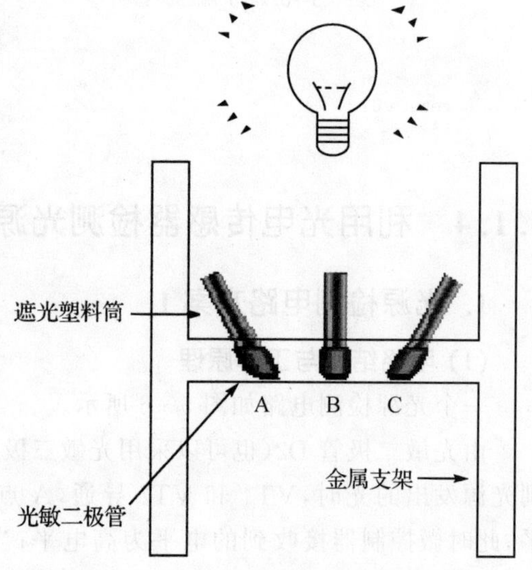

图 5-7 光源检测实物示意图

微控制器可根据这三个光敏二极管的状态,控制小车动作,寻找光源。小车的动作和传感器的对应关系如表 5-2所列。

下面给出方案 1 的程序流程图(见图 5-8)和程序清单 5.1。此处以查询法判断是否检测到光源为例，也可以接控制器的 3 个中断引脚，通过中断法来判断哪个方向检测到光源。

表 5-2　状态-动作对应表

状　态	光敏二极管 A	光敏二极管 B	光敏二极管 C	光源与车之间的位置	小车动作
1	0	0	0	非常远	—
2	0	0	1	在车的右端	右转(幅度大)
3	0	1	0	正对车	直线行驶
4	0	1	1	在车的右端	右转(幅度小)
5	1	0	0	在车的左端	左转(幅度大)
6	1	0	1	—	—
7	1	1	0	在车的左端	左转(幅度小)
8	1	1	1	非常近	减速直线行驶

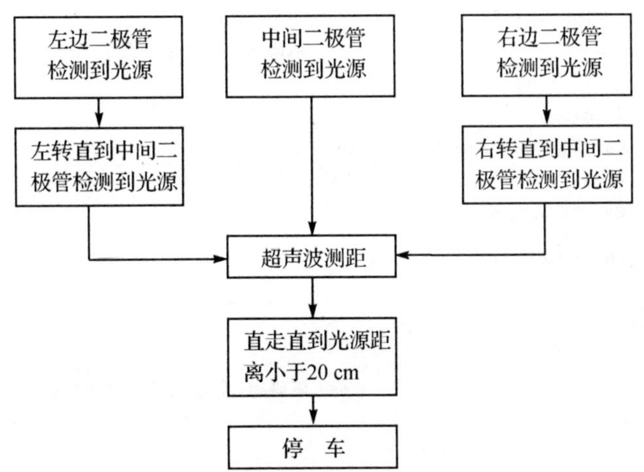

图 5-8　方案 1 程序流程图

程序清单 5.1　方案 1 程序示例

```
#include "config.h"
#define left_gy      1<<7;              // 定义 P0.7 作为左边光源检测引脚
#define middle_gy    1<<8;              // 定义 P0.8 作为中间光源检测引脚
#define right_gy     1<<9;              //定义 P0.9 作为右边光源检测引脚
uint8 GuangYuan_check (void)
{
    uint16 data = 0;
    if((IO0PIN & left_gy) == 0)         // 检测到左边有光源
    {
        while(IO0PIN & middle_gy)       // 小车左转直到中间检测到光源
```

```
            MOTO_TL();
        data = Distance();                          // 超声波测距
        if(data>20)                                 // 障碍物(光源)距离大于 20 cm 时直走
        {
            if(data > 30)
            Car_Go_Forward_mm((data - 20) / 2);
            else
            Car_Go_Forward_mm(50);

            while(MOTO_done_Flag == 0);
        }
        else
        {
            STOP();                                 // 障碍物(光源)距离小于 20 cm 时停车
        }
    }
    if((IO0PIN & middle_gy) == 0)                   // 检测到中间有光源
    {
        data = Distance();                          // 超声波测距
        if(data>20)                                 // 障碍物(光源)距离大于 20 cm 时直走
        {
            if(data > 30)
            Car_Go_Forward_mm((data - 20) / 2);
            else
            Car_Go_Forward_mm(50);

            while(MOTO_done_Flag == 0);
        }
        else
        {
            STOP();                                 // 障碍物(光源)距离小于 20 cm 时停车
        }
    }
    if((IO0PIN & right_gy) == 0)                    // 检测到右边有光源
    {
        while(IO0PIN & middle_gy)
            MOTO_TR();                              // 小车右转直到中间检测到光源
        data = Distance();                          // 超声波测距
        if(data>20)                                 // 障碍物(光源)距离大于 20 cm 时直走
        {
            if(data > 30)
            Car_Go_Forward_mm((data - 20) / 2);
            else
            Car_Go_Forward_mm(50);

            while(MOTO_done_Flag == 0);
        }
        else
```

```
            {
                STOP();                         // 障碍物(光源)距离小于 20 cm 时停车
            }
        }
        return 0;
    }
```

2. 光源检测电路方案 2

(1) 电路结构与工作原理

一个利用 3DU5C 硅光敏三极管构成的光源检测电路如图 5－9 所示。3DU5C 硅光敏三极管传感器在有光照的情况下将会导通，且光敏三极管导通后的输出电压随光源强度的变化而变化，利用 LPC2148 微控制器上的 ADC，通过 ADC 采样光敏三极管的输出电压，从而判断光敏三极管探测到的光源的强度。使用两个这样的电路，来分别判断光敏三极管探测到的光源的强度，并进行比较，便可以准确地判断光源的位置。

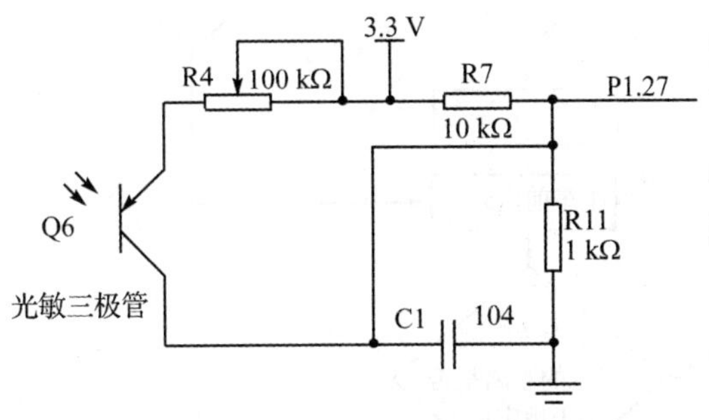

图 5－9　光源检测电路 2

(2) 应用实例

与光源检测电路方案 1 一样，同样用于在 2003 年 E 题“简易智能电动车”中，小车在光源的引导下，通过障碍区进入停车区并到达车库。

光源检测电路程序流程图如图 5－10 所示，程序代码如程序清单 5.2 所示。

程序清单 5.2　方案 2 程序示例

```
/*********************************************************************
*  文件名:GuangYuan_Check.C
*  功能:光源查找程序包,包括对光敏三极管电压值 A/D 采样、控制小车朝向光源,并
*        逐步向光源运动
*********************************************************************/
#include  "config.h"
uint32 Right_gy_Date;
uint32 Left_gy_Date;
uint32 battery_rate;                    // 检测到的电源值
uint32 right_data;
```

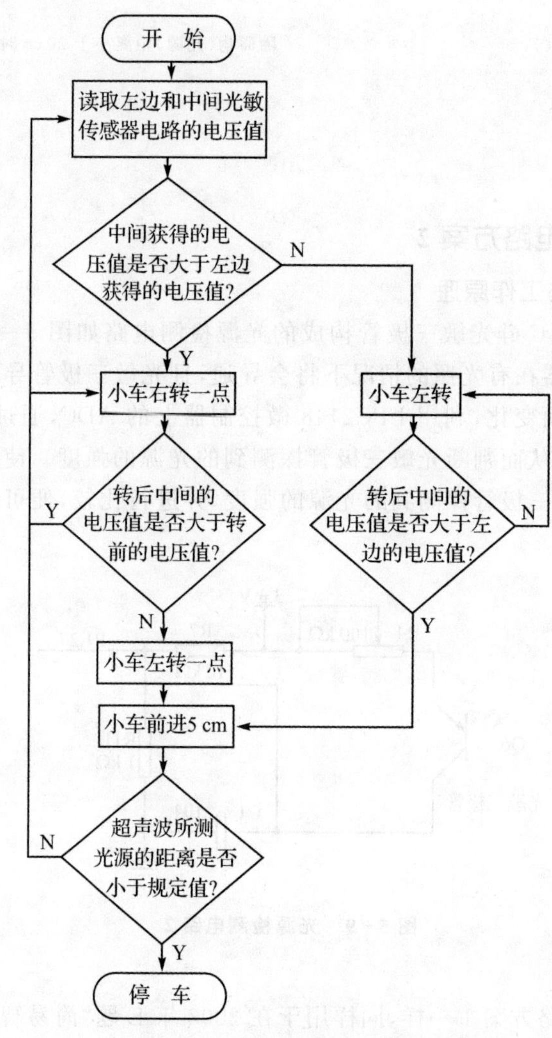

图 5-10 光源检测程序流程图

```
extern volatile uint8 MOTO_done_Flag;          // 小车运转完成标志
/***********************************************************************
 * 文件名:Left_gy_sure
 * 功能:  用于读取左光源转换后的确定值,读 5 次值,排序后,读取中间值
 * 说明:  调用
***********************************************************************/
uint32 Left_gy_sure(void)
{
    uint8 times ;                              // 探测光源次数
    uint32 Left_table[5]  ;
    uint32 data;
    uint8 i,j;
    /********************* 读取 5 次光源的电压值 *************************/
```

```
    for(times = 0;times < 5;times ++ )
    {
        Left_table[times]  = Left_ADC_Read_Date();DelayNS(10);
    }
    /******************** 将 5 次光源的电压值按升序排列 *****************/
    for(i = 0; i < 4;i ++ )
    {
        for(j = i + 1;j < 5;j ++ )
            if(Left_table[i] > Left_table[j])
            {
                data = Left_table[i];
                Left_table[i] = Left_table[j];
                Left_table[j] = data;
            }
    }
    Left_gy_Date   = Left_table[2];          // 取数组中间值
    return (Left_gy_Date) ;
}
/****************************************************************************
*  文件名:Right_gy_sure
*  功能:  用于读取右光源转换后的确定值,读 5 次值,排序后,读取中间值
*  说明:  调用
****************************************************************************/
uint32  Right_gy_sure(void)
{
    uint8   times ;                         // 探测光源次数
    uint32 Right_table[5] ;
    uint32 data;
    uint8   i,j;
    /******************** 读取 5 次光源的电压值 *************************/
    for(times = 0;times < 5;times ++ )
    {
        Right_table[times] = Right_ADC_Read_Date();
    }
    /******************* 将 5 次光源的电压值按升序排列 *******************/
    for(i = 0; i < 4;i ++ )
    {
        for(j = i + 1;j < 5;j ++ )
        if(Right_table[i] > Right_table[j])
        {
            data = Right_table[i];
            Right_table[i] = Right_table[j];
            Right_table[j] = data;
        }
    }
    Right_gy_Date = Right_table[2];                          // 取数组中间值
    return (Right_gy_Date);
```

```
}
/**************************************************************************
 * 文件名:gy_decide_MOTO
 * 功能:  用于寻找光源,控制小车朝光源方向前进,直到超声波测距小于 Guangyuan_Safe 的值
 * 入口参数:无
 * 出口参数:0 表示未到达光源,1 表示已经结束光源寻找
 **************************************************************************/
uint8  Seek_END(void)
{
   uint32 data;
   gy_decide_MOTO();                                        // 检测光源,前进
   data = Distance();
   if(data > Guangyuan_Safe)
   {
      gy_decide_MOTO();                                     // 检测光源,前进
      data = Distance();
      if(data > Guangyuan_Safe)
      {
         if( (500<data) & (data<3000) )                     // 小于 3 m
         {
            Car_Go_Forward_mm((data - Guangyuan_Safe)/2);   // 前进
            while(MOTO_done_Flag == 0);
         }
         else if( (Guangyuan_Safe + 10) <data)
         {
            Car_Go_Forward_mm(100);                         // 前进 100 mm
            while(MOTO_done_Flag == 0);
         }
         else
         {
            Car_Go_Forward_mm(50);                          // 前进 50 mm
            while(MOTO_done_Flag == 0);
         }
      }
     return 0;
   }
   else
   {
      STOP();
      Beep_Time1(80);                                       // 蜂鸣器鸣响
      return 1;                                             // 完成光源查找任务
   }
}
/**************************************************************************
 * 文件名:gy_decide_MOTO
 * 功能:  用于检测光源,控制小车朝光源方向转动
```

```
* 入口参数：无
* 出口参数：无
*****************************************************************/
void  gy_decide_MOTO(void)
{
    Right_gy_Date = Right_gy_sure();                // 读取右边光源强度值
    Left_gy_Date  = Left_gy_sure();                 // 读取左边光源强度值
    if(Right_gy_Date  > Left_gy_Date)               // 若右边光源强于左边
    {
        right_data = Right_gy_Date;                 // 将右边光源强度值保存在 right_data
        MOTO_Turn_Right_Angle(5);                   // 小车右转一点
        while(MOTO_done_Flag == 0);                 // 等待车轮转完
        Right_gy_Date = Right_gy_sure();            // 读取转动后的右边的强度值
        if(Right_gy_Date  > right_data)             // 如果向右转一点后,强度变强
        {
            /* 直到检测到右转后的光强度比不转时弱才跳出循环 */
            while( (Right_gy_Date > right_data))
            {
                right_data = Right_gy_Date;         // 将最大的数据放入 right_data 中
                MOTO_Turn_Right_Angle(5);           // 继续右转一点
                while(MOTO_done_Flag == 0);
                Right_gy_Date = Right_gy_sure();
            }
            right_data = Right_gy_Date;
            MOTO_Turn_Left_Angle(5);  // 左转一点(和之前右转的时间相同)此时小车对准光源
            while(MOTO_done_Flag == 0);
            Right_gy_Date = Right_gy_sure();
        }
        else
        {
            MOTO_Turn_Left_Angle(5);
            while(MOTO_done_Flag == 0);
        }
    }
    else                                            //若左边光源强于右边
    {
    /* 左转直到右边强于左边 */
            while( (Left_gy_Date > Right_gy_Date))
            {
                MOTO_Turn_Left_Angle(5);
                while(MOTO_done_Flag == 0);
                Left_gy_Date = Left_gy_sure();
                Right_gy_Date = Right_gy_sure();
            }
    }
```

```
}
/****************************************************************************
*  文件名:ADC.C
*  功能:   LPC2148 的 ADC 软件包,包括 ADC 引脚初始化和数据读取等,用于对电池电压采
*          样和对光敏三极管电压值采样
*  说明:   采用直接读取方式,未使用中断
****************************************************************************/
#include   "config.h"
/****************************************************************************
*  函数名:ADC_Init
*  功能:   ADC 引脚初始化,清除所用 ADC 通道的 DONE 标志位
*  说明:   调用
****************************************************************************/
void ADC_Init(void)
{
   uint32     ADC_Data;
   PINSEL1 = (PINSEL1 & 0xFcFFFFFF )| 0x01000000;        // 设置 P0.28 连接到 AD0.1 电池
/* 设置 P0.5  连接到 AD0.7(左光源) 设置 P0.12 连接到 AD1.3(右光源) */
  PINSEL0 = (PINSEL0 & 0xFcFFF3FF )| 0x03000c00;

/*********************** 清零 AD0DR7 中标志位 ***************************/
   ADCR =   ( 1 << 7 )                       | // 选择 AD0.7
           ((Fpclk / 1000000 - 1) << 8)      | // 即转换时钟为 1 MHz
           (0 << 16)                         | // BURST = 0,软件控制转换操作
           (0 << 17)                         | // CLKS = 0,使用 11 clock 转换
           (1 << 21)                         | // PDN = 1,正常工作模式
           (0 << 22)                         | // TEST1:0 = 00,正常工作模式
           (1 << 24)                         | // START = 1,直接启动 ADC 转换
           (0 << 27);                          // EDGE = 0 (CAP/MAT 引脚下降沿触发 ADC 转换)
   DelayNS(10);
   ADC_Data = ADDR;                            // 读取 ADC 结果,并清除 DONE 标志位
  /********************** 清零 AD1GDR 中标志位 **************************/
   AD1CR = ( 1 << 3 )                        | // 选择 AD1.3
           ((Fpclk / 1000000 - 1) << 8)      | // 转换时钟为 1 MHz
           (0 << 16)                         | // BURST = 0,软件控制转换操作
           (0 << 17)                         | // CLKS = 0,使用 11 clock 转换
           (1 << 21)                         | // PDN = 1,正常工作模式
           (0 << 22)                         | // TEST1:0 = 00,正常工作模式
           (1 << 24)                         | // START = 1,直接启动 ADC 转换
           (0 << 27);                          // EDGE = 0 (CAP/MAT 引脚下降沿触发 ADC 转换)
   DelayNS(10);
   ADC_Data   =   AD1GDR;
 /********************** 清零 AD0DR1 中标志位 ***************************/
   ADCR = ( 1 << 1 )                         | // SEL.1 = 1,选择 AD0.1
           ((Fpclk / 1000000 - 1) << 8)      | // CLKDIV = Fpclk / 1 000 000 - 1,即转换时钟为 1 MHz
           (0 << 16)                         | // BURST = 0,软件控制转换操作
           (0 << 17)                         | // CLKS = 0,使用 11 clock 转换
```

```
            (1 ≪ 21)                    | // PDN = 1,正常工作模式
            (0 ≪ 22)                    | // TEST1:0 = 00,正常工作模式
            (1 ≪ 24)                    | // START = 1,直接启动 ADC 转换
            (0 ≪ 27);                     // EDGE = 0 (CAP/MAT 引脚下降沿触发 ADC 转换)
    DelayNS(10);
    ADC_Data = ADDR;                      // 读取 ADC 结果,并清除 DONE 标志位
}
/*********************************************************************
*  函数名称: Left_ADC_Read_Date
*  函数功能: 读取左边光源 ADC 转换结果
*  入口参数: 无
*  出口参数: 左边光源强度
*  说    明: 调用
*********************************************************************/
uint32 Left_ADC_Read_Date(void)
{
    uint32   Left_ADC_Data;
    ADCR = (ADCR & 0xFFFFFF00)| (1≪7) | (1 ≪ 24);        // 切换通道并进行第一次转换
    while( (ADDR & 0x80000000) == 0 );                   // 等待转换结束
    ADCR = ADCR | (1 ≪ 24);                              // 再次启动转换
    while( (ADDR & 0x80000000) == 0 );
    Left_ADC_Data = ADDR;                                // 读取 ADC 结果
    Left_ADC_Data = (Left_ADC_Data≫6) & 0x3FF;
    Left_ADC_Data = Left_ADC_Data * 2048;                // 基准电压为 2 048 mV
    Left_ADC_Data = Left_ADC_Data / 1024;

    return   Left_ADC_Data;
}
/*********************************************************************
*  函数名称: Right_ADC_Read_Date
*  函数功能: 读取右边光源 ADC 转换结果
*  入口参数: 无
*  出口参数: 右边光源强度
*  说    明: 调用
*********************************************************************/
uint32 Right_ADC_Read_Date(void)
{
    uint32   Right_ADC_Data;
    AD1CR = (AD1CR & 0xFFFFFF00)|(1≪3)|(1 ≪ 24);         // 切换通道并进行第一次转换
    while( (AD1GDR & 0x80000000) == 0 );                 // 等待转换结束
    AD1CR = AD1CR | (1 ≪ 24);                            // 再次启动转换
    while( (AD1GDR & 0x80000000) == 0 );
    Right_ADC_Data = AD1GDR;                             // 读取 ADC 结果

    Right_ADC_Data = (Right_ADC_Data≫6) & 0x3FF;
    Right_ADC_Data = Right_ADC_Data * 2048;              // 基准电压为 2 048 mV
    Right_ADC_Data = Right_ADC_Data / 1024;

    return   Right_ADC_Data;
```

```
}
/***********************************************************************
* 函数名称：ADC_Read_PowerDate
* 函数功能：读取 A/D 转换后的值
* 入口参数：无
* 出口参数：电源电压值
* 说    明：调用
***********************************************************************/
uint32 ADC_Read_PowerDate(void)
{
    uint32  ADC_Power_Data;
    ADCR = (ADCR&0xFFFFFF00)|(1<<1)|(1 << 24);            // 切换通道并进行第一次转换
    while( (ADDR&0x80000000) == 0 );                       // 等待转换结束
    ADCR = ADCR | (1 << 24);                               // 再次启动转换
    while( (ADDR&0x80000000) == 0 );
    ADC_Power_Data = ADDR;                                 // 读取 ADC 结果
    ADC_Power_Data = (ADC_Power_Data>>6) & 0x3FF;
    ADC_Power_Data = ADC_Power_Data * 2048;                // 基准电压为 2 048 mV
    ADC_Power_Data = ADC_Power_Data / 1024;
    return ADC_Power_Data;
}
```

5.2 超声波传感器及其应用

5.2.1 超声波传感器的基本特性与选型

超声波传感器可以用来测量距离，探测障碍物，区分被测物体的大小。超声波传感器的种类繁多，生产厂商也很多，以"超声波传感器"为关键词，可以在相关网站查询到大量的资料。

1. 超声波的基本特性

超声波检测装置包含有一个发射器和一个接收器。发射器向外发射一个固定频率的声波信号，当遇到障碍物时，声波返回被接收器接收。

超声波探头可由压电晶片制成，超声波探头可以发射超声波，也可以接收超声波。小功率超声探头多作探测用，有多种不同的结构。

超声探头构成晶片的材料可以有许多种。晶片的大小，如直径和厚度各不相同，因此每个探头的性能也不同。超声波传感器的主要性能指标包括：工作频率、灵敏度和工作温度。

工作频率就是压电晶片的共振频率。当加到晶片两端的交流电压的频率和晶片的共振频率相等时，输出的能量最大，灵敏度也最高。

灵敏度主要取决于制造晶片本身。机电耦合系数大,灵敏度高;反之,灵敏度低。

2. 超声波传感器的类型

(1) T/R－40－XX 系列通用型超声波发射/接收传感器

T/R－40－XX 系列超声波传感器分为发射和接收两种,发射器型号为 T－40－XX,接收器型号为 R－40－XX。T/R－40－XX 系列超声波传感器的振子是用压电陶瓷制成的,利用压电效应工作的传感器,加上谐振喇叭可提高动作灵敏度。当处于发射状态时,外加共振频率的电压能产生超声波,将电能转化为机械能;当处于接收状态时,又能很灵敏地探测到共振频率的超声波,将机械能转化为电能。T/R－40－XX 系列超声波传感常在以空气作为传播媒介的遥控发射和接收电路中使用。

T/R－40－XX 系列超声波传感器在实际应用中分为:用于遥控及报警电路的直射型;用于测距、料位测量等电路的分离反射型;用于材料的探伤、测厚等电路的反射型。

(2) MA40EIS/EIR 系列密封式超声波发送/接收传感器

MA40EIS/EIR 系列密封式超声波发射/接收传感器是一种具有防水功能的超声波传感器(但不能放入水中),适用于物位监测及遥控开关电路。其外形、尺寸和符号与 T/R－40－XX 系列类似。

(3) UCM－40－T/R 超声波发射/接收传感器

UCM－40－T/R 超声波发射/接收传感器的外形、尺寸及特性与 T/R－40－XX 系列基本相同。

(4) 美国 Honeywell 生产的 900 系列超声波精密接近传感器

900 系列超声波精密接近传感器的发射距离为 100～6 000 mm,声波发射角为 5.9°～236°,重复精度为±1～±5 mm 或测量距离的 0.3%,工作电压范围为 10～30 V,具有数字和模拟输出形式,可在液位或料位控制、装瓶或装罐时的液位控制、工件的有/无检测、高度/宽度测量、距离测量等领域应用。

5.2.2　超声波传感器用于障碍物检测与测距

1. 超声波障碍物检测原理

超声波发射器向某一方向发射超声波,在发射时刻的同时开始计时,超声波在空气中传播,途中碰到障碍物就立即返回来,超声波接收器收到反射波就立即停止计时。超声波在空气中的传播速度为 340 m/s,根据计时器记录的时间 t,就可以计算出发射点距障碍物的距离 s,即

$$s = 340t/2 \qquad (5.2.1)$$

障碍物检测电路中的超声波传感器可以采用 T/R－40、UCM40 等压电陶瓷超声波传感器,它的工作电压是 40 kHz 的脉冲信号。

超声波传感器构成的障碍物检测与测距电路由发射和接收两部分组成。发射部分包括 40 kHz PWM 脉冲的产生和输出电路，接收部分包括接收信号的放大与比较电路。

2. 超声波发射电路

超声波发射电路包括超声波发射器、40 kHz 超音频振荡器、驱动(或激励)电路，有时还包括编码调制电路，设计时应注意以下两点：

① 普通用的超声波发射器所需电流小，只有几毫安到十几毫安，但激励电压要求在 4 V 以上。

② 激励交流电压的频率必须调整在发射器中心频率 f_0 上，才能得到高发射功率和高效率。

一个超声波发射电路如图 5-11 所示，PWM 脉冲由微控制器 LPC2148 产生。例如：PWM 脉冲可以通过控制 LPC2148 内部的 PWM 模块中的匹配寄存器 0 (PWMR0)来控制 PWM 的周期和频率，选用 PWM5 引脚即 LPC2148 的 P0.21 引脚输出，其输出频率为

$$f = \frac{f_{\mathrm{PCLK}}}{\mathrm{PWMR0}(P_{\mathrm{WMPR}} + 1)} \tag{5.2.2}$$

式中，f_{PLCK} 为晶振频率；P_{WMPR} 为 PWM 预分频的值，若要达到频率 $f=40$ kHz，则取 $P_{\mathrm{WMPR}}=99$，匹配寄存器 PWMR0 的值为 3。因此，取 $P_{\mathrm{WMPR}}=99$，PWMR0＝3，即可满足要求。

产生的 PWM 脉冲通过三极管提高其电压值后再输入 CD4049 的 3 引脚，经 CD4049 增强驱动能力后由 CD4049 的 10 引脚、12 引脚输出，10 引脚和 12 引脚的输出信号需分别串联一个 224 的电容，以滤除输出信号中的直流信号。最后通过超声波发射传感器将信号输出。

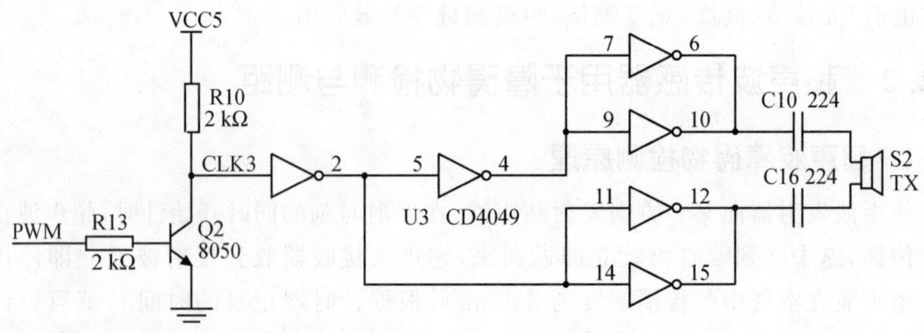

图 5-11 超声波发射电路

3. 超声波接收电路

配套的超声波接收电路如图 5-12 所示。

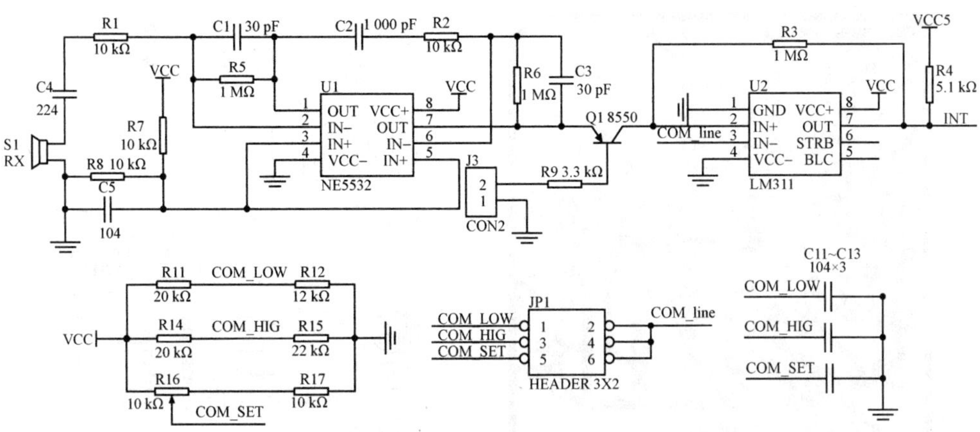

图 5-12　超声波接收电路

当超声波接收传感器接收到信号时，由超声波接收器的两脚输出，2 引脚信号接地，1 引脚信号经 224 电容滤除直流后接 10 kΩ 电阻 R1，再将 R1 接 NE5532 的 2 引脚反相输入端，3 引脚正相输入端通过 10 kΩ 阻值的 R2、R3 分得 VCC 的 2.5 V 电压，保证了 3 引脚工作所需的正常电位(这是因为输入的交流信号电压太小了，要加个直流偏置电压，放大器才能正常放大)，输出端经阻值为 1 MΩ 的电阻 R5 和容量为 30 pF 的电容 C3 反馈到反相输入端；经 NE5532 的一级放大后的交流信号电压由 1 引脚输出，再将 1 引脚输出信号接 NE5532 的 6 引脚反相输入端，用相同电路进行第二级的同倍放大，最后将最终放大了 10 000 倍的超声波信号由 7 引脚输出。

由 $\frac{U_{\text{out1}}}{U_{\text{in}}}=\frac{\text{R5}}{\text{R1}}$ 和 $\frac{U_{\text{out2}}}{U_{\text{out1}}}=\frac{\text{R6}}{\text{R2}}$ 得出放大倍数计算公式如下：

$$\frac{U_{\text{out2}}}{U_{\text{in}}}=\frac{\text{R6}\cdot\text{R5}}{\text{R2}\cdot\text{R1}} \tag{5.2.3}$$

式中，U_{in} 为信号的输入电压，U_{out1} 为一级放大后的信号电压，U_{out2} 为二级放大后的信号电压，将 R6＝R5＝1 MΩ，R2＝R1＝10 kΩ 代入上式得出放大后的信号电压与放大信号的输入电压之比为 10 000。

将放大 10 000 倍后的信号，输入到 LM311 电压比较器的 2 引脚，3 引脚输入比较电压，对接收信号进行调整；将调整后的超声波信号通过 7 引脚输出到 LCP2148 的外部中断引脚，若放大后的信号电压高于比较电压，则 7 引脚输出高电平 5 V，反之，则为低电平。可通过 3 引脚输入的比较电压的不同，选择不同的测距模式，距离越远，信号电压越弱，这时需降低 3 引脚输入的比较电压。在放大器和比较器之间用 8550PNP 三极管作为通路选择，使用时，将 J1 跳线帽短接，固定时三极管导通便可。

超声波传感器构成的障碍物检测电路的 PCB 图和元器件布局图如图 5-13 所示。

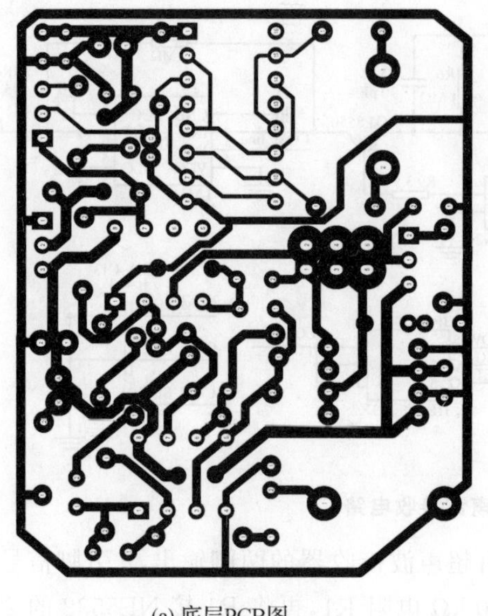

(a) 底层PCB图

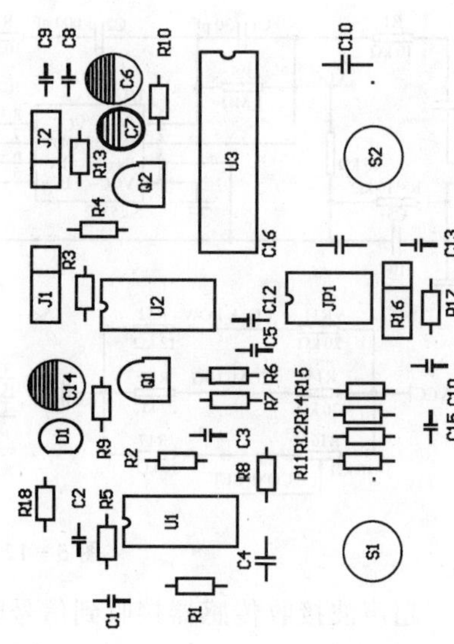

(b) 元器件布局图

图 5-13　超声波传感器构成的障碍物检测电路的 PCB 图和元器件布局图

4. 超声波障碍物检测与测距的注意事项

超声波障碍物检测与测距时应注意避免余波信号的干扰，如图 5-14 所示。图中显示了通过示波器采集的超声波发射头和接收头的两个波形。

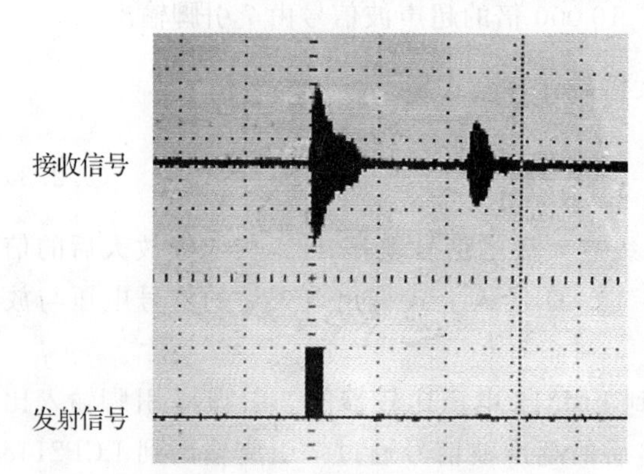

图 5-14　余波信号的干扰

从图 5-14 中可以看到，当发射头发出一组 40 kHz 的脉冲后（图中下面的波形），接收头几乎在同一时间就收到了超声波信号，这个波束是余波信号。持续一段时间后，才可以看到超声波接收头又收到了一组波束，这个才是经过被测物表面反射的回波信号。

超声波障碍物检测与测距，需要检测的是开始发射到接收到信号的时间差，由图 5-14 中可以看出，需要检测的有效信号为反射物反射的回波信号，故要尽量避免检测到余波信号。这就要求对接收头收到的波束进行处理，这也是超声波检测中存在最小测量盲区的主要原因。

在软件中的处理方法就是，当发射头发出脉冲后，计时器同时开始计时。在计时器开始计时一段时间后再开启检测回波信号，以避免余波信号的干扰。等待的时间可以为 1 ms 左右。更精确的等待时间可以减小最小测量盲区。

5. 超声波测距模块

为了方便使用，目前有一些生产厂商能够提供成品的超声波测距模块，以“超声波测距模块”为关键词，可以在相关网站查询到超声波测距模块的资料。

例如：凌阳公司的超声波测距模组如图 5-15 所示。该超声波测距模组的超声波传感器谐振频率为 40 kHz；模组传感器工作电压为 4.5～9 V；模组接口电压为 4.5～5.5 V；三种测距模式选择跳线 J1（短距 10～80 cm，中距 80～400 cm，可调距，其范围由可调节参数确定）。一般应用时，只需要用 10PIN 排线把 J8 与 SPCE061A 的 IOB 低 8 位接口接起来，同时设置好 J7、J1、J2 跳线就可完成硬件的连接了。不同测距模式的选择只需改变测距模式跳线 J1 的连接方法。

图 5-15 超声波测距模组

5.2.3 超声波传感器用于障碍物检测与测距编程示例

以 LPC2148 控制器为例，给出使用超声波传感器进行障碍物测距的软件流程图和编程示例。流程图如图 5-16 所示，编程示例如程序清单 5.3 所示。

启动 ADS 1.2，使用 ARM Executable Image for lpc2148 工程模板创建一个工程。使用过程中，需要 Startup.s 修改系统模式，堆栈设置为 0x5f，开启中断。本示例是将所测距离使用 MP3 语音模块进行语音播报，用户可根据实际情况进行修改，如采用液晶显示等。

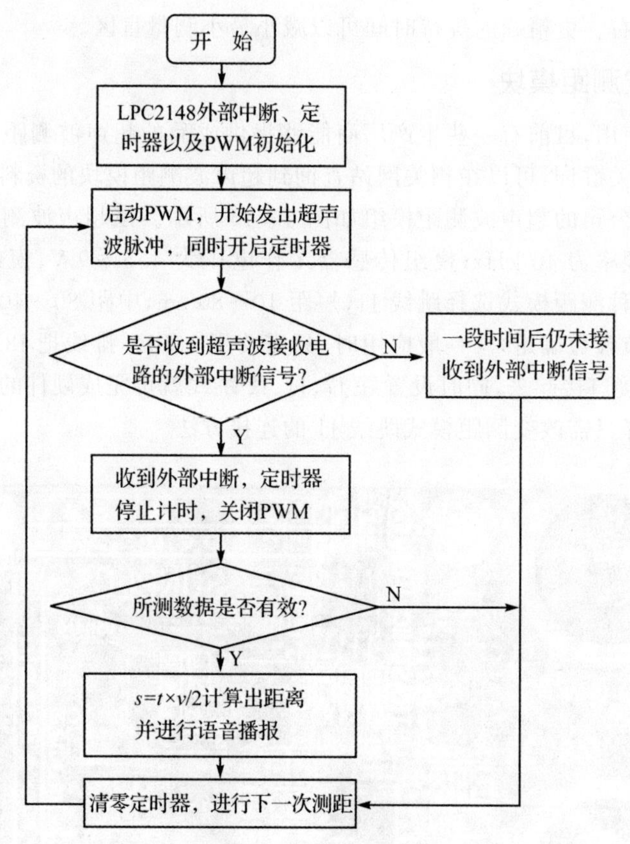

图 5－16　程序流程图

程序清单 5.3　超声波测距示例程序

```
/************************************************************************
*  文件名:ChaoShengbo_Ceju.C
*  功　能:LPC2148 控制超声波测距模块进行距离测量,使用 LCD12864 显示
*  说　明:将 P0.21、P0.7 分别连接到超声波模块的 CLK 引脚和 INT 外部中断引脚
************************************************************************/
#include   "config.h"

char  ChaoShengFlag = 0;                          // 定义超声波接收标志
uint32 Time0_times = 0;                           // 定时器 0 的 TC 计数值
uint16 distance = 0;                              // 障碍物距小车距离
fp64  middle_v = 0;                               // 中间转换变量
/************************************************************************
*  函数名称:mDelaymS ()
*  函数功能:软件延时
*  入口参数:ms  延时参数,值越大,延时越久
*  出口参数:无
```

```
*********************************************************************/
void    mDelaymS( uint32 ms )
{
    uint32     i;
    while ( ms -- ) for ( i = 25000; i != 0; i -- );
}
/*********************************************************************
* 函数名称:Time0Init()
* 函数功能:初始化定时器0,用于超声波测距计时
* 入口参数:无
* 出口参数:无

*********************************************************************/
void Time0Init(void)
{
    T0PR   = 0;                         // 设置定时器0不分频,得12 MHz
    T0MCR = 0x02;                       // 匹配通道0匹配不中断复位T0TC
    T0MR0 = 0xffffffff;                 // 匹配值为最大
    T0TCR = 0x03;                       // 启动并复位T0TC
    T0TCR = 0x01;                       // 启动T0TC
}
/*********************************************************************
* 函数名称:IRQ_Eint1()
* 函数功能:外部中断EINT1服务函数,关闭定时器0,用于超声波测距
* 入口参数:无
* 出口参数:无
*********************************************************************/
void __irq IRQ_Eint1(void)
{
    T0TCR   = 0x00;                     // 关闭定时器
    PWMTCR = 0x00;                      // 关闭PWM定时器,停止PWM
    Time0_times = T0TC;                 // 读取定时器0的值
    ChaoShengFlag = 1;                  // 超声波接收标志置1
    EXTINT = 0x02;                      // 清零EINT1中断标志
    VICIntEnClr = 1<<15;                // 关闭EINT1中断,在play()函数再开启中断
    VICVectAddr = 0;                    // 结束中断
}
/*********************************************************************
* 函数名称:Eint1_Init()
* 函数功能:初始化外部中断1(EINT1)为向量中断,并设置为上升沿触发模式,用于超声波
* 说    明:在STARTUP.S文件中使能IRQ中断(清零CPSR中的I位)
*********************************************************************/
```

```
void Eint1_Init(void)
{
    /* 设置P0.14为外部中断1(超声波接收中断) */
    PINSEL0 = (PINSEL0 & 0xcfffffff) | 0x20000000;
    EXTMODE |= (1 << 1);                          // 设置外部中断为边沿触发模式
    EXTPOLAR &=  ~(1 << 1);                       // 设置为下降沿触发
     /*采用向量IRQ开启EXINT1*/
    VICIntSelect = 0x00000000;                    // 设置所有中断为IRQ中断
    VICVectCntl0 = 0x2f;
    VICVectAddr0 = (int)IRQ_Eint1;                // 设置中断服务程序地址
    EXTINT = 0x02;                                // 清零EINT1中断标志
    VICIntEnable = 1<<15;                         // 1<<15使能EINT1中断
}
/*********************************************************************
* 函数名称: PWM5_Init()
* 函数功能: 初始化PWM6连接到P0.9引脚
* 入口参数: freq 初始化PWM频率
*********************************************************************/
void PWM5_Init(uint32 freq)                       // 所需的freq
{
    PINSEL1 = (PINSEL1 & 0xffff3ff) | 0x00000400;  // 将PWM5连接到P0.21引脚
    PWMPCR |= (1<<13);                            // 允许PWM5输出,单边PWM
    PWMPR  = 0x00;                                // 不分频,计数频率为Fpclk
    PWMMCR  = 0x02;                               // 设置PWMMR0匹配时复位PWMTC
    PWMMR0 = Fpclk/freq;                          // PWM频率
    PWMMR5 = PWMMR0/2;                            // 设置PWM占空比1/2
    PWMLER |= (0x01 | 1<<5);                      // PWMMR0、PWMMR5锁存
    PWMTCR = 0x09;                                // 启动定时器,PWM使能
}
/*********************************************************************
* 函数名称:Play()
* 函数功能:超声波测距初始化,清零标志和数据变量,开启定时器和外部中断,进行下一次测距
* 入口参数:无
* 出口参数:无
*********************************************************************/
void Play(void)
{
    ChaoShengFlag = 0;
    Time0_times = 0;
    T0TCR  = 0x03;                                // 使能定时器并复位定时器
```

```
    T0TCR   = 0x01;                          // 使能定时器
    PWMTCR = 0x09;                           // 启动定时器,PWM 使能

    VICIntEnable = 1≪15;                     // 使能外部中断 EINT1
}

/*********************************************************************
*  函数名称:main()
*  函数功能:测量障碍物,语音播报障碍物距离
*  入口参数:无
*  出口参数:无
*********************************************************************/
int main(void)
{
    fp64 x = 0;

    UART1_Ini(9600);
    PWM5_Init(40000);                        // 调用 PWM 初始化函数
    Eint1_Init();                            // 调用外部中断初始化函数
    Time0Init();                             // 调用 Time0 初始化函数

    mDelaymS(500);

    while(1)
    {
        if(ChaoShengFlag == 1)               // 接收到外部中断
        {
            if(Time0_times > 30)             // 由时间判断数据是否有效
            {
                x = (fp64)Time0_times;
                middle_v =  x * 0.345 /24.0; // 由声速和时间计算障碍物距离
                distance = (uint32)middle_v;

                MP3_distance ();             // 语音播报障碍物距离
            }
                Play();
        }
        else
        {
            if(Time0_times > 300)            // 若定时器时间超过 300,则数据无效
                Play();                      // 重新开启
        }

    } return(0);
}
```

5.3 图像识别传感器及其应用

5.3.1 图像识别模组的内部结构

图像识别技术目前在人们的生活中应用得越来越普遍,如在数码相机、摄像头、具有摄像功能的手机,以及一些智能玩具上都具有图像识别的功能。在电子设计竞赛中,通常采用图像识别模组。图像识别模组由光学镜头、CMOS传感器、图像处理芯片组成。目前有一些厂商生产图像识别模组成品,以"图像识别模组"为关键词,可以在相关网站查询到有关资料。

凌阳公司生产的图像识别模组的内部结构如图5-17所示,由光学镜头、SPCA561A CMOS传感器、SPCA563A图像处理芯片组成。

SPCA563A是凌阳科技生产的图像识别控制器,此芯片主要应用于具有图像识别处理功能的交互式智能玩具中,芯片内置AE/AWB功能,能够把来自于CMOS传感器的数据处理成CIF/QVGA格式。芯片内部主要嵌入了图像捕获单元、特征识别单元、unSP内核的16位CPU单元、ROM单元等,使其具有颜色识别、形状识别等功能。用户能够使用这些辨识结果去控制一些交互式人机接口。SPCA563A分析和处理SPCA561A传过来的图像信号,并得出颜色、形状等相应的信息,通过模组接口与微控制器连接。

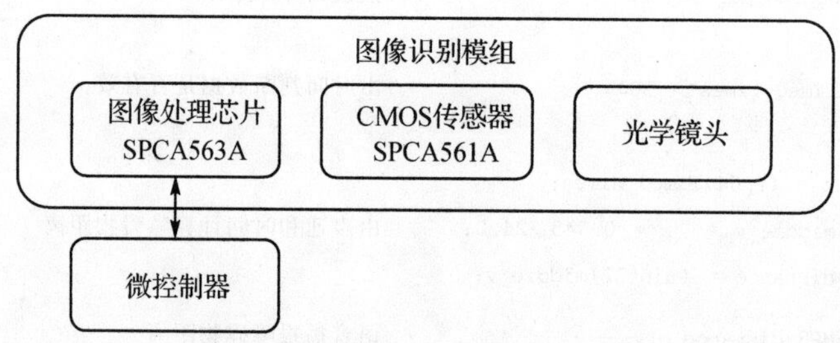

图5-17 图像识别模组的内部结构方框图

图像识别模组的接口共有6个引脚,依次接在SPCA563A的VCC、SCK、SD、RDY、图像识别3_RESET和GND上。

5.3.2 图像识别模组的电路

图像识别传感器电路图如图5-18所示。

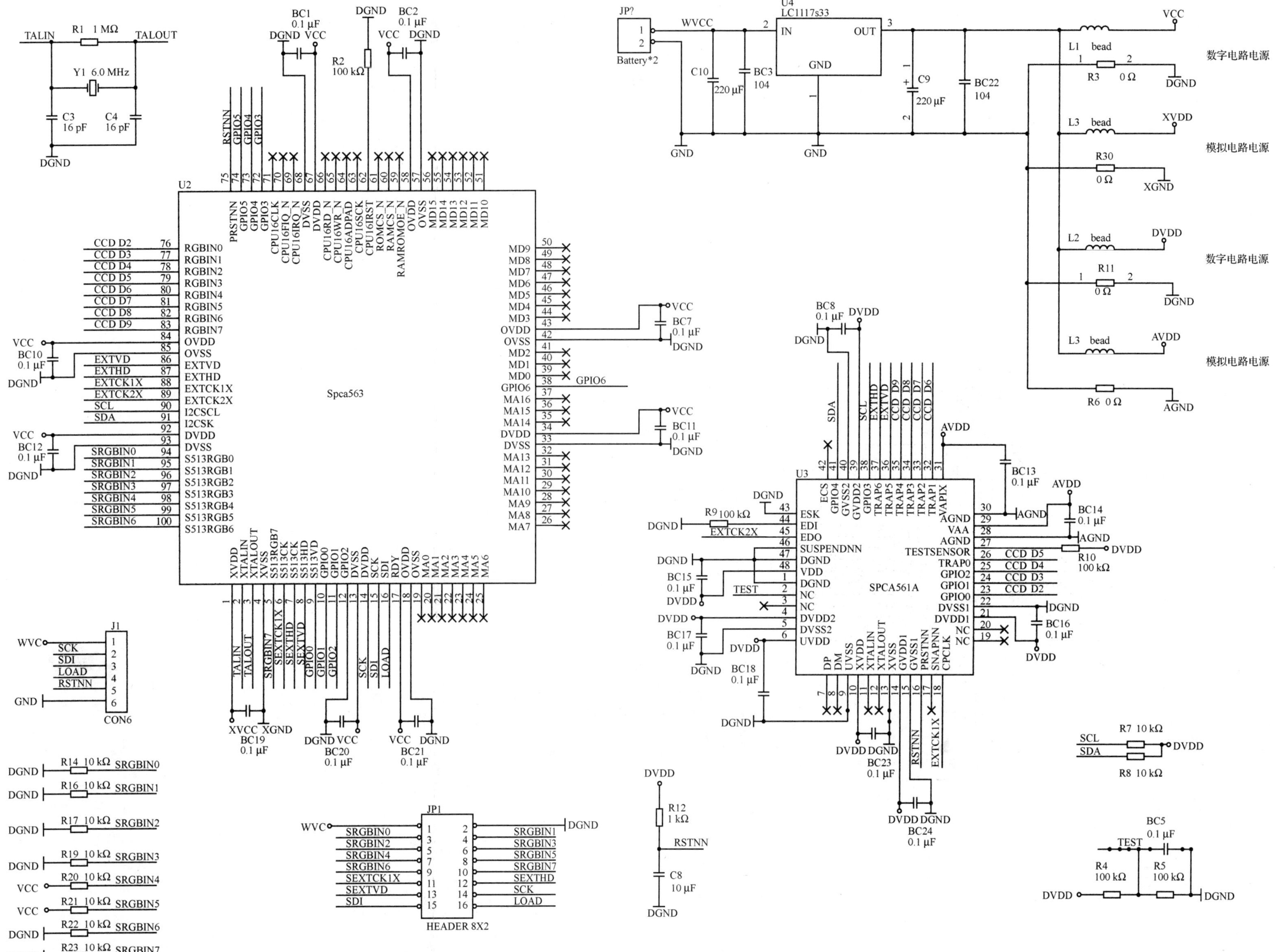

图 5-18　图像识别传感器电路图

5.3.3 图像识别模组的应用

1. 图像识别模组与微控制器的连接

图像识别模组以主、从模式与微控制器连接,微控制器为主机(master),SPCA563A为从机(slave),如图 5-19 所示。

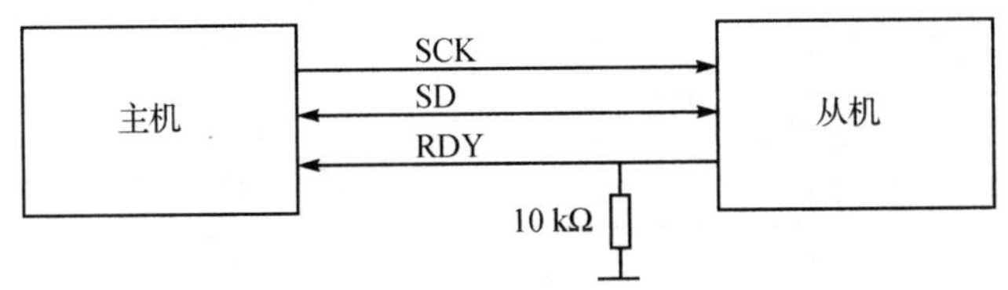

图 5-19 主、从机之间的连接

2. 微控制器与 SPCA563A 的串行通信

微控制器与 SPCA563A 串行通信的时序图和流程图如图 5-20 至图 5-22 所示。凌阳公司可以提供 SPCE061A 单片机读取 SPCA563A 内容的有关程序,更多内容请登录 http://www.unsp.com.cn 查询。

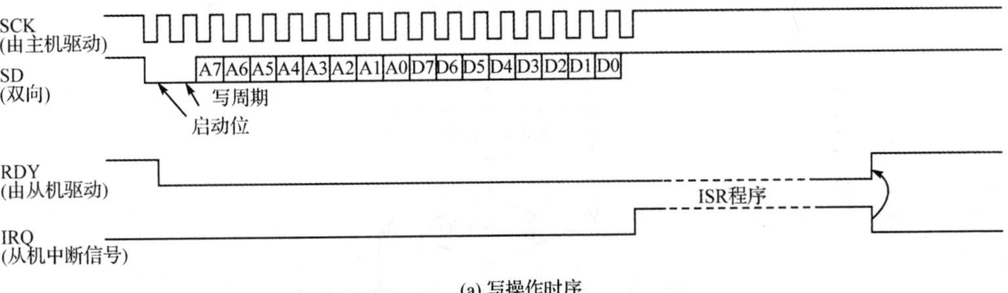

(a) 写操作时序

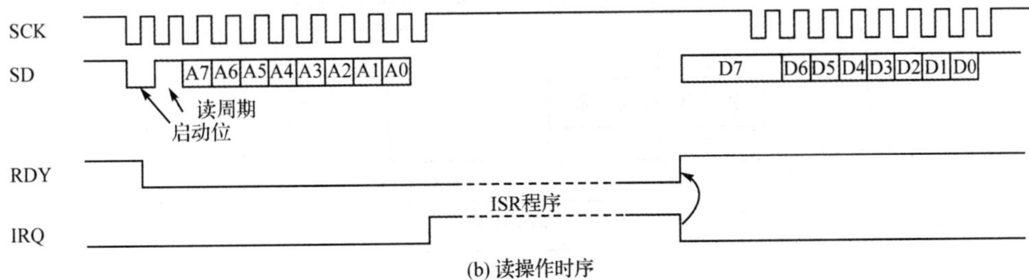

(b) 读操作时序

图 5-20 主、从机之间串行通信时序图

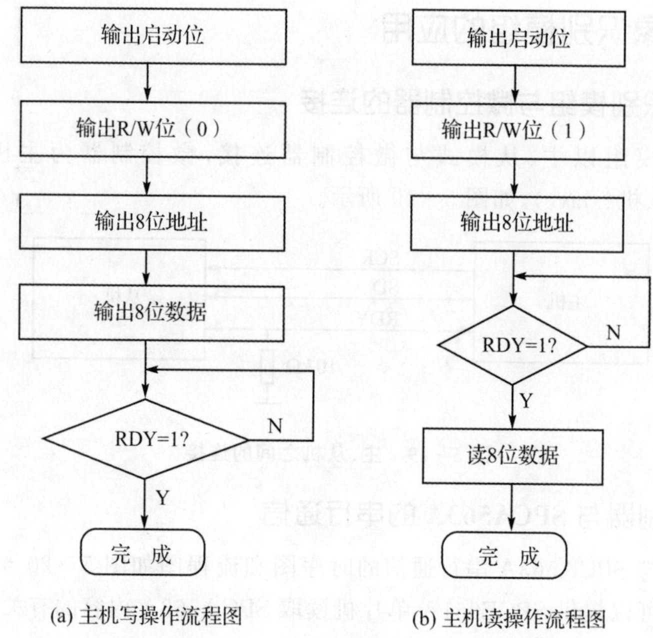

图 5－21 主机读/写操作流程图

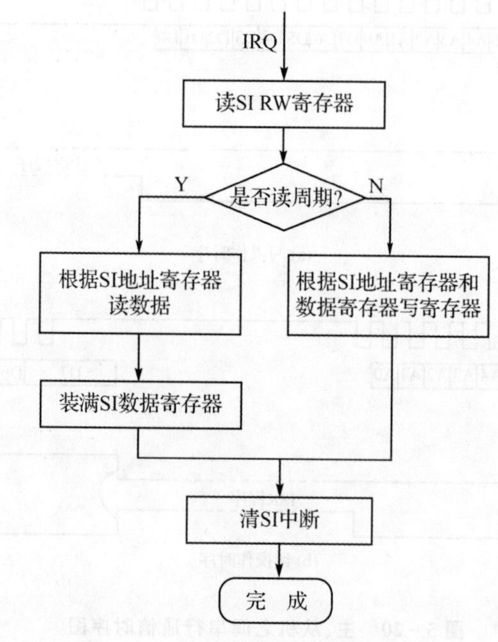

图 5－22 从机读/写操作流程图

5.3.4　SPCA563A 图像识别模块编程示例

利用 LPC2148 控制 SPCA563A 图像识别模块编程示例如程序清单 5.4 所示。

程序清单 5.4　图像识别示例程序

(1) 图像识别模块主程序

```
/************************************************************
* 文件名:TXSB_master.C
* 功能: SPCA563A 图像识别模块数据读/写软件包,供颜色、形状处理子函数调用
************************************************************/
#include  "config.h"

extern    int     VR_PrevResult_Color;
extern    int     VR_PrevResult_Shape;
extern    int     VR_Result;
extern    int     VR_OverFlag;
extern    int     VR_PrevResult_CenterX;
extern    int     VR_PrevResult_CenterY;
extern    int     VR_PrevResult_AreaH;
extern    int     VR_PrevResult_AreaL;
extern    int     gActivated;
extern    int     PlayFlag ;
extern    int     VR_TimeDeldy;
extern    int     VR_TimeFlag;
/************************************************************
* 函数名称:umDelay()
* 函数功能:软件延时
* 入口参数:ms   延时时长参数
* 出口参数:无
************************************************************/
void  umDelay( uint32 ms )
{
    uint32    i;
    while ( ms -- ) for ( i = 250; i != 0; i -- );
}
/************************************************************
* 函数名称:Set_SDA_Output()
* 函数功能:设置 SPCA563A_DATA 引脚为输出,用于向 SPC563 写数据
* 入口参数:无
* 出口参数:无
************************************************************/
void Set_SDA_Output(void)
{
    IO1DIR |= SPCA563A_DATA;
}
```

```
/**********************************************************************
 * 函数名称:Set_SDA_INput()
 * 函数功能:设置 SPCA563A_DATA 引脚为输入,用于从 SPC563 读数据
 * 入口参数:无
 * 出口参数:无
**********************************************************************/
void Set_SDA_INput(void)
{
    IO1DIR &= ~(SPCA563A_DATA);
}
/**********************************************************************
 * 函数名称:F_WriteOneByte()
 * 函数功能:CPU 向 SPCA563A 写一个字节的数据
 * 入口参数:data   要写的数据
 * 出口参数:无
**********************************************************************/
void F_WriteOneByte(uint8 data)
{
    uint8 i;
    Set_SDA_Output();
    for(i = 0;i<8;i++)
    {
        IO1CLR = SPCA563A_CLK;
        if( (data & 0x80) != 0 ) IO1SET = SPCA563A_DATA;
        else IO1CLR = SPCA563A_DATA;
        data = data << 1;
        IO1SET = SPCA563A_CLK;
    }
}
/**********************************************************************
 * 函数名称:F_WriteOneByte()
 * 函数功能:CPU 向 SPCA563A 读一个字节的数据
 * 入口参数:无
 * 出口参数:data   读出的数据
**********************************************************************/
uint8 F_ReadOneByte(void)
{
    uint8 i = 0,data = 0;
    Set_SDA_INput();
    for(i = 0;i<7;i++)
    {
        IO1CLR = SPCA563A_CLK;
        IO1SET = SPCA563A_CLK;
        if((IO1PIN & SPCA563A_DATA) != 0)
        {
            data += 1;
        }
```

```
        else
        {
            data + = 0;
        }
        data = data << 1;
    }
    IO1CLR = SPCA563A_CLK;
    IO1SET = SPCA563A_CLK;
    if((IO1PIN & SPCA563A_DATA) ! = 0)
    {
        data + = 1;
    }
    else
    {
        data + = 0;;
    }
    Set_SDA_Output();
    return data;
}
/*****************************************************************
* 函数名称:F_WriteOper()
* 函数功能:CPU 向 SPCA563A 的一个寄存器写数据
* 入口参数:addr   要写入的寄存器地址低 8 位
*          data   要写入的数据
* 出口参数:1 表明写入成功,0 表明写入失败,SPCA563 将会被复位
*****************************************************************/
uint8 F_WriteOper(uint8 addr, uint8 data)
{
    Set_SDA_Output();
    if((IO1PIN & SPCA563A_RDY) == 0)          // 若没准备好
    {
        mDelaymS(50);
        if ((IO1PIN & SPCA563A_RDY) == 0)
        {
          F_ResetEagle3Again();               // 复位 SPC563
          return 0;
        }
    }
    IO1SET = SPCA563A_CLK;
    IO1SET = SPCA563A_DATA;
    IO1CLR = SPCA563A_CLK;
    IO1CLR = SPCA563A_DATA;
    IO1SET = SPCA563A_CLK;
    IO1CLR = SPCA563A_DATA;
    IO1CLR = SPCA563A_CLK;
    IO1SET = SPCA563A_CLK;
```

```
    IO1CLR = SPCA563A_DATA;
    F_WriteOneByte(addr);
    F_WriteOneByte(data);
    IO1CLR = SPCA563A_CLK;
    IO1SET = SPCA563A_DATA;
    IO1SET = SPCA563A_CLK;
    mDelaymS(1);
    if((IO1PIN & SPCA563A_RDY) == 0)
    if((IO1PIN & SPCA563A_RDY) == 0)          // 若 RDY 为 0,说明还没有写完
    {
        mDelaymS(10);
        if ((IO1PIN & SPCA563A_RDY) == 0)
        {
            F_ResetEagle3Again();             // 复位 SPC563
            return 0;
        }
    }

  return 1;
}
/****************************************************************************
* 函数名称:F_ReadOper()
* 函数功能:CPU 从 SPCA563A 的一个寄存器读取数据
* 入口参数:addr   要读的地址低 8 位
* 出口参数:data   读出的数据
****************************************************************************/
uint8 F_ReadOper (uint8 addr)
{
    uint8 data,i = 0;
    Set_SDA_Output();
    if((IO1PIN & SPCA563A_RDY) == 0)          // 若没准备好
    {
        mDelaymS(10);
        if ((IO1PIN & SPCA563A_RDY) == 0)
        {
            F_ResetEagle3Again();             // 复位 SPC563
            return 0;
        }
    }
    IO1SET = SPCA563A_DATA;
    IO1SET = SPCA563A_CLK;
    IO1CLR = SPCA563A_DATA;
    IO1CLR = SPCA563A_CLK;
    IO1SET = SPCA563A_CLK;
    IO1CLR = SPCA563A_DATA;
```

```
    IO1CLR = SPCA563A_CLK;
    IO1SET = SPCA563A_DATA;
    IO1SET = SPCA563A_CLK;
    F_WriteOneByte(addr);
    IO1CLR = SPCA563A_CLK;
    IO1SET = SPCA563A_CLK;
    mDelaymS(1);
    if((IO1PIN & SPCA563A_RDY) == 0)          // 若没准备好
    {
        mDelaymS(10);
        if ((IO1PIN & SPCA563A_RDY) == 0)
        {
            F_ResetEagle3Again();             // 复位 SPC563
            return 0;
        }
    }
    i = 6;
    while(i--);                               // 延时
    IO1SET = SPCA563A_CLK;
    data = F_ReadOneByte();
    IO1CLR = SPCA563A_CLK;
    IO1SET = SPCA563A_DATA;
    IO1SET = SPCA563A_CLK;
    return data;
}
/*************************************************************
* 函数名称:F_HighAddr70()
* 函数功能:允许中心控制寄存器使用
* 入口参数:无
* 出口参数:无
*************************************************************/
void F_HighAddr70(void)
{
    F_WriteOper(0x70,0x0f);
}
/*************************************************************
* 函数名称:F_HighAddr74()
* 函数功能:允许中心控制寄存器使用
* 入口参数:无
* 出口参数:无
*************************************************************/
void F_HighAddr74(void)
{
    F_WriteOper(0x0f,0x74);
}
```

```
/************************************************************************
*  函数名称:F_HighAddr75()
*  函数功能:允许中心控制寄存器使用
*  入口参数:无
*  出口参数:无
************************************************************************/
void F_HighAddr75(void)
{
    F_WriteOper(0x0f,0x75);
}
/************************ 数据子函数结束 ****************************/
```

(2) 图像识别主机控制程序

```
/************************************************************************
*  文件名:TXSB_master_control.C
*  功能:  SPCA563A 图像识别模块工作模式配置软件包,供图像识别模块初始化函数调用
************************************************************************/
#include   "config.h"
int     VR_PrevResult_Color;
int     VR_PrevResult_Shape;
int     VR_Result;
int     VR_OverFlag;
int     VR_PrevResult_CenterX;
int     VR_PrevResult_CenterY;
int     VR_PrevResult_AreaH;
int     VR_PrevResult_AreaL;
int     gActivated;
int     PlayFlag ;
int     VR_TimeDeldy;
int     VR_TimeFlag;
/************************************************************************
*  函数名称:IntialToEagle3()
*  函数功能:用来初始化 Eagle3(摄像头)
*  入口参数:无
*  出口参数:无
************************************************************************/
void IntialToEagle3(uint8 R_Temp)
{
    uint8 i;
    for(i = 0;i<R_Temp;i ++ )                          // 初始化次数
    {
        mDelaymS(10);                                  // 延时 ms 毫秒
        Set50Hz();                                     // 设置光源为 50 Hz
        SetAWBON();                                    // 自动色彩调整功能
        SetClassDataMode();                            // 采集数据分类模式
        SetAEDefault();                                // 自动曝光的功能
    }
```

```
}
/************************************************************
*  函数名称:Set50Hz()
*  函数功能:光源选择是 50 Hz
*  入口参数:无
*  出口参数:无
************************************************************/
void Set50Hz(void)
{
    uint8 databuf = 0;
    F_HighAddr70();
    databuf = F_ReadOper(0xe1);
    databuf |= 0x18;
    F_HighAddr70();
    F_WriteOper(0xe1,databuf);
    F_FeatureEngine();
}
/************************************************************
*  函数名称:F_FeatureEngine()
*  函数功能:允许特征寄存器使用
*  入口参数:无
*  出口参数:无
************************************************************/
void F_FeatureEngine(void)
{
    F_WriteOper(0x0f,0x74);
}
/************************************************************
*  函数名称:SetAWBON()
*  函数功能:自动色彩调整使用
*  入口参数:无
*  出口参数:无
************************************************************/
void SetAWBON(void)
{
    uint8 databuf = 0;
    F_HighAddr70();
    databuf = F_ReadOper(0xe2);
    databuf |= 0x80;
    F_WriteOper(0xe2,databuf);
    F_FeatureEngine();
}
/************************************************************
*  函数名称:SetAWBOFF()
*  函数功能:选择黑白增益匹配不使用
*  入口参数:无
```

```
 * 出口参数:无
*****************************************************************/
void SetAWBOFF(void)
{
    uint8 data = 0;
    F_HighAddr70();
    data = F_ReadOper(0xe2);
    data &= 0x7f;
    F_WriteOper(0xe2,data);
    F_FeatureEngine();
}
/*****************************************************************
 * 函数名称:SetClassDataMode()
 * 函数功能:用来选择 SPCA561 模式,选择:分类数据
 * 入口参数:无
 * 出口参数:无
*****************************************************************/
void SetClassDataMode(void)
{
    F_HighAddr74();
    F_WriteOper(0x08,0x03);
}
/*****************************************************************
 * 函数名称:SetAEDefault()
 * 函数功能:用来设置默认的自动曝光设置
 * 入口参数:无
 * 出口参数:无
*****************************************************************/
void SetAEDefault(void)
{
    uint8 T_AETargetTable[16] = {0x90,0xa0,0xb0,0xc0,0xd0,
    0xe0,0xf0,0x00,0x10,0x20,0x30,0x40,0x50,0x60,0x70,0x7f};
    F_HighAddr70();
    F_WriteOper(0xe3,T_AETargetTable[7]);
    F_FeatureEngine();
}
/*****************************************************************
 * 函数名称:ResetEagle3()
 * 函数功能:长时间复位 SPCE563A、SPCE561A
 * 入口参数:无
 * 出口参数:无
*****************************************************************/
void ResetEagle3(void)
{
    IO0CLR = SPCA563A_RST;                         // 低电平进行复位
    mDelaymS(25);
```

```
    IO0SET = SPCA563A_RST;
}
/****************************************************************
*  函数名称:F_ResetEagle3Again()
*  函数功能:复位 SPCE563A、SPCE561A
*  入口参数:无
*  出口参数:无
****************************************************************/
void F_ResetEagle3Again(void)
{
    IO0CLR = SPCA563A_RST;                        // 低电平进行复位
    mDelaymS(15);
    IO0SET = SPCA563A_RST;
}
```

(3) 读取 SPC563 图像识别模块的颜色、形状等数据程序

```
/****************************************************************
*  文件名:Find_Color_Shape.C
*  功能:读取 SPCA563A 图像识别模块的数据,并将颜色、形状等存入各全局变量供其他函数使用
*  说明:需要调用 TXSB_master.c 中向 SPC563A 各寄存器读/写数据的子函数
****************************************************************/
#include   "config.h"
/************************* 定义全局标号 ****************************/
uint8     R_AETarget;                      // AE 设置
uint8     R_SeekFlag;                      // 功能标志
uint8     R_Flag;                          // 标志
uint8     R_Offset;                        // 偏移量

uint8     R_AddrBuffer;                    // 读地址
uint8     R_WriteBuffer;                   // 写地址
uint8     R_ReadBuffer;                    // 读地址
uint8     R_WriteDataBuffer;               // 写的内容
uint8     R_ReadDataBuffer;                // 读的内容

uint8     R_ObjNum;                        // 物体数目
uint8     R_Shape;                         // 物体外形
uint8     R_Color;                         // 物体颜色
uint8     R_AreaL;                         // 物体距离低位
uint8     R_AreaH;                         // 物体距离高位
uint8     R_CenterX;                       // 中心水平距离
uint8     R_CenterY;                       // 中心垂直距离
uint8     R_CompX;                         // 物体空间水平尺寸
uint8     R_CompY;                         // 物体空间垂直尺寸
uint8     R_StaX;                          // 水平距离的初始值
uint8     R_StaY;                          // 垂直距离的初始值
uint8     R_EndX;                          // 水平距离的结束值
uint8     R_EndY;                          // 垂直距离的结束值

uint8     R_DelayTime;                     // 延时时间
```

```
uint8    R_WaitRDYTime;              // 数据准备好的时间
uint8    R_CmpColor;                 // 分配的颜色
uint8    R_ColorIndex;               // 颜色索引
uint8    R_PreColor;                 // 校正后的物体颜色
uint8    R_PreShape;                 // 校正后的物体外形
uint8    R_PrevArea;                 // 校正后的物体空间尺寸低位
uint8    R_PreAreaH;                 // 校正后的物体空间尺寸高位
uint8    R_PreCompX;                 // 校正后的物体的水平尺寸
uint8    R_PreCompY;                 // 校正后的物体的垂直尺寸
uint8   SeekFlag;
uint8    R_Temp;                     // 数组
uint8    R_Temp0;                    // 数组 0
uint8    R_Temp1;                    // 数组 1
uint8    R_Temp2;                    // 数组 2
/********************** 定义全局标号结束 ******************************/

/****************************************************************************
 * 函数名称:F_Clear70E8()
 * 函数功能:初始化 70E8 为 0
 * 入口参数:无
 * 出口参数:无
****************************************************************************/
void F_Clear70E8(void)
{
    F_HighAddr70();
    F_WriteOper(0xe8,0x00);
}
/****************************************************************************
 * 函数名称:ShapeAnaly()
 * 函数功能:允许外形分解功能的使用
 * 入口参数:无
 * 出口参数:无
****************************************************************************/
void ShapeAnaly(void)
{
    F_HighAddr70();
    F_WriteOper(0xe0,0x08);
}
/****************************************************************************
 * 函数名称:NormalOperMode()
 * 函数功能:允许地址 75 选用,地址为 75 + 00 选择手动设置
 * 入口参数:无
 * 出口参数:无
****************************************************************************/
void NormalOperMode (void)
{
```

```
    F_HighAddr75();
    F_WriteOper(0x00,0x00);
}
/*****************************************************************
*  函数名称:OperMode()
*  函数功能:允许中心控制寄存器使用
*  入口参数:无
*  出口参数:无
*****************************************************************/
void  OperMode(void)
{
    F_WriteOper(0x0f,0x70);
}
/*****************************************************************
*  函数名称:GetObjNum()
*  函数功能:获得物体的数目
*  入口参数:无
*  出口参数:无
*****************************************************************/
uint8   GetObjNum(void)
{
    uint8   read_data = 0, ObjNum ;
    uint8   SeekFlag;
    F_HighAddr74();
    read_data  =  F_ReadOper(0x10) ;
    ObjNum  =  ( read_data & 0x0f);              // 判断低 4 位的值,得到物体数目
    if (ObjNum & 0x08 )    SeekFlag | = 0x80 ;
    else  SeekFlag &= 0x7f;
    return SeekFlag;
}
/*****************************************************************
*  函数名称:GetObjDataOne()
*  函数功能:获得在空间的远距离
*  入口参数:无
*  出口参数:无
*****************************************************************/
uint8 GetObjDataOne(void)
{
    if(R_SeekFlag & 0x01 !  = 0)                 // 若标志位不是 1,则……
        return 0;
    else
        return GetObj1stData();
}
/*****************************************************************
*  函数名称:GetObj1stData()
*  函数功能:获得数据 0x7421~0x7427 单元的内容,包括颜色寄存器、形状寄存器、物体数
```

```
*           目寄存器等
*  入口参数:无
*  出口参数:无
*******************************************************************/
uint8 GetObj1stData(void)
{
    F_HighAddr74();                          // 设置高 8 位地址为 0x74,准备读取数据

    GetShapeAndColor();
    GetObjStaX();                            // Addr 0x22
    GetObjStaY();                            // Addr 0x23
    GetObjEndX();                            // Addr 0x24
    GetObjEndY();                            // Addr 0x25
    GetObjACG();                             // StaX + End = CenterX;StaY + endY = CenterY;
    GetAreaL();                              // Addr 0x26
    GetAreaH();                              // Addr 0x27
    return 0;
}
/*******************************************************************
*  函数名称:GetShapeAndColor()
*  函数功能:获得颜色和外形内容
*  入口参数:无
*  出口参数:无
*******************************************************************/
uint8 GetShapeAndColor(void)
{
    uint8 data = 0;
    uint8 bufer = 0;
    data = F_ReadOper(0x21);
    bufer = data;                            // 保存这个值
    data &= 0x07;
    if(data == 0x07)
    {
        data -= 1;
        R_Color = data;
    }
    else                                     // 若不是 Yellow
    {
        R_Color = data;
        data = bufer;
        data &= 0x70;
        data = data >> 4;
        R_Shape = data;
    }
    return R_Color;
}
/*******************************************************************
```

```
*  函数名称:GetObjStaX()
*  函数功能:获得物体水平方向的近距离内容
*  入口参数:无
*  出口参数:无
*********************************************************************/
uint8 GetObjStaX(void)
{
    R_StaX = 0;
    R_StaX = F_ReadOper(0x22);
    return R_StaX;
}
/*********************************************************************
*  函数名称:GetObjStaY()
*  函数功能:获得物体垂直方向的近距离内容
*  入口参数:无
*  出口参数:无
*********************************************************************/
uint8 GetObjStaY(void)
{
    R_StaY = 0;
    R_StaY = F_ReadOper(0x23);
    return R_StaY;
}
/*********************************************************************
*  函数名称:GetObjEndX()
*  函数功能:获得物体水平方向的远距离内容
*  入口参数:无
*  出口参数:无
*********************************************************************/
uint8 GetObjEndX(void)
{
    R_EndX = 0;
    R_EndX = F_ReadOper(0x24);
    return R_EndX;
}
/*********************************************************************
*  函数名称:GetObjEndY()
*  函数功能:获得物体垂直方向的远距离内容
*  入口参数:无
*  出口参数:无
*********************************************************************/
uint8 GetObjEndY()
{
    R_EndY = 0;
    R_EndY = F_ReadOper(0x25);
    return R_EndY;
}
```

```
/************************************************************************
* 函数名称:GetObjACG()
* 函数功能:获得物体垂直方向、水平方向的中心距离和水平、垂直方向物体的实际距离
* 入口参数:无
* 出口参数:无
************************************************************************/
uint8 GetObjACG(void)
{
    R_CenterX = R_StaX + R_EndX;
    R_CenterX = R_CenterX >> 1;
    R_CenterY = R_StaY + R_EndY;
    R_CenterY = R_CenterY >> 1;
    R_CompX = R_EndX - R_StaX;
    R_CompY = R_EndY - R_StaY;
    return R_CompY;
}
/************************************************************************
* 函数名称:GetAreaL()
* 函数功能:获得在空间的近距离
* 入口参数:无
* 出口参数:无
************************************************************************/
uint8 GetAreaL(void)
{
    R_AreaL = 0;
    R_AreaL   = F_ReadOper(0x26);
    return R_AreaL;
}
/************************************************************************
* 函数名称:GetAreaH()
* 函数功能:获得在空间的远距离
* 入口参数:无
* 出口参数:无
************************************************************************/
uint8 GetAreaH(void)
{
    R_AreaH = 0;
    R_AreaH   = F_ReadOper(0x27);
    return R_AreaH;
}
/************************************************************************
* 函数名称:Find_Color_Shape()
* 函数功能:获得物体的颜色和外形并播报
* 入口参数:无
* 出口参数:无
************************************************************************/
```

```
void Find_Color_Shape (void)
{
    NormalOperMode();
    OperMode();
    ShapeAnaly();
    F_Clear70E8();
    F_FeatureEngine();
    F_HighAddr74();                         // 使能 0x74xx
    F_WriteOper (0x05,0x08);                // 选择辨别蓝色、红色、绿色、黄色
    mDelaymS(65);
    SeekFlag = GetObjNum();
    while( SeekFlag & 0x80 )
    {
        if(SeekFlag == 0x08)
        {
            F_HighAddr74();                 // 使能 0x74xx
            F_WriteOper (0x05,0x08);        // 选择辨别蓝色、红色、绿色、黄色
            mDelaymS(65);
            SeekFlag   = GetObjNum();
        }
        else break;
    }
   SeekFlag &= 0xfe;
   GetObjDataOne();
}
```

重要说明：使用该图像模块识别颜色还可以，识别形状不是很准确。

5.4 色彩传感器及其应用

5.4.1 常用的几种色彩传感器的解决方案

色彩传感器在终端设备中起着极其重要的作用，比如色彩监视器的校准装置，彩色打印机和绘图仪，涂料、纺织品和化妆品制造，以及医疗方面的应用，如血液诊断、尿样分析和牙齿整形等。色彩传感器系统的复杂性在很大程度上取决于其用于确定色彩的波长谱带或信号通道的数量。此类系统从相对简单的三通道色度计到多频带频谱仪种类繁多。

目前，常用的几种色彩传感器的解决方案如下。

1. 具有滤色器的分立型光电二极管

感测色彩的传统做法是把 3～4 个光电二极管组合在一块芯片上，而将红、绿、蓝滤色器置于光电二极管的表面(通常将 2 个蓝滤色器组合在一起以补偿硅片对于蓝

光的低灵敏度)。独立的跨阻抗放大器将每个光电二极管的输出馈送到具有 8～12 位典型分辨率的 A/D 转换器中。A/D 转换器的输出随后被馈送至一个微控制器或其他类型的数字处理器中。

这种方法的主要优点是灵活性高,因为能够使放大器的增益和带宽以及 A/D 转换器的速度和分辨率适合具体应用的要求,从而可以对设计进行调整以实现性能与成本的折中。为获得这种灵活性所付出的代价是增加了设计的复杂性,另外也使模拟电路的电路板布局变得非常苛刻。该方案的主要应用包括:工业控制中需要短暂响应时间的高速过程检验,或因光照条件不定而要求随意调节增益和速度的应用。

2. 集成光-电压转换器

另一种方法是将用于单一色彩谱带的一个光电二极管、滤色器和跨阻抗放大器组合在一块芯片上。与分立型实现方案一样,三个元件的输出被馈送到一个外部三通道 A/D 转换器中,接着进行数字处理。Texas Advanced Optoelectronic Solutions (TAOS)公司推出的 TSLR257、TSLG257 和 TSLB257 就是这些元件的实例。

这种方法所需的元件数量比分立型光电二极管的要少,由于对噪声敏感的模拟电路位于芯片之上,因此压缩了电路板的占用空间,降低了安装成本,并且简化了设计和电路板布局。缺点是传感器的增益和灵敏度不能动态地改变。该方法的应用实例有:具有定义明确的光照条件、空间约束条件、灵敏度要求的系统或那些对面市时间或设计周期有着较高要求的系统。

3. 集成光-频率转换器

第三种方法是将光强度直接转换为频率分别与每个红、绿、蓝通道的红、绿、蓝光分量的强度成正比的一个脉冲序列。脉冲序列可以直接提供给微处理器,而无需增设 A/D 转换器。TAOS 公司的 TCS230 就是此类器件的一个实例。它把红、绿和蓝传感器-滤波器组合(和一个没有滤波器的额外"干净"传感器)划分为栅格状,从而将元素扩散到整个感测区域,因此不再需要光扩散器。将每种颜色的光电二极管并联起来最终可使任何不均匀的照度达到平衡。该方案取消了跨阻抗放大器和 A/D 转换器,处理器只是简单地测量周期或计算一个周期内来自传感器的脉冲数。传感器和微处理器之间的直接连接具有较高水平的抗噪声度,为将传感器放置在远处创造了条件。

RGB 频率转换法的局限性会在光强度较低的应用中显现出来。光强度较低,产生的频率也会随之降低,从而增加了转换时间。该方案的应用实例包括:空间因素至关重要的便携式系统和需要以低成本来实现更高分辨率的系统。

4. 集成的数字颜色光传感器

集成的数字颜色光传感器 TCS3404CS/TCS3414CS 具有可编程的中断功能和用户可设置阈值功能,芯片内部集成有光滤波器,采用 SMBus 100 kHz 或者采用 I^2C

400 kHz 输出16 位数字信号，可编程的模拟增益和集成的定时器支持 1～1 000 000 的动态范围，工作温度范围为－40～85 ℃，采用单电源供电，电压范围为 2.7～3.6 V。

有关 TCS3404CS/TCS3414CS 的使用和更多内容请登录 http://www.taosinc.com 查询。

5.4.2　TCS230 可编程颜色光-频率转换器

1. TCS230 简介

TCS230 是 TAOS 公司推出的可编程彩色光到频率的转换器。它把可配置的硅光电二极管与电流频率转换器集成在一个单一的 CMOS 电路上，同时在单一芯片上集成了红绿蓝（RGB）三种滤光器，是业界第一个有数字兼容接口的 RGB 彩色传感器。

TCS230 的输出信号是数字量，可以驱动标准的 TTL 或 CMOS 逻辑输入，因此可直接与微处理器或其他逻辑电路相连接。由于输出的是数字量，并且能够实现每个彩色信道 10 位以上的转换精度，因而不再需要 A/D 转换电路，使电路变得更简单。

2. TCS230 的内部结构与工作原理

TCS230 的封装形式与内部结构方框图如图 5－23 所示，图中，TCS230 采用8 引脚的 SOIC 表面贴装式封装，在单一芯片上集成有 64 个光电二极管。这些二极管共

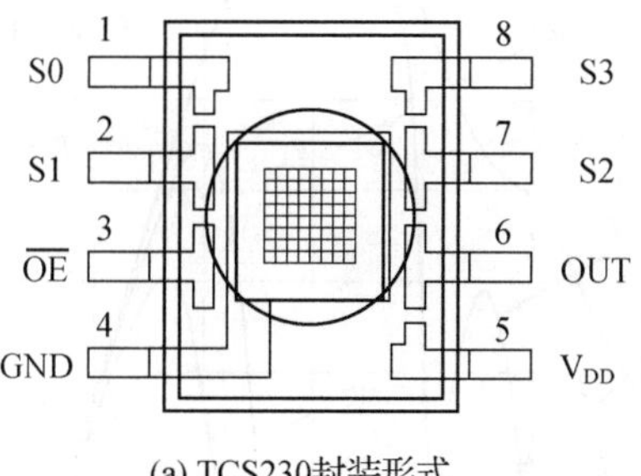

(a) TCS230封装形式

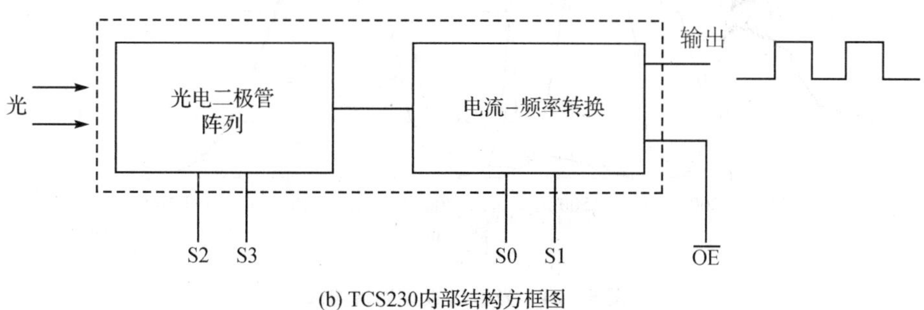

(b) TCS230内部结构方框图

图 5－23　TCS230 的封装形式与内部结构方框图

分为4种类型。其中16个光电二极管带有红色滤波器;16个光电二极管带有绿色滤波器;16个光电二极管带有蓝色滤波器;其余16个不带有任何滤波器,可以透过全部的光信息。这些光电二极管在芯片内是交叉排列的,能够最大限度地减少入射光辐射的不均匀性,从而增加颜色识别的精确度;另一方面,相同颜色的16个光电二极管是并联连接的,均匀分布在二极管阵列中,可以消除颜色的位置误差。

工作时,通过两个可编程的引脚来动态选择所需要的滤波器,如表5-3所列。该传感器的典型输出频率范围从2 Hz～500 kHz,用户还可以通过两个可编程引脚来选择100%、20%或2%的输出比例因子,或低功耗模式。输出比例因子使传感器的输出能够适应不同的测量范围,提高了它的适应能力。例如,当使用低速的频率计数器时,就可以选择小的定标值,使TCS230的输出频率和计数器相匹配。

从图5-23可知:当入射光投射到TCS230上时,通过光电二极管控制引脚S2、S3的不同组合,可以选择不同的滤波器;经过电流到频率转换器后输出不同频率的方波(占空比是50%),不同的颜色和光强对应不同频率的方波;还可以通过输出定标控制引脚S0、S1,选择不同的输出比例因子,对输出频率范围进行调整,以适应不同的需求。

TCS230的光敏二极管光谱响应如图5-24所示。有关TCS230的更多内容请登录http://www.taosinc.com查询。

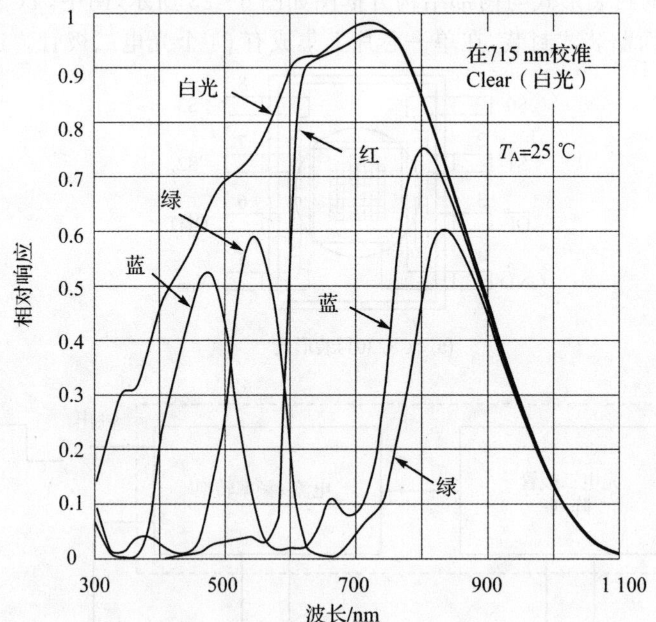

图5-24　光敏二极管光谱响应

表 5－3　S0～S3 选择

S0	S1	模　式	S2	S3	滤波器类型
低电平	低电平	低功耗模式	低电平	低电平	红色
低电平	高电平	2%输出比例因子	低电平	高电平	蓝色
高电平	低电平	20%输出比例因子	高电平	低电平	无
高电平	高电平	100%输出比例因子	高电平	高电平	绿色

3. 基于 TCS230 的颜色识别模块

基于 TCS230 的颜色识别模块如图 5－25 所示，尺寸为 72×16×12(mm^3)，配备 4 只大功率白色 LED 灯，可识别 R(红色)，G(绿色)，B(蓝色)三原色的分量，通过这三原色分量的混合，可以识别颜色。工作电压 5 V，工作电流 0.12 A。输出引脚可直接与微控制器的 I/O 口连接。

图 5－25　颜色识别模块

TCS230 模组具有 3 个颜色滤波器，由德国物理学家赫姆霍兹的三原色理论可知：各种颜色是由不同比例的三原色(红、绿、蓝)混合而成的。该模组正是利用了这个原理，通过获得物体的三原色比例来判断出物体的颜色。该模组具有 3 个颜色滤波器，当选择其中的一个颜色滤波器时，它只允许某个特定的原色通过，阻止其他原色的通过。例如：当选择蓝色滤波器时，入射光中只有蓝色可以通过，红色和绿色都被阻止，而此时 TCS230 的脉冲输出引脚会输出一定频率的脉冲，由脉冲数就可以得到蓝色光的光强；同理，选择其他的滤波器，就可以得到红色光和绿色光的光强，通过这三个值，就可以分析投射到 TCS230 传感器上。而颜色滤波器的选择由模组 S2 引脚和 S3 引脚决定，如表 5－3 所列。

因此，将模组的 S2 和 S3 与控制器的 I/O 口相连接，便可方便地控制对滤波器的选择；同时，将模组的脉冲输出引脚与控制器的外部中断引脚相连接，控制器便可通过外部中断的方式获得模组的脉冲输出个数，从而获得三原色的成分，分析出物体的颜色。模组与控制器 LPC2148 之间的连接示意图如图 5－26 所示。

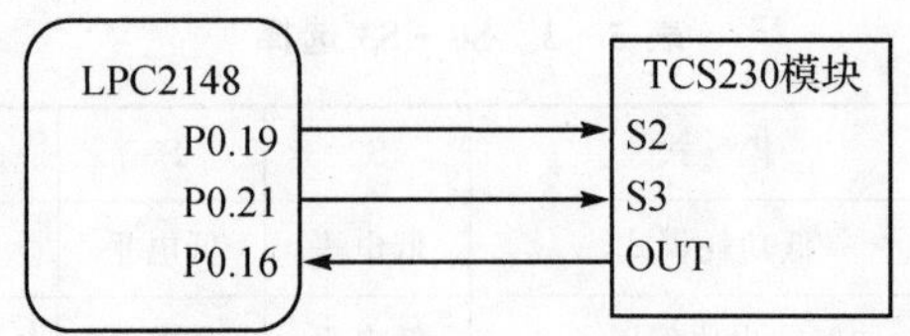

图 5-26 TCS230 与 LPC2148 的连接示意图

5.4.3 颜色识别模块的编程示例

颜色识别模块的编程示例流程图如图 5-27 所示，程序如程序清单 5.5 所示。

启动 ADS 1.2，使用 ARM Executable Image for lpc2148 工程模板创建一个工程。在使用过程中，需要 Startup.s 修改系统模式，堆栈设置为 0x5f，开启中断。本示例是将颜色识别结果通过彩屏液晶显示，用户可根据实际情况做修改。下面只给出了能正确进行颜色识别的程序代码。

程序清单 5.5 颜色识别示例程序

```
#include "config.h"
#define  MAX_Uart1buf_Size   200                         // 定义串口数据长度
#define Red        1
#define Blue       2
#define Green      3
#define Yellow     4
#define TCS230_S2       (1<<19)                          // TCS230 控制端口 S2
#define TCS230_S3       (1<<21)                          // TCS230 控制端口 S3
uint32 Colour_num = 0;                                   // 用于颜色传感器脉冲计数
uint8   rcv_buf[50] = {0},num = 0;                       // 定义存放串口接收数据的数组
uint8   rcv_flag = 0;                                    // 定义表示串口收到数据的标志
uint8   rcv_buf1[MAX_Uart1buf_Size] = {0},num1 = 0;      // 定义存放串口接收数据的数组
uint8   rcv_flag1 = 0;                                   // 定义表示串口收到数据的标志
uint8   RTC_flag = 0;                                    // 设置中断标志,用于定时 1 s
volatile  uint8  Time0_Flag = 0;                         // 定时器 0 中断标志
/****************************************************************************
 * 函数名称:mDelaymS()
 * 函数功能:延时 ms 毫秒
 * 入口参数:ms  延时参数,值越大,延时越久
 * 出口参数:无
****************************************************************************/
void mDelaymS( uint32 ms )
{
    uint32     i;
    while ( ms -- ) for ( i = 25000; i ! = 0; i -- );
```

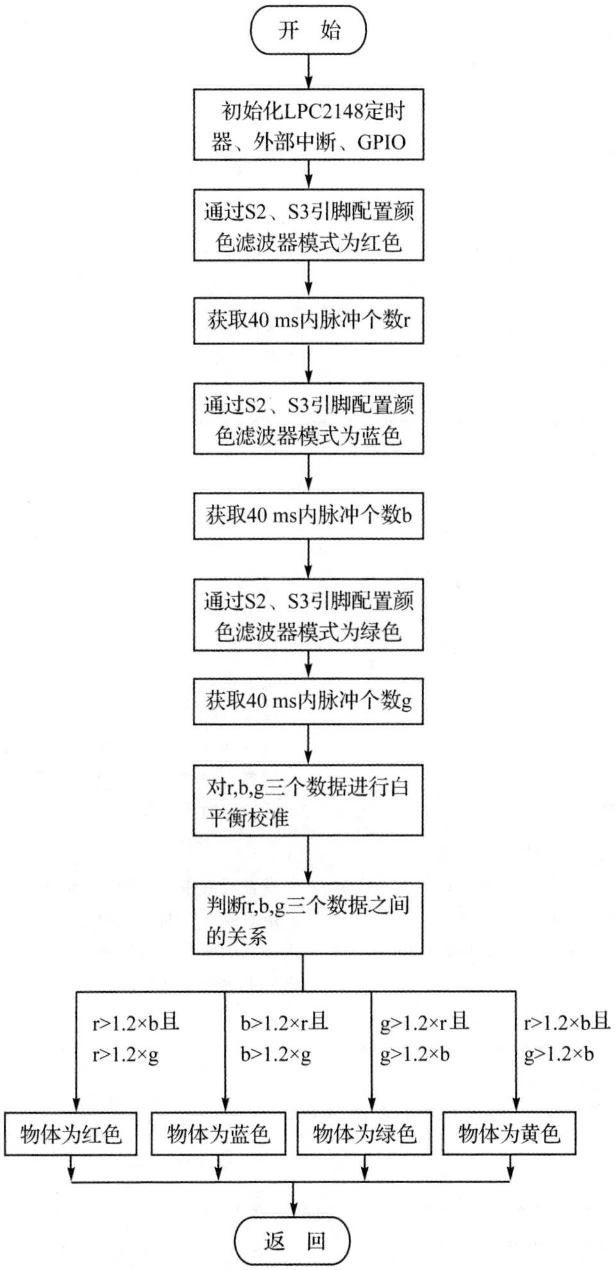

图 5－27 TCS230 颜色识别程序流程图

```
}
/************************************************************************
 *函数名称:IRQ_Time0()
 *函数功能:定时器 0 中断服务程序
 *入口参数:无
 *出口参数:无
************************************************************************/
void __irq IRQ_Time0(void)
{
    T0IR = 0x01;                              // 清除中断标志
    Time0_Flag = 1;
    VICIntEnClr = 1≪14;                       // 关闭 EINT1 中断
    VICVectAddr = 0;                          // 退出中断
}
/************************************************************************
 *函数名称:Time0Init()
 *函数功能:初始化定时器 0
 *入口参数:无
 *出口参数:无
************************************************************************/
void Time0Init(void)
{
    T0PR = 11;                                // 设定定时器分频为 12 分频,得 1 MHz
    T0MCR = 0x03;                             // 设置匹配通道 0 匹配中断并复位 T0TC
    T0MR0 = 50000;                            // 设置预分频寄存器值
    T0TCR = 0x03;                             // 设置定时器/计数器和预分频计数器启动并复位
    T0TCR = 0x01;                             // 开启定时器工作
    /*设置定时器 0 向量中断 IRQ*/
    VICIntSelect = 0x00000000;                // 设置所有中断分配为 IRQ 中断
    VICVectCntl0 = 0x24;                      // 定时器 0 中断分配向量中断 0,开启 IRQ 中断使能
    VICVectAddr0 = (uint32)IRQ_Time0;         // 设置中断服务程序地址
    VICIntEnable = 1≪4;                       // 使能定时器 0 中断
}
/************************************************************************
 * 函数名称:IRQ_Eint0()
 * 函数功能:外部中断 EINT0 服务函数,取反 B1 控制口
 * 入口参数:无
 * 出口参数:无
************************************************************************/
void __irq IRQ_Eint0(void)
{
    Eint0_flag = 1;
    Colour_num++;

    EXTINT = 0x01;                                    // 清零 EINT0 中断标志
    VICVectAddr = 0;                                  // 结束中断
```

```
}
/*******************************************************************
*  函数名称:Eint0_init()
*  函数功能:初始化外部中断 0(EINT1)为向量中断,并设置为下降沿触发模式
*  函数说明:在 STARTUP.S 文件中使能 IRQ 中断(清零 CPSR 中的 I 位)
*******************************************************************/
void Eint0_init(void)
{

    PINSEL1 = (PINSEL1 & 0xfffffffc) | 0x00000001;    // 设置 P0.16 为外部中断 0
    EXTMODE |= (0x01);                                 // 设置外部中断为边沿触发模式
    EXTPOLAR &= ~0x01;                                 // 设置为下降沿触发
    EXTPOLAR |= 0x01;                                  // 设置为上升沿触发
    /* 采用向量 IRQ 开启 EXINT1 */
    VICIntSelect    = 0x00000000;                      // 设置所有中断为 IRQ 中断
    VICVectCntl4 = 0x2e;
    VICVectAddr4 = (int)IRQ_Eint0;                     // 设置中断服务程序地址
    EXTINT = 0x01;                                     // 清零 EINT0 中断标志
    VICIntEnable = 1<<14;                              // 使能 EINT0 中断
}
/*******************************************************************
 *函数名称:Colour_init()
 *函数功能:TCS230 内部颜色滤波器配置,通过控制 TCS230 的 S2、S3 引脚配置为各种模式
 *入口参数:Colour  选择配置的模式
 *出口参数:无
*******************************************************************/
void Colour_init(uint8 Colour)
{
    switch(Colour)
    {
        case Red:
                  IO0CLR = TCS230_S2;        // S2 = 0
                  IO0CLR = TCS230_S3;        // S3 = 0
                  break;
        case Blue:
                  IO0CLR = TCS230_S2;        // S2 = 0
                  IO0SET = TCS230_S3;        // S3 = 1
                  break;
        case Green:
                  IO0SET = TCS230_S2;        // S2 = 1
                  IO0SET = TCS230_S3;        // S3 = 1
                  break;
        default:
                  break;
    }
```

```
}
/***********************************************************************
 *函数名称:TCS230_Play()
 *函数功能:初始化定时器0和外部中断,用于脉冲计数
 *入口参数:无
 *出口参数:无
***********************************************************************/
void TCS230_Play(void)
{
    T0TCR = 0x03;                       //使能定时器并复位定时器
    T0TCR = 0x01;                       //使能定时器
    PWMTCR = 0x09;                      //启动定时器,PWM使能

    Colour_num = 0;                     //清零计数变量
    Time0_Flag = 0;                     //清零定时器中断标志

    VICIntEnable = 1≪14;               //使能外部中断EINT0
}
/***********************************************************************
 *函数名称:TCS320_Read_Clour()
 *函数功能:设置颜色滤波器并读出相应频率参数
 *入口参数:Colour      颜色滤波器配置参数
 *出口参数:Colour_num  测得的脉冲个数
***********************************************************************/
uint32 TCS230_Read_Clour(uint8 Colour)
{
    Colour_init(Colour);                //配置TCS230端口S2,S3
    mDelaymS(5);
    TCS230_Play();                      //初始化各变量,定时器和外部中断
    while(Time0_Flag == 0);             //等待计时完毕
    return Colour_num;                  //返回计数个数
}
/***********************************************************************
 *函数名称:Disp_Colour()
 *函数功能:在DMT32240S035彩屏液晶上显示识别到的颜色
 *入口参数:dat      颜色种类,红、蓝、绿、黄
 *          x       显示字符所在液晶的X坐标
 *          y       显示字符所在液晶的Y坐标
 *出口参数:无
***********************************************************************/
void Disp_Colour(uint16 x,uint8 y,uint32 dat)
{
    uint8 buf_Red[20] = "红色";
    uint8 buf_Blue[20] = "蓝色";
    uint8 buf_Green[20] = "绿色";
    uint8 buf_Yellow[20] = "黄色";
    if(dat == Red) Disp_words(0x55,x,y,buf_Red);
```

```
    if(dat == Blue)Disp_words(0x55,x,y,buf_Blue);
    if(dat == Green)Disp_words(0x55,x,y,buf_Green);
    if(dat == Yellow)Disp_words(0x55,x,y,buf_Yellow);
}
/*****************************************************************
 * 函数名称:TCS230_See()
 * 函数功能:通过各种颜色滤波器得到的频率参数判断识别到的颜色
 * 入口参数:无
 * 出口参数:颜色种类,红、蓝、绿、黄
*****************************************************************/
uint8 TCS230_See(void)
{
    uint8   i;
    uint32 r = 0,b = 0,g = 0;
    fp32 k  =  1.2;                         // 比例系数
    for(i = 0;i<3;i ++ )                    // 读取三次数据
    {
        r + = TCS230_Read_Clour(Red);       // 读取 40 ms 内的脉冲个数
        b + = TCS230_Read_Clour(Blue);
        g + = TCS230_Read_Clour(Green);
    }
    r  =  r/3;                              // 取三次数据的平均值
    b  =  b/3;
    g  =  g/3;
    if( (r>k * b) & (r>k * g) )
    {
        return Red;                         // 红色
    }
    if( (b>k * r) & (b>k * g) )
    {
        return Blue;                        // 蓝色
    }
    if( (g>k * r) & (g>k * b) )
    {
        return Green;                       // 绿色
    }
    if( (r>k * b) & (g>k * b) )
    {
        return Yellow;                      // 黄色
    }
    return 0;
}
/*****************************************************************
*  函数名称:main()
*  函数功能:TCS230 颜色识别示例程序主函数
*****************************************************************/
```

```
int main(void)
{
    uint32 a;
    PINSEL0 = 0;
    PINSEL1 = 0;
    PINSEL2 &= ~(0x00000006);
    IO0DIR = 0;
    IO1DIR = 0;
    IO0DIR |= TCS230_S2 | TCS230_S3;              // 设置 TCS230 控制端口为输出引脚
    UART1_Ini(115200);                            // UART1 初始化,用于与彩屏液晶通信
    Time0Init();                                  // 定时器 0 初始化,用于计时
    Eint0_init();                                 // 外部中断 0 初始化,用于脉冲计数
    mDelaymS(20);
    LCD_Clear();                                  // 清屏
    mDelaymS(20);
    Disp_System_time();                           // 液晶显示系统时间
    while(1)
    {
        a = TCS230_See();                         // 控制 TCS230 识别颜色
        Disp_Colour(0,120,a);                     // 在彩屏液晶上显示颜色
    }
    return 0;
}
/**********************************************************************
*                              END                                   *
**********************************************************************/
```


5.5 电子罗盘及其应用

5.5.1 电子罗盘简介

电子罗盘,也叫数字罗盘,是利用地磁场来确定北极的一种方法。一般采用磁阻传感器和磁通门加工而成。电子罗盘可以分为平面电子罗盘和三维电子罗盘。平面电子罗盘要求用户在使用时必须保持罗盘的水平,否则当罗盘发生倾斜时,也会给出航向的变化,而实际上航向并没有变化。虽然平面电子罗盘对使用要求很高,但如果能保证罗盘所附载体始终水平的话,平面罗盘是一种性价比很好的选择。

三维电子罗盘在其内部加入了倾角传感器,可以克服平面电子罗盘在使用中的严格限制。如果罗盘发生倾斜时可以对罗盘进行倾斜补偿,这样即使罗盘发生倾斜,航向数据依然准确无误。有时为了克服温度漂移,罗盘也可内置温度补偿,以最大限度地减少倾斜角和指向角的温度漂移。

BQ-CA80-TTL电子罗盘模块是一款低成本的数字罗盘模块,其工作原理是通过磁阻传感器感应地球磁场的磁分量,从而检测出方位角度。该罗盘的分辨率为1°,精度小于3°,测量范围为0°~360°,响应频率为10 Hz,工作温度为-40~85 ℃,支持两种电源电压模式,直流6~9 V或5 V可选,工作电流小于30 mA;同时具有标定功能,安装角和磁偏角补偿功能,可适应不同的工作环境;尺寸为40 mm×40 mm×12 mm。

BQ-CA80-TTL电子罗盘模块的方位角度检测数据以TTL电平十六进制方式与微控制器进行半双工的通信方式,只需要占用微控制器的2个串口引脚便可实现对其的控制。对其的读出和写入数据都是先高字节,再低字节。单字节传送格式为:1位开始位,8位数据位,1位结束位。其输出波特率可调(4.8 kbit/s,9.6 kbit/s,19.2 kbit/s),默认波特率为9.6 kbit/s,有连续输出和询问输出两种工作模式。

注意: 如果在使用的环境中有除了地球以外的磁场且这些磁场无法有效屏蔽时,那么电子罗盘计算输出的角度信息将是不准确的,为了消除周围固有磁场变化带来的影响,需要对罗盘进行标定,对罗盘因周围磁场改变所产生的影响进行校正。

BQ-CA80-TTL电子罗盘的引脚端:

- 引脚端1(GND)为电源地;
- 引脚端2(VCC)为电源输入正端 V_{DC}(+5 V);
- 引脚端3(VDD)为电源输入正端 V_{DC}(+6~+9 V);
- 引脚端4(RX)为数据接收端;
- 引脚端5(TX)为数据发送端。

5.5.2 BQ-CA80-TTL电子罗盘与微控制器的连接

BQ-CA80-TTL电子罗盘模块与微控制器的连接示意图如图5-28所示。

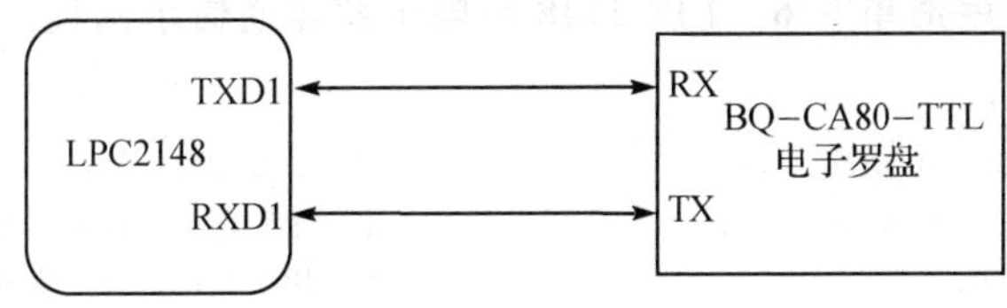

图5-28 电子罗盘模块与控制器的连接示意图

5.5.3 BQ-CA80-TTL电子罗盘模块的编程示例

BQ-CA80-TTL电子罗盘模块的程序流程图如图5-29所示。实现1 s读取一次电子罗盘的数据,然后将结果送数码管显示角度,通过串口送PC显示方向。

在设置好罗盘模块为单次采集或连续采集模式(74H或76H)后,主机向罗盘发送命令77H,此时罗盘模块将返回一组双字节的信息,包含方向、角度、工作状态。

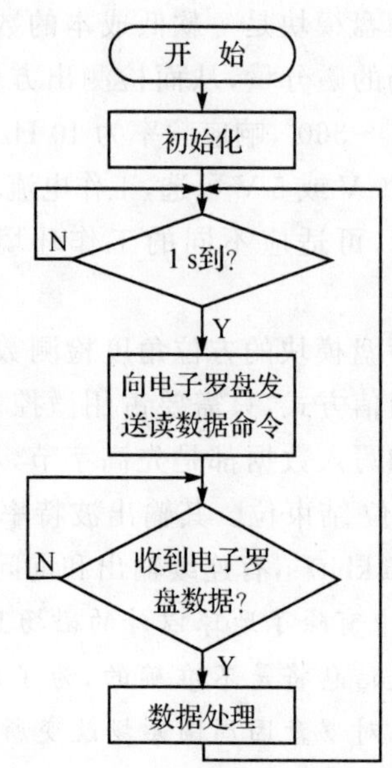

图 5-29 电子罗盘程序流程图

返回数据的第15、14位表示罗盘的工作状态,定义为:01——查询状态,10——正常状态,11——标定状态。第13、12位保留。第11、10、9位表示罗盘的方向,定义为000——北,001——东北,010——东,011——东南,100——南,101——西南,110——西,111——西北。第8位至第0位表示罗盘的角度(0°~360°),为二进制数形式。

启动ADS 1.2,使用ARM Executable Image for lpc2148工程模板创建一个工程。使用过程中,需要Startup.s修改系统模式,堆栈设置为0x5f,开启中断。下面只给出LPC2148与电子罗盘的通信程序,如程序清单5.6所示。

程序清单 5.6 LPC2148 与电子罗盘通信示例程序

```
#include"config.h"
#define  MAX_Uart1buf_Size   50                         // 定义串口数据长度
uint8   rcv_buf[100] = {0},num = 0;                     // 定义用于存放串口接收数据的数组
uint8   rcv_flag = 0;                                   // 定义用于表示串口接收到数据的标志变量
uint8   RTC_flag;                                       // RTC 中断标志,用于定时 1 s
uint8   rcv_buf1[50] = {0},num1 = 0;                    // 定义用于存放串口接收数据的数组
uint8   rcv_flag1 = 0;                                  // 定义用于表示串口接收到数据的标志变量
/*********************************************************************
*  函数名称:IRQ_UART()
*  函数功能:串口 UART1 接收中断
*  入口参数:无
*  出口参数:无
*********************************************************************/
void     __irq IRQ_UART1(void)
{
```

```
    uint8 i;
    switch(U1IIR&0x0f)                              // U0IIR 为中断标志,低 4 位有效
    {
        // 发生 RDA(有效数据达到触发点)中断,读 U0RBR 中断复位
        case 0x04:
            for(i = 0; i<8; i++)                    // 触发点为 8
                rcv_buf1[num1++] = U1RBR;
            rcv_flag1 = 1;
            break;
        case 0x0c:
            while((U1LSR&0x01) == 1)
            {
                rcv_buf1[num1++] = U1RBR;
            }
            rcv_flag1 = 1;
            break;
        default:
            break;
    }

   if(num1 >= MAX_Uart1buf_Size) num1 = 0;
   VICVectAddr = 0x00;                              // 中断处理结束
}
/************************************************************************
* 函数名称:UART1_Ini()
* 函数功能:初始化串口 0,设置为 8 位数据位,1 位停止位,无奇偶校验
* 入口参数:bps     UART 通信波特率
* 出口参数:无
************************************************************************/
void UART1_Ini(uint32 bps)
{
    uint16 Fdiv;
    uint32 tmp;
    PINSEL0 = (PINSEL0 & 0xFFFCFFFF)|0x50000;     // 设置 I/O 为 UATR1
    if(bps == 115200)
    {
        U1FDR    = 1 | 12<<4;                       // LPC2148 串口除数校准波特率
    }

    tmp = bps;

    U1LCR = 0x83;     // DLAB = 1,允许访问除数锁存寄存器,可设置波特率 Fdiv = (Fpclk/16)/tmp;
    U1DLM = Fdiv/256;                               // 根据波特率设定 U0DLM 和 U0DLL 的值
    U1DLL = Fdiv % 256;

    U1LCR = 0x03;                                   // 禁止访问除数锁存寄存器

    U1FCR = 0x81;                                   // 使能 FIFO,并设置触发点为 8 字节
    U1IER = 0x01;                                   // 允许 RBR 中断,即接收中断
```

```
    /* 设置中断允许 */
    VICIntSelect = 0x00000000;                       // 设置所有通道为 IRQ 中断
    VICVectCntl1 = 0x27;                             // UART0 中断通道分配到 IRQ slot 0
    VICVectAddr1 = (int)IRQ_UART1;                   // 设置 UART0 向量地址
    VICIntEnable = 1≪7;                              // 使能 UART0 中断
}
/******************************************************************************
* 函数名称:UART1_SendByte()
* 函数功能:向串口发送字节数据,并等待发送完毕
* 入口参数:data  要发送的数据
* 出口参数:无
******************************************************************************/
void UART1_SendByte(uint8 data)
{
    U1THR = data;                                    // 将要发送的数据传给发送器保持寄存器
    while( (U1LSR&0x40 == 0) ) ;                     // 等待数据发送完毕
}
/******************************************************************************
* 函数名称:UART1_SendBuf ()
* 函数功能:向串口发送数组,并等待发送完毕
* 入口参数:*str      要发送的数组
           *no       发送数据的个数
* 出口参数:无
******************************************************************************/
void UART1_SendBuf(uint8 *str,uint16 no)             // 串口 0 发送字符串
{
     uint8 i;
     for(i = 0;i<no;i++)
      UART1_SendByte(str[i]);                        // 发送数据
}
/******************************************************************************
* 函数名称:main()
* 函数功能:电子罗盘示例程序主函数,显示测得的数据并由串口返回方向值
******************************************************************************/
int main(void)
{
    uint16 a = 0,b = 0,u8_tempa,u8_tempb;
    uint8 temp;
    PINSEL0 = 0;
    PINSEL1 = 0;
    UART0_Init(9600);
    UART1_Ini(9600);
    I2C_Init(40000);
    RTCIni();
    DelayNS(1000);                                          // 罗盘上电比较慢,要延时
    for (;;)
    {
```

```
        if(RTC_flag)                                    // RTC 秒中断标志
        {
            RTC_flag = 0;
            UART1_SendByte(0x77);                       // 控制电子罗盘返回数据
        }

        if(rcv_flag1)
        {
            DelayNS(10);
            rcv_flag1 = 0;
            num1 = 0;
            /* 读取 2 字节数据,包含当前方位及角度信息 */
            u8_tempa = rcv_buf1[0];
            u8_tempb = rcv_buf1[1];
            b = (u8_tempa&0x01) * 0xff + u8_tempb;      // 取出角度,范围 0°~360°
            Display(b);                                 // 在数码管上显示角度
            temp = (u8_tempa>>1)&0x07;                  // 在串口上显示当前方位
            switch(temp)
            {
                case 0x00: UART0_SendStr("北");     break;
                case 0x01: UART0_SendStr("东北");   break;
                case 0x02: UART0_SendStr("东");     break;
                case 0x03: UART0_SendStr("东南");   break;
                case 0x04: UART0_SendStr("南");     break;
                case 0x05: UART0_SendStr("西南");   break;
                case 0x06: UART0_SendStr("西");     break;
                case 0x07: UART0_SendStr("西北");   break;
            }
        }
    }
    return 0;
}
```


5.6 倾角传感器及其应用

5.6.1 倾角传感器简介

倾角传感器可以用来测量相对于水平面的倾角变化量。理论基础就是牛顿第二定律,根据基本的物理原理,在一个系统内部,速度是无法测量的,但却可以测量其加速度。如果初始速度已知,就可以通过积分计算出线速度,进而可以计算出直线位移。所以它其实是运用惯性原理的一种加速度传感器。

当倾角传感器静止时也就是侧面和垂直方向没有加速度作用,那么作用在它上面的只有重力加速度。重力垂直轴与加速度传感器灵敏轴之间的夹角就是倾斜角。

倾角传感器把MCU、MEMS加速度计、模/数转换电路及通信单元全都集成在一块非常小的电路板上,可以直接输出角度等倾斜数据。

MSIN-LD60是上海麦游电子OEM的MSIN-LDxx系列低成本双轴倾角传感器,此集成模块检测量程为±60°;精度可达0.1°;其工作原理是利用测量重力加速度的分量,通过计算将其转为绝对倾角。可根据需要选择输出RS232串行数字信号和TTL电平串行信号,选择方法是直接连接TTL输出引脚或RS232输出引脚。模块提供了9 600 bit/s、19 200 bit/s、4 800 bit/s这3种波特率可供用户选择,但校验位、数据位、停止位不提供调整,必须设定为无校验位、数据位为8、停止位为1。此模块还具有零点自动设定功能,精确度可调整,输出频率可调,工作模式可调,波特率可调等功能,参数具有断电保护,工作电压范围V_{DC}值为3.0~5 V。外观尺寸为:边长25 mm,安装孔直径3 mm,安装孔圆心距离19 mm。

MSIN-LD60倾角传感器的引脚端:

- 引脚端1(VCC)为电源正极(V_{DC}值+5 V);
- 引脚端2(GND)为电源地;
- 引脚端3(TXD)为数据发送端(TTL电平);
- 引脚端4(RXD)为数据接收端(TTL电平);
- 引脚端5(TXD)为数据发送端(RS232电平);
- 引脚端6(RXD)为数据接收端(RS232电平)。

5.6.2 LPC214x开发板与MSIN-LD60倾角传感器的连接

本设计选择MSIN-LD60,其与控制器LPC2148之间采用TTL电平相连,连接示意图如图5-30所示。

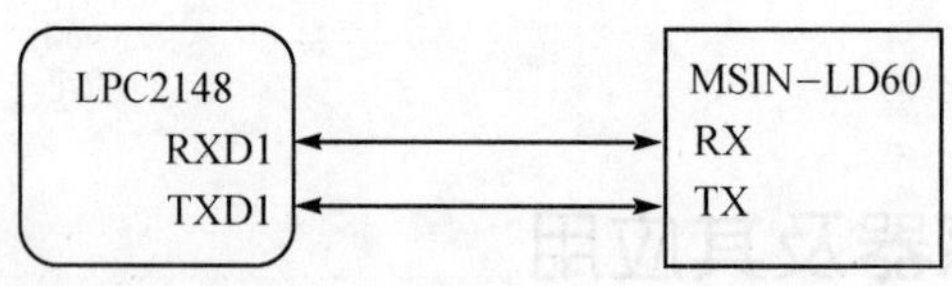

图5-30 MSIN-LD60与LPC2148连接示意图

5.6.3 MSIN-LD60倾角传感器编程示例

利用LPC2148串口1与MSIN-LD60通信,当收到MSIN-LD60数据后,控制数码管显示X轴和Y轴的角度数据。程序流程图如图5-31所示。

启动ADS 1.2,使用ARM Executable Image for lpc2148工程模板创建一个工程。使用过程中,需要Startup.s修改系统模式,堆栈设置为0x5f,开启中断。下面只给出LPC2148与MSIN-LD60的通信程序,详细代码如程序清单5.7所示。

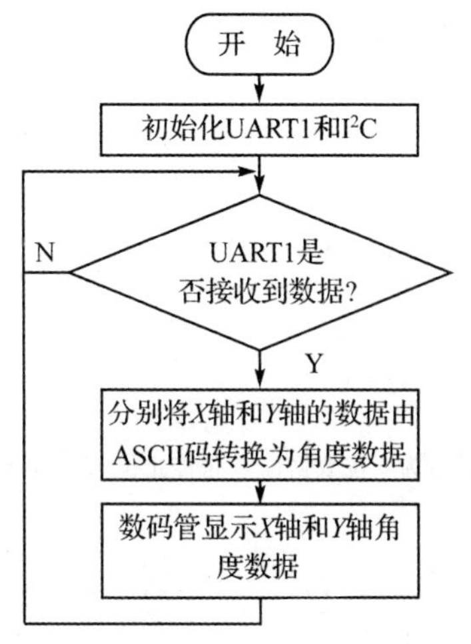

图 5-31　MSIN-LD60 程序流程图

程序清单 5.7　LPC2148 与 MSIN-LD60 的通信示例程序

(1) 串口 1 程序

```
#include   "config.h"
#include   "uart1.h"
#define  MAX_Uart1buf_Size   200                             // 定义串口数据长度
uint8    rcv_buf1[MAX_Uart1buf_Size] = {0},num1 = 0;         // 存放串口接收数据的数组
uint8    rcv_flag1  =  0;                                    // 串口接收到数据的标志变量
/*****************************************************************
*  函数名称:IRQ_UART1()
*  函数功能:串口 UART1 接收中断
*  入口参数:无
*  出口参数:无
*****************************************************************/
void    __irq IRQ_UART1(void)
{
    uint8 i;
    switch(U1IIR&0x0f)                                       // U0IIR 低 4 位为中断标志
    {
        case 0x04:
            for(i = 0; i<8; i++)                             // 触发点为 8,需修改
                rcv_buf1[num1++]  =  U1RBR;
            rcv_flag1  =  1;
            break;
        case 0x0c:                                           // FIFO 接收数据少于触发个数
            while((U1LSR&0x01) == 1)                         // 0x01 表示接收数据就绪
            {
                rcv_buf1[num1++]  =  U1RBR;
            }
```

```
            rcv_flag1 = 1;
            break;                                  // 发送空,则触发 THER 中断
        default:
            break;
    }
    if(num1 >= MAX_Uart1buf_Size) num1 = 0;
    VICVectAddr = 0x00;                             // 中断处理结束
}
/********************************************************************
* 函数名称:UART1_Ini()
* 函数功能:初始化串口 1。设置为 8 位数据位,1 位停止位,无奇偶校验
* 入口参数:bps      波特率
* 出口参数:无
********************************************************************/
void UART1_Ini(uint32 bps)
{
    uint16 Fdiv;
    uint32 tmp;
    /*  设置 P0.8,P0.9 为 UATR1 */
    PINSEL0 = (PINSEL0 & 0xFFFCFFFF)|0x50000;
    if(bps == 115200)
    {
        U1FDR    = 1 | 12<<4;                       // LPC2148 串口除数校准波特率
    }
    tmp = bps;
    U1LCR = 0x83;                                   // 一个停止位,无奇偶校验位
    Fdiv = (Fpclk/16)/tmp;
    U1DLM = Fdiv/256;                               // 根据波特率设定 U0DLM 和 U0DLL 的值
    U1DLL = Fdiv % 256;
    U1LCR = 0x03;                                   // 禁止访问除数锁存寄存器
    U1FCR = 0x81;                                   // 使能 FIFO,并设置触发点为 8 字节
    U1IER = 0x01;                                   // 允许 RBR 中断,即接收中断
      /* 设置中断允许 */
    VICIntSelect = 0x00000000;                      // 设置所有通道为 IRQ 中断
    VICVectCntl1 = 0x27;                            // UART1 中断通道分配到 IRQ
    VICVectAddr1 = (int)IRQ_UART1;                  // 设置 UART1 向量地址
    VICIntEnable = 1<<7;                            // 使能 UART1 中断
}
/********************************************************************
* 函数名称:UART1_SendByte()
* 函数功能:向串口发送字节数据,并等待发送完毕
* 入口参数:data  要发送的数据
* 出口参数:无
********************************************************************/
void UART1_SendByte(uint8 data)
```

```
{
    U1THR = data;                                   // 数据传给发送器保持寄存器
    while( (U1LSR&0x40 == 0) ) ;                    // 等待数据发送完毕
}

/*************************************************************************
*  函数名称:UART1_SendBuf()
*  函数功能:向串口发送数组,并等待发送完毕
*  入口参数:*str    要发送的数据
*            no     发送数据的个数
*  出口参数:无
*************************************************************************/
void UART1_SendBuf(uint8 *str,uint16 no)
{
     uint8 i;
     for(i = 0;i<no;i++)
      UART1_SendByte(str[i]);                       // 发送数据
}
```

(2) 主函数程序

```
#include"config.h"
extern uint8    rcv_buf1[50],num1;                 // 存放串口接收数据的数组
extern uint8    rcv_flag1;                         // 串口接收到数据的标志
uint8 tmp_buf[200];                                // 倾角传感器数据缓存数组
/*************************************************************************
*  函数名称:Display_Angle()
*  函数功能:控制 ZLG7290 数码管驱动,显示测得的角度
*  入口参数:*buf_x     X 轴角度数据
*            *buf_y     Y 轴角度数据
*  出口参数:无
*************************************************************************/
void Display_Angle(uint8 *buf_x,uint8 *buf_y)
{
    uint8 i;
    uint8 LED_Buf[8];
    uint8 LED_tab[10] = {0xfc,0x60,0xda,0xf2,0x66,0xb6,0xbe,0xe0,0xfe,0xf6};
    /* Y 轴倾角显示,数码管译码 */
    if(buf_y[0] == 1) LED_Buf[0] = 0x02;           // 若是负角度,显示'-'号
    else LED_Buf[0] = 0x00;                        // 若是正角度,不显示
    LED_Buf[1] = LED_tab[buf_y[1]/10%10];
    LED_Buf[2] = LED_tab[buf_y[1]%10] | 0x01;      // "|0x01"为小数点显示
    LED_Buf[3] = LED_tab[buf_y[2]%10];             // 小数位显示
    /* X 轴倾角显示,数码管译码 */
    if(buf_x[0] == 1)LED_Buf[4] = 0x02;            // 若是负角度,显示'-'号
    else LED_Buf[4] = 0x00;
    LED_Buf[5] = LED_tab[buf_x[1]/10%10];
```

```
    LED_Buf[6] = LED_tab[buf_x[1] % 10] | 0x01;         // "|0x01"为小数点显示
    LED_Buf[7] = LED_tab[buf_x[2] % 10];                // 小数位显示
    for(i = 0;i<8;i++)
    {
        ZLG7290_SendData(0x10 + i,LED_Buf[i]);          // 直接显示法显示在 ZLG7290 上
    }
}
/****************************************************************************
* 功能:LPC2148 控制倾角传感器测量角度主函数
* 说明:倾角传感器通过串口返回的数据为 ASCII 码数据,需要转为可用的角度数据
****************************************************************************/
int main(void)
{
    uint8 temp,buf_x[100],buf_y[100];
    PINSEL0 = 0;
    PINSEL1 = 0;
    UART1_Ini(9600);                                    // UART1 初始化,用于接收倾角传感器数据
    I2C_Init(40000);                                    // I²C 初始化,用于 ZLG7290 数码管显示
    DelayNS(300);                                       // 延时,等待传感器就绪
    for (;;)
    {
        if(rcv_flag1)                                   // UART1 接收数据标志
        {
            DelayNS(5);
            rcv_flag1 = 0;
            for(temp = 0;temp<num1;temp++)
               tmp_buf[temp] = rcv_buf1[temp];
               num1 = 0;
            /* 若收到的数据为"X = ±xx.x",处理 X 轴数据 */
            if(tmp_buf[0] == 'X')
            {
                if(tmp_buf[2] =='-')                    // 若角度为负
                {
                    buf_x[0] = 1;                       // 1 表示角度为负
                }
                else if(tmp_buf[2] =='+')
                {
                    buf_x[0] = 0;                       // 0 表示角度为正
                }
                /* buf_x[1],buf_x[2]分别存放角度整数部分和小数部分 */
                buf_x[1] = (tmp_buf[3] - '0') * 10 + (tmp_buf[4]-'0');
                buf_x[2] = tmp_buf[6]-'0';
            }
            /* 若收到的数据为"Y = ±xx.x",处理 Y 轴数据 */
            else if(tmp_buf[0] == 'Y')
```

```
            {
                if(tmp_buf[2]=='-')
                {
                    buf_y[0] = 1;
                }
                else if(tmp_buf[2]=='+')
                {
                    buf_y[0] = 0;
                }
                buf_y[1] = (tmp_buf[3]- '0')*10 + (tmp_buf[4]-'0');
                buf_y[2] = tmp_buf[6]-'0';
            }
            /* 若收到的数据为"X=±xx.x  Y=±xx.x",处理Y轴数据 */
            else if(tmp_buf[9] == 'Y')
            {
                if(tmp_buf[11]=='-')
                {
                    buf_y[0] = 1;
                }
                else if(tmp_buf[11]=='+')
                {
                    buf_y[0] = 0;
                }
                buf_y[1] = (tmp_buf[12]-'0')*10 + (tmp_buf[13]-'0');
                buf_y[2] = tmp_buf[15]-'0';
            }
            Display_Angle(buf_x,buf_y);                    //控制数码管显示X轴、Y轴角度
        }
    }
    return 0;
}
```


5.7 角度传感器及其应用

5.7.1 WDD35D-4角度传感器简介

WDD35D-4角度传感器实际是一个高精度、标称阻值为5 kΩ的电位器,独立线性度达到0.1%,机械转角可达360°,理论电旋转角为345°,故WDD35D-4测量角度的最大偏差为0.345°。其结构示意图如图5-32所示。旋转角度传感器转轴时,其电阻值随之改变,当转轴转动360°后,电阻值与旋转前的相等。因此可通过读取电阻值的大小来计算旋转的角度,也可利用ADC采样将电阻值转换为电压值来判断旋转角度。

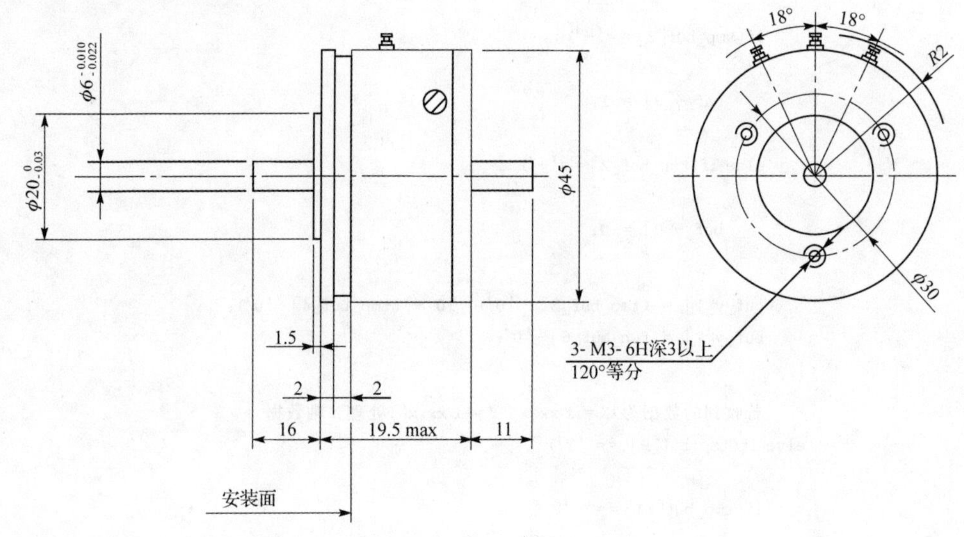

图 5-32　WDD35D-4 角度传感器结构示意图

5.7.2　LPC214x 开发板与 WDD35D-4 角度传感器的连接

WDD35D-4 角度传感器的应用示意图如图 5-33 所示。下面以 LPC2148 微控制器使用其自带的 10 位 ADC 采样来判断旋转角度为例，说明该传感器的用法。WDD35D-4 角度传感器的引脚端连接基准电压 V_{REF}，引脚端 2 连接到地(GND)，中心头(引脚端 3)与 LPC2148 微控制器的 AD0.3 引脚端连接。当角度传感器从 0°～345°变化时，引脚端 3 的电压从 0 V 到 V_{REF} 线性变化，首先找到引脚端 3 输出电压为 0 V 时动臂的位置，以该处作为起始位置 0°，然后旋转动臂到任意位置，通过 A/D 采样读取该点处的电压值 V，便可计算出该点角度 $\Phi=(345/V_{REF})\times V(°)$。

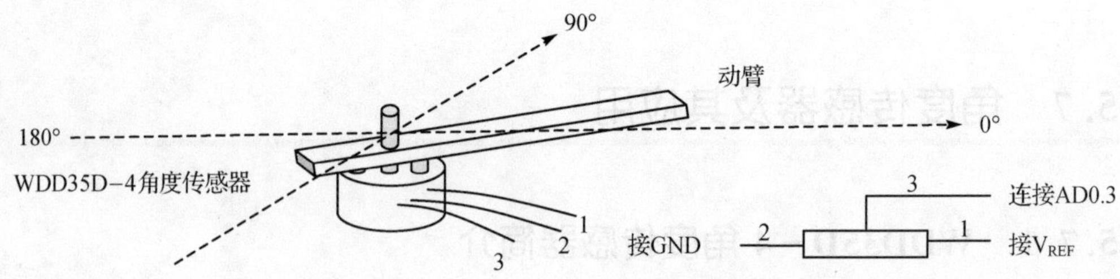

图 5-33　WDD35D-4 角度传感器应用示意图

5.7.3　WDD35D-4 角度传感器编程示例

启动 ADS 1.2，使用 ARM Executable Image for lpc2148 工程模板创建一个工程。使用 LPC2148 的 AD0.3 通道采样电压值，最后换算为角度值后显示出来，软件流程图如图 5-34 所示,程序代码如程序清单 5.8 所示,注:清单中没有给出显示程序代码。

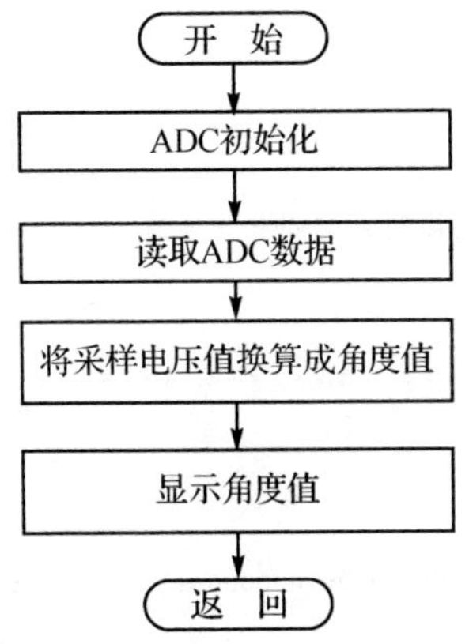

图 5-34　计算动臂旋转角度程序流程图

程序清单 5.8　角度传感器应用示例程序

```
#include "config.h"
fp64 Per_Angle = 345.0/1024.0;                    // 单位 ADC 值的角度
/******************************************************************
* 函数名称:DelayNS()
* 函数功能:长软件延时
* 入口参数:dly  延时参数,值越大,延时越久
* 出口参数:无
******************************************************************/
void  DelayNS(uint32  dly)
{
    uint32  i;
    for(; dly>0; dly--)
        for(i=0; i<5000; i++);
}
/******************************************************************
* 函数名称:AD0_3_P0_30_Read()
* 函数功能:读出 ADC0 的第 3 通道,P0.30 引脚 ADC 数据
* 入口参数:无
* 出口参数:读出的 ADC 数据
******************************************************************/
uint32  AD0_3_P0_30_Read(void)
{
    uint32  ADC_Data;
    ADCR = (ADCR&0xFFFFFF00)| (1<<3) | (1 << 24);       // 切换通道
    while( (ADDR&0x80000000)==0 );                      // 等待转换结束
    ADCR = ADCR | (1 << 24);                            // 再次启动转换
```

```
    while( (ADDR&0x80000000) == 0 );
    ADC_Data = ADDR;
    ADC_Data = (ADC_Data≫6) & 0x3ff;
    return ADC_Data;
}
/*************************************************************************
 * 函数名称:AD0_3_P0_30()
 * 函数功能:初始化 P0.30 引脚连接到 ADC0 的第 3 通道
 * 入口参数:无
 * 出口参数:无
*************************************************************************/
uint32  AD0_3_P0_30(void)
{
     uint32  ADC_Data;
     PINSEL1 = PINSEL1 & 0x3FFFFFFF | 0x10000000;        // 设置 P0.30 连接到 AD0.3
     ADCR = ( 1 ≪ 3 )                          |         // SEL.1 = 1,选择 AD0.1
            ((Fpclk / 1000000 - 1) ≪ 8)        |         // 转换时钟为 1 MHz
            (0 ≪ 16)                           |         // 软件控制转换操作
            (0 ≪ 17)                           |         // 使用 11 clock 转换
            (1 ≪ 21)                           |         // 非掉电转换模式
            (0 ≪ 22)                           |         // 非测试模式
            (1 ≪ 24)                           |         // 直接启动 ADC 转换
            (0 ≪ 27);                                    // EDGE = 0
    DelayNS(10);
    ADC_Data = ADDR;
    ADCR = (ADCR&0xFFFFFF00)| (1≪3) | (1 ≪ 24);         // 切换通道
    while( (ADDR&0x80000000) == 0 );                     // 等待转换结束
    ADCR = ADCR | (1 ≪ 24);                              // 再次启动转换
    while( (ADDR&0x80000000) == 0 );
    ADC_Data = ADDR≫6;
    return ADC_Data;
}
/*************************************************************************
 * 函数名称:Read_Angle()
 * 函数功能:通过角度传感器数据计算角度
 * 入口参数:无
 * 出口参数:无
*************************************************************************/
fp64 Read_Angle(void)
{
    uint32 temp;
    temp = AD0_3_P0_30_Read();
    return Per_Angle * temp;                             // 由 A/D 值计算角度
}
/*************************************************************************
 * 名称:main ()
```

全国大学生电子设计竞赛ARM嵌入式系统应用设计与实践(第2版)

```
* 功能:角度传感器测量主函数
*********************************************************************/
int main(void)
{
    fp64   a;
    PINSEL0 = 0;
    PINSEL1 = 0;
    PINSEL2 &= ~(0x00000006);
    IO0DIR = 0;
    IO1DIR = 0;
    AD0_3_P0_30();                              // A/D引脚初始化
    while(1)
    {
        a = Read_Angle();                       // 读取并计算角度
        Display(a);                             // 显示角度
    }
}
```

第 6 章

数据存储

6.1 E^2PROM 24LC256

6.1.1 E^2PROM 24LC256 简介

24LC256 是 Microchip 公司生产的一种串行的 E^2PROM 存储器，存储容量为 32K×8 bit(256 Kbit)；工作电压为 1.8～5.5 V，芯片的传输速率与工作电压的大小有关，当工作电压为 1.8 V 时，最大传输速率为 100 kHz，而当工作电压不低于 2.5 V 时，最大传输速率可达400 kHz；与外界通信采用 I^2C 标准总线接口，仅需 2 根串行接口总线；具有硬件写保护、10 万次的擦写和 200 年以上的数据保护功能。

24LC256 采用 8 引脚的 DIP 封装，引脚端 A0～A2 为用户可配置的芯片选择引脚，所接的电平高低决定了 24LC256 的 I^2C 通信地址；SCL 和 SDA 分别为 I^2C 通信的串行时钟引脚和串行数据引脚；WP 为芯片写保护输入引脚，当其接低电平时可对芯片写入数据，而接高电平时，芯片为只读，VDD 和 VSS 为电源引脚。

6.1.2 24LC256 的典型应用电路

图 6-1 为 24LC256 的典型应用电路。

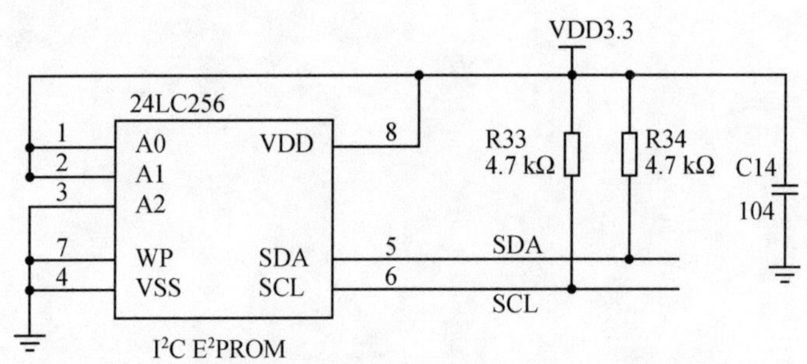

图 6-1　24LC256 典型应用电路原理图

图 6-1 中的 R33、R34 是上拉电阻，与 I^2C 通信速度有关，传输速度越高上拉电阻阻值越小，典型值是 1～10 kΩ，在此选用 4.7 kΩ。将 A0、A1 配置成高电平，器件从机地址是 0xA6。将微控制器系统板上的 SDA 和 SCL 引脚端、电源(3.3 V)和地

分别与 24LC256 的对应引脚连接即可。使用时应注意:为避免与系统中的其他 I^2C 器件地址冲突,应正确设置其地址位。

6.1.3 24LC256 读/写操作编程示例

微控制器可以使用 I^2C 主模式对 24LC256 进行读/写操作。

1. 24LC256 读/写流程图

下面给出 LPC2148 使用 I^2C 主模式对 24LC256 进行读/写并校验操作的示例程序,程序流程如图 6-2 所示。

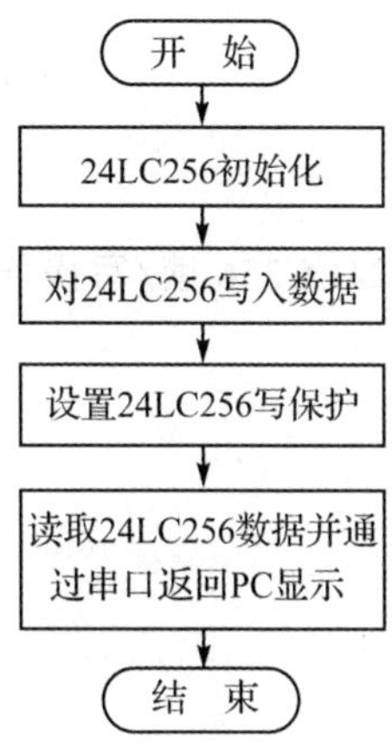

图 6-2 对 24LC256 进行读/写的流程图

2. 示例程序

启动 ADS 1.2,使用 ARM Executable Image for lpc2148 工程模板创建一个工程。使用过程中,需要将 I^2C 软件包文件 I2CINT.c、I2CINT.h 包含进工程,并且在 config.h 中将 I2CINT.h 包括进去,且还需要 Startup.s 修改系统模式,堆栈设置为 0x5f,开启中断。主模式 I^2C 初始化示例如程序清单 6.1 所示。该示例程序操作调用 I^2C 软件包,服务程序中向有子地址器件写入和读取多字节数据的函数,程序清单 6.2 给出了读/写操作主要函数代码。I^2C 软件包可以在周立功公司网站(http://www.zlgmcu.com)上下载。

程序清单 6.1 主模式 I^2C 初始化示例程序

```
/*******************************************************************
** 函数名称:I2C_Init
** 函数功能:主模式 I²C 初始化,包括初始化其中断为向量 IRQ 中断
** 入口参数:fi2c          初始化 I²C 总线速率,最大值为 400K
** 出口参数:无
*******************************************************************/
void  I2C_Init(uint32 fi2c)
{
```

```
    if (fi2c > 400000)
    {
        fi2c = 400000;
    }
    PINSEL0 = (PINSEL0&0xFFFFFF0F) | 0x50;             // 设置 I²C 控制口有效
    I2C0SCLH = (Fpclk / fi2c + 1) / 2;                  // 设置 I²C 时钟为 fi2c
    I2C0SCLL = (Fpclk / fi2c) / 2;
    I2C0CONCLR = 0x2C;
    I2C0CONSET = 0x40;                                  // 使能主 I²C
    /* 设置 I²C 中断允许 */
    VICIntSelect = 0x00;                                // 所有中断通道设置为 IRQ 中断
    VICVectCntl0   = 0x29;                              // 设 I²C 中断最高优先级
    VICVectAddr0 = (uint32)IRQ_I2C;                     // 设置中断服务程序地址
    VICIntEnable = 1 << 0x09;                           // 使能 I²C 中断
}
```

程序清单 6.2　24LC256 读/写操作主要函数代码

```
#include "config.h"
#include "I2CINT.H"
#include  "uart0.h"
#define  Address   0xa6                          // 10100110
uint8     reciver[1000] = {0};
/****************************************************************
* 函数名称:DelayNS()
* 函数功能:长软件延时
* 入口参数:dly  延时参数,值越大,延时越久
* 出口参数:无
****************************************************************/
void  DelayNS(uint32  dly)
{  uint32  i;

   for(; dly > 0; dly--)
       for(i = 0; i < 5000; i++);
}
/****************************************************************
** 函数名称:storage_data()
** 函数功能:将需要存放的数据依次写入 24LC256 中
** 入口参数:data      所需存放的数组名
            Long      数组长度,若不为 10 的整数倍,可根据需要更改函数
** 出口参数:无
****************************************************************/
void storage_data(uint8 *data, uint8 Long)
{
    uint16 add = 0;
    uint8 k = Long/10;
    for(add = 0;add<k;add++)
```

```
    {
        I2C_WriteNByte(Address,TWO_BYTE_SUBA,add * 10,data + add * 10,10);
        DelayNS(10);
        I2C_ReadNByte(Address,TWO_BYTE_SUBA,add * 10,reciver + add * 10,10);
    }
}
/**********************************************************************
*  名称:main()
*  功能:24LC256 存储器应用实例程序主函数
**********************************************************************/
int main(void)
{
    uint8 data[1000] = {0};                    // 用于存放需要发送的数据
    uint8 Long  =  80;                         // Long 为自己定义的数组长度,根据需要更改大小
    uint8 i = 0;

    I2C_Init(400000);                          // 设置 I²C 波特率为最大 400 kbit/s
    UART0_Init(9600);
    DelayNS(300);
 /* 给需要存放的数组赋值 */
    for(i = 0;i<100;i + + )
    {
        data[i]  =  i;
    }

    storage_data(data,Long);
    for(i = 0;i<Long;i + + )
    UART0_SendByte(reciver[i]);
    while(1);
    return 0;
}
```

注意:I^2C 初始化可参照程序清单 6.1(主模式 I^2C 初始化示例),通信速率不大于 400 kbit/s,本示例程序将 I^2C 中断分配为 IRQ 模式。

6.2　SK - SDMP3 语音模块及其应用

6.2.1　SK - SDMP3 模块简介

SK - SDMP3 语音模块是广州市苏凯电子有限公司生产的产品,其直接支持 MP3 语音文件,文件来源广泛、占据容量小、容易制作、音质优美、通用性好;将 SD 卡作为存储媒体,存储容量大、容易复制保存、更新十分方便;存储内容按文件夹的形式编排,按名称分段存储,易存、易改;支持 4 种工作模式:标准模式、按键模式、并口模式、串口模式;可以播放背景音乐,广告语;可以进行任意段语音的播放;模块按照工业级设计,模块尺寸为 44 mm×40 mm×8 mm,电源电压为直流 5～9 V(内部稳压成

3.3 V 工作电压),可以适用于各种复杂的场合。

1. 模块结构与引脚端功能

SK - SDMP3 语音模块如图 6 - 3 所示,模块内部含 MCU、MP3 解码芯片,AMS1117 - 3.3 稳压管等电路,模块的 I/O 口封装形式如图 6 - 4 所示,I/O 口电平为 3.3 V,各引脚端功能如表 6 - 1 所列。微控制器仅需通过串口就可以控制该模块播放 SD 卡中 MP3 格式的语音数据。

图 6 - 3 SK - SDMP3 语音模块

表 6 - 1 I/O 口引脚端功能

引脚号	符　号	功　能	引脚号	符　号	功　能
1	＋VCC	直流 5～9 V	12	BUSY	忙信号
2	GND	电源地	13	P01	I/O 口
3	GND	音频地	14	P02	I/O 口
4	TXD	串口数据发送端	15	P03	I/O 口
5	RXD	串口数据接收端	16	P04	I/O 口
6	系统备用	悬空	17	P05	I/O 口
7	系统备用	悬空	18	P06	I/O 口
8	系统备用	悬空	19	GBUF	音频输出缓冲地
9	系统备用	悬空	20	R	音频右声道输出
10	RST	低电平复位	21	L	音频左声道输出
11	GND	电源地	22	GND	数字地

2. 工作模式

SK - SDMP3 语音模块支持标准模式、按键模式、并口模式、串口模式 4 种工作模式。

(1) 标准模式

该工作模式和一般的 MP3 十分类似，P01～P06 端口平时为高电平，低电平时触发。

注意：模块上电后，要先给 P03 一个负脉冲信号，这样模块才能开始播放，然后其他 I/O 口才能起作用。同时，文件夹里面 MP3 文件名称必须是连续的。如有不连续，那么在断点之后的歌曲将不能被识别。

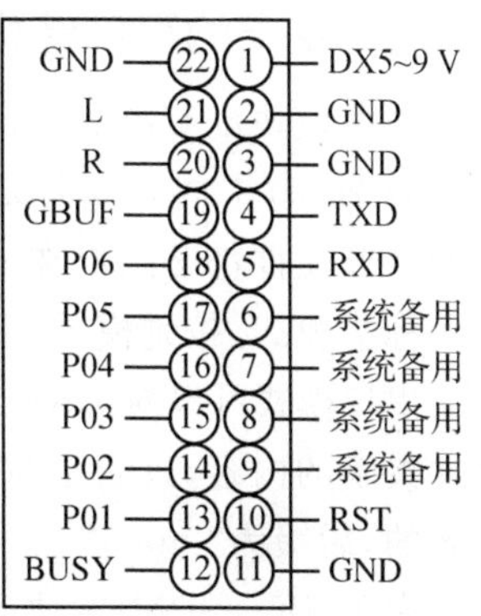

图 6-4　I/O 口封装形式

(2) 按键模式

6 个 I/O 口平时为高电平，低电平触发。一个 I/O 对应触发一段语音，只能放 6 段语音，而且文件名称必须为 001.mp3～006.mp3。

(3) 并口模式

从 00H～1FH 一共有 32 个地址，正好对应 32 段语音。首先将 P01～P06 设置成地址，再将 P01 从高电平拉为低电平，这样就能触发语音播放了。

(4) 串口模式

以标准的 RS232 串口通信时序为基础，波特率 9 600，自行定制了一个通信协议，该通信协议数据格式如表 6-2 所列。在表 6-3 中，操作码共 9 个，其中 A0、A4 和 B0 后面必须带数据位，其余的不用。

表 6-2　通信协议数据格式

起始码	数据长度	操作码	文件夹		曲　目			结束码
			十位	个位	百位	十位	个位	
7E	07	xx	xx	xx	xx	xx	xx	7E

表 6-3　操作码功能与数据格式

操作码功能	操作码	数　据
播放广告(重新播放)	A0H	xx xx xx xx xx
暂停广告	A1H	
从暂停处播放广告	A2H	
停止广告	A3H	
调节音量	A4H	0～8 (代表八级音量)
播放背景(重新播放)	B0H	xx xx xx xx xx
暂停背景	B1H	
从暂停处播放背景	B2H	
停止背景	B3H	

3. SD卡的存储结构

与本模块配套的SD卡的容量必须为32 MB～1 GB,卡的格式为FAT。本模块只能识别SD卡内名称为advert00～advert99的文件夹,其余的名称都不能识别,所有的语音文件都必须放在advert00～advert99其中的一个文件夹里面。

标准模式、按键模式、并口模式都只能读/写advert01文件夹内的内容。

串口模式可以对advert00～advert99共100个文件夹的内容进行读取。advertxx文件夹下的内容,只能是000.mp3～999.mp3共1 000个文件,都是以数字000～999命名的,后缀为“.mp3”。

有关SK-SDMP3语音模块的更多内容请登录http://www.dzkf.com,查询《SK-SDMP3模块应用手册V1.5》。

6.2.2 音频功率放大器电路

1. 采用LM386的音频功率放大器电路

SK-SDMP3语音模块的音频功率放大器电路可以采用LM386音频功率放大器芯片。

LM386是美国国家半导体公司生产的音频功率放大器芯片,电压增益可调,范围为20～200,静态功耗低,电流约为4 mA,工作电压范围为4～12 V或者5～18 V,主要应用于低电压消费类产品。

LM386为使外围元件最少,电压增益内置为20。但在引脚1和引脚8之间增加了一只外接电阻和电容,可在20～200范围内调节电压增益。输入端以地位参考,同时输出端被自动偏置到电源电压的一半,在6 V电源电压下,它的静态功耗仅为24 mW,使得LM386特别适用于电池供电的场合。LM386采用塑封8引线双列直插式和贴片式的封装形式。

一个采用LM386的音频功率放大器电路如图6-5所示,调节电位器R2的阻值可以控制输出信号的大小。使用两个相同的电路分别用于放大SK-SDMP3语音模块的左右音频信号,通过扬声器便可获得很好的音频信号输出。

2. 采用TEA2025B的音频功率放大器电路

一个采用TEA2025B的音频功率放大器电路原理图和印制板图如图6-6所示。

TEA2025B是ST公司生产的双声道音频功率放大器芯片,该电路具有声道分离度高,电源接通时冲击噪声小,外接元件少,最大电压增益可由外接电阻调节等特点,在袖珍式或便携式立体声音响系统中做音频功率放大。TEA2025的工作电源电压范围为3～12 V,典型工作电压为6～9 V。当工作电压为12 V时,TEA2025B用于双声道,其最大可输出每声道2.3 W;当工作电压为9 V时,用于单声道最大可输出功率为4.7 W。

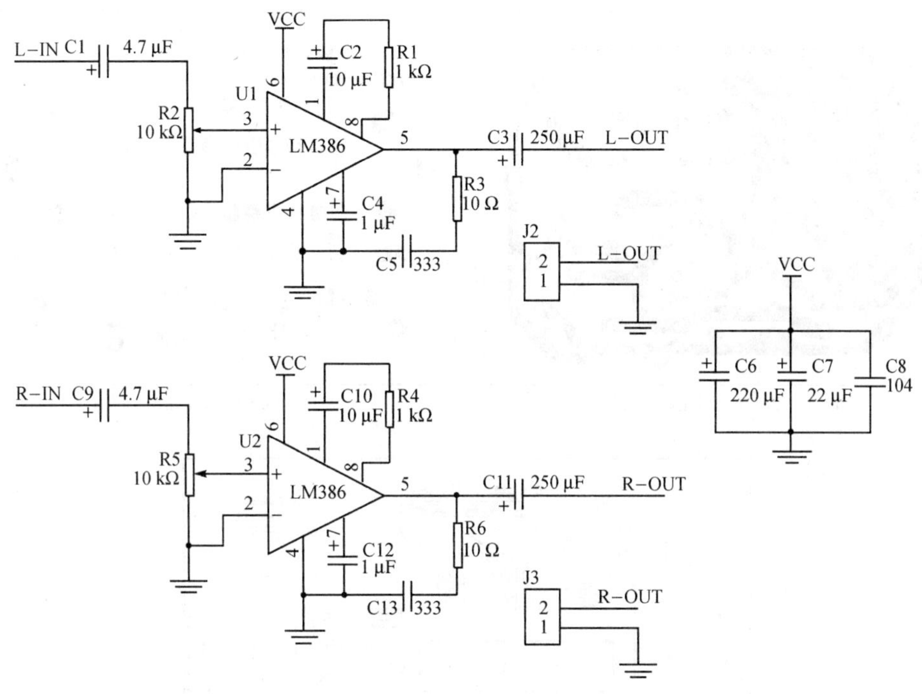

图 6-5 LM386 双声道音频放大器电路

将 SK-SDMP3 语音模块的左右音频信号分别通过两个 20 kΩ 可调电位器接在放大器的输入端 10 引脚和 7 引脚，经过 TEA2025 放大的音频信号，通过扬声器便可获得很好的音频信号输出，在音频信号输入端和输出端都需要加上滤波电容器。

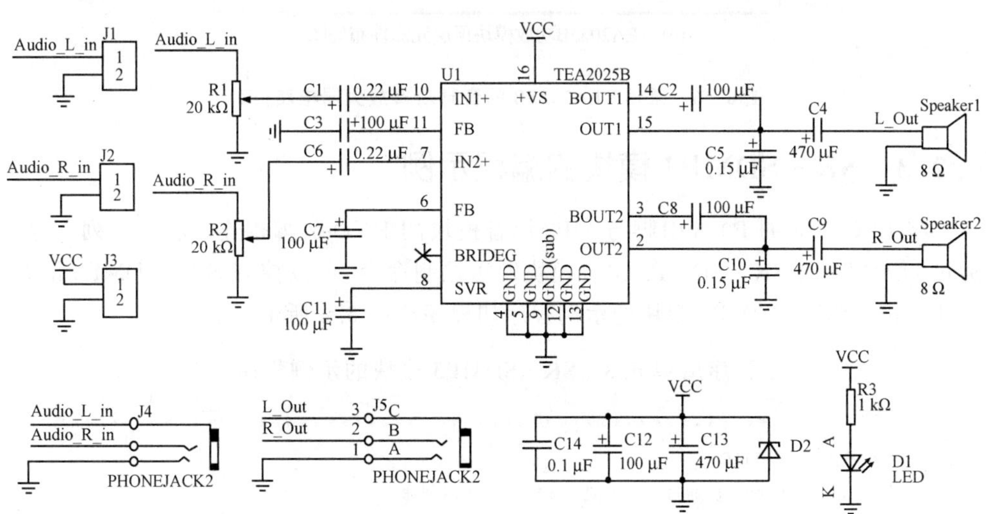

(a) TEA2025B构成的双通道音频功率放大器电路原理图

图 6-6 采用 TEA2025B 的音频功率放大器电路

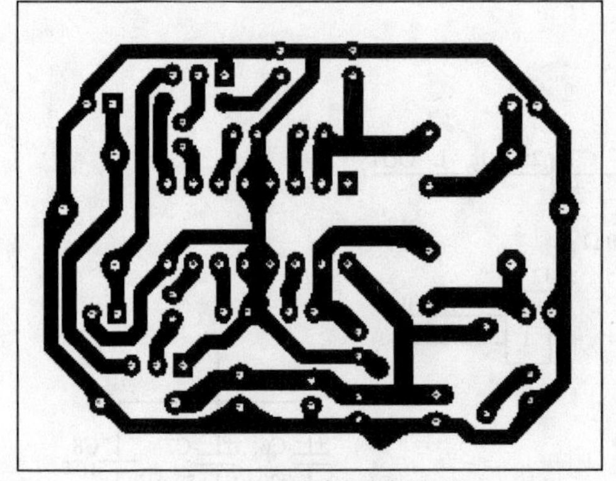

(b) TEA2025B功放电路底层PCB

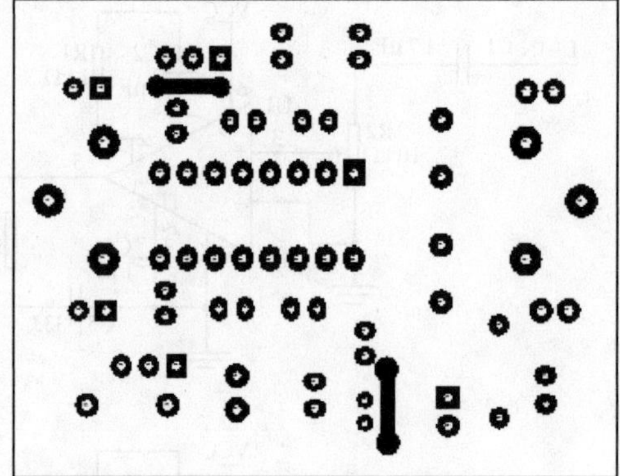

(c) TEA2025B功放电路顶层PCB

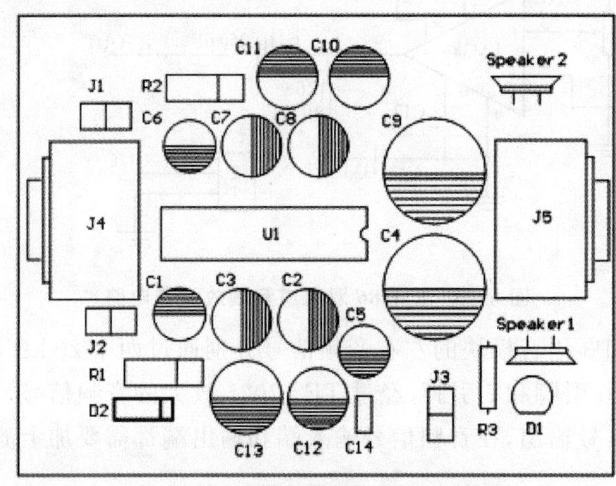

(d) TEA2025B功放模块顶层元器件布局图

图 6-6 采用 TEA2025B 的音频功率放大器电路(续)

6.2.3 SK-SDMP3 模块的编程示例

将 LPC2148 的 P0.8 引脚与 MP3 语音模块的 RXD 引脚相连。利用下列程序，通过 UART1 向 SK-SDMP3 语音模块发送控制命令，可以控制 SK-SDMP3 语音模块发声。SK-SDMP3 模块的示例程序如程序清单 6.3 所示。

程序清单 6.3 SK-SDMP3 模块的示例程序

```
/****************************************************************************
* 文件名:MP3_UART1.C
* 功能:MP3 控制软件包,通过 UART1 向 MP3 模块发送控制命令
* 说明:将 LPC2148 的 P0.8 引脚与 MP3 语音模块的 RXD 引脚相连
****************************************************************************/
#include  "config.h"
```

```
/**********************************************************************
*  函数名称:UART1_Ini()
*  函数功能:初始化串口 1,设置为 8 位数据位,1 位停止位,无奇偶校验
*  入口参数:bps      UART 通信波特率
*  出口参数:无
**********************************************************************/
void UART1_Ini(uint32 bps)
{
    uint16 Fdiv;
    uint32 tmp;
    PINSEL0 = (PINSEL0 &0xFFFeFFFF)|0x10000;          // 设置 P0.8 为 UART1_TXD
     tmp = bps;
    U1LCR = 0x83;                                    // DLAB = 1
    Fdiv = (Fpclk/16)/tmp;                           // UART_BPS
    U1DLM = Fdiv/256;                                // 根据波特率设定 U0DLM 和 U0DLL 的值
    U1DLL = Fdiv % 256;
    U1FDR = 1<<4;                                    // 设置小数波特率的值
    U1LCR = 0x03;                                    // 禁止访问除数锁存寄存器
    UART1_MP3_voice(8);                              // MP3 语音模块音量选择为最大声
}
/**********************************************************************
*  函数名称:UART1_SendByte()
*  函数功能:向串口发送字节数据,并等待发送完毕
*  入口参数:data   要发送的数据
*  出口参数:无
**********************************************************************/
void UART1_SendByte(uint8 data)
{
    U1THR = data;                                    // 将要发送的数据传给发送器保持寄存器
    while( (U1LSR&0x40) = = 0 ) ;                    // 等待数据发送完毕
}
/**********************************************************************
*  函数名称:UART1_MP3display_Folder_song()
*  函数功能:向串口发送播放指令
*  入口参数:需要播放的曲目及其所在文件夹
            Folder     文件夹数
            song       歌曲数
*  出口参数:无
*  说明:MP3 模块收到本命令后,开始播放指定文件夹的指定 MP3 格式文件
**********************************************************************/
void UART1_MP3display_Folder_song(uint8   Folder,uint16 song)
```

```
{
    UART1_SendByte(0x7e);
    UART1_SendByte(0x07);
    UART1_SendByte(0xb0);

    UART1_SendByte(0x30 + Folder/10 % 10);
    UART1_SendByte(0x30 + Folder % 10);

    UART1_SendByte(0x30 + song/100 % 10);
    UART1_SendByte(0x30 + song/10 % 10);
    UART1_SendByte(0x30 + song % 10);

    UART1_SendByte(0x7e);
}
/**********************************************************************
* 函数名称:UART1_MP3_voice()
* 函数功能:控制 MP3 模块,音量
* 入口参数:voice  音量等级,值为 0~8,共 9 个等级,0 为关闭声音,8 为最大
* 出口参数:无
**********************************************************************/
void UART1_MP3_voice(uint8 voice)
{
    UART1_SendByte(0x7e);
    UART1_SendByte(0x03);
    UART1_SendByte(0xa4);
    UART1_SendByte(voice);
    UART1_SendByte(0x7e);
}
/**********************************************************************
* 函数名称:UART1_MP3_Pause_play_stop()
* 函数功能:向 MP3 发送暂停/播放/停止指令
* 入口参数:code    暂停或播放操作码  0xb1 表示暂停,0xb2 表示恢复,0xb3 表示停止
* 出口参数:无
**********************************************************************/
void UART1_MP3_Pause_play_stop(uint8 code)
{
    UART1_SendByte(0x7e);
    UART1_SendByte(0x07);
    UART1_SendByte(code);

    UART1_SendByte(0);
    UART1_SendByte(0);
    UART1_SendByte(0x30);
    UART1_SendByte(0);
    UART1_SendByte(0);
```

```
    UART1_SendByte(0x7e);
}
/*****************************************************************
*  名称:main ()
*  功能:使用最大音量播报文件夹中的第 15 首歌曲
*****************************************************************/
int main(void)
{
    UART1_Ini(9600);                              // 串口初始化
    UART1_MP3_voice(8);                           // MP3 音量选择最大
    UART1_MP3display_Folder_song(20,15);          // 播报第 20 个文件夹中的第 15 首歌曲
    while(1);
    return 0;
}
```

第 7 章 数据传输

7.1 无线数据传输

7.1.1 基于 nRF905 的无线收发器电路模块

1. nRF905 简介

nRF905 是一个工作在 433 MHz、868 MHz、915 MHz 的 ISM 频段，完全集成的单片无线收发器芯片。

nRF905 芯片内部包含有一个完全集成的调制器、接收解调器、功率放大器、晶体振荡器等电路。其采用"DSS+PLL"频率合成技术和 GMSK 调制，频率稳定性非常好，抗干扰能力强；可以很容易通过 SPI 接口编程配置其工作模式；信道数最多可达 170 个，能够满足需要多信道工作的特殊场合；传输距离最远可达 1 000 m；数据速率最高为 50 kbit/s；工作电压为 1.9～3.6 V；在发射功率为 10 dBm 时，电流消耗为 11 mA；在接收状态时，电流消耗为12.5 mA，具有低功耗模式；采用 32 引脚端的 QFN 5 mm×5 mm 封装。

2. nRF905 的工作模式

nRF905 通过对 TRX_CE、TX_EN、PWM_UP 的设置来实现不同的工作模式，nRF905 工作模式如表 7-1 所列，模式设置如表 7-2 所列。

表 7-1 nRF905 工作模式

工作模式	工作状态
活动模式	ShockBurst RX（突发接收）
	ShockBurst TX（突发发射）
节电模式	掉电和 SPI 编程
	STANDBY（待机）和 SPI 编程

表7-2　nRF905工作模式设置

PWM_UP	TRX_CE	TX_EN	工作模式
0	X	X	掉电和SPI编程
1	0	X	Standby(待机)和SPI编程
1	1	0	ShockBurst RX(突发接收)
1	1	1	ShockBurst TX(突发发射)

3. nRF905的配置

nRF905的工作模式通过SPI接口对芯片内部的寄存器进行配置实现。所有配置字都是通过SPI接口传送给nRF905。SIP接口的工作方式可通过SPI指令进行设置。当nRF905处于空闲模式或关机模式时,SPI接口可以保持在工作状态。

(1) SPI接口寄存器配置

SPI接口由状态寄存器、射频配置寄存器、发送地址寄存器、发送数据寄存器和接收数据寄存器5个寄存器组成。状态寄存器包含数据准备好的引脚状态信息和地址匹配引脚状态信息;射频配置寄存器包含收发器配置信息,如频率和输出功能等;发送地址寄存器包含接收机的地址和数据的字节数;发送数据寄存器包含待发送的数据包的信息,如字节数等;接收数据寄存器包含要接收的数据的字节数等信息。

(2) SPI接口指令与操作

当CSN为低时,SPI接口开始等待一条指令。任何一条新指令均由CSN的由高到低的转换开始。用于SPI接口的命令如表7-3所列。

表7-3　SPI接口指令

指令名称	指令格式	操　作
W_CONFIG (WC)	0000AAAA	写配置寄存器AAAA指出写操作的开始字节,字节数量取决于AAAA指出的开始地址
R_CONFIG (RC)	0001AAAA	读配置寄存器AAAA指出读操作的开始字节,字节数量取决于AAAA指出的开始地址
W_TX_PAYLOA D (WTP)	00100000	写TX有效数据1～32字节,写操作全部从字节0开始
R_TX_PAYLOA D (RTP)	00100001	读TX有效数据1～32字节,读操作全部从字节0开始
W_TX_ADDRES S (WTA)	00100010	写TX地址1～4字节,写操作全部从字节0开始
R_TX_ADDRES S (RTA)	00100011	读TX地址1～4字节,读操作全部从字节0开始

续表 7-3

指令名称	指令格式	操 作
R_RX_PAYLOA D (RRP)	00100100	读 RX 有效数据 1~32 字节,读操作全部从字节 0 开始
CHANNEL_CON FIG (CC)	1000pphc cccccccc	快速设置配置寄存器中 CH_NO HFREQ_PLL 和 PA_PWR 的专用命令 CH_NO=cccccccc HFREQ_PLL=h PA_PWR=pp

(3) SPI 接口操作时序

SPI 接口读操作和写操作时序图分别如图 7-1 和图 7-2 所示。

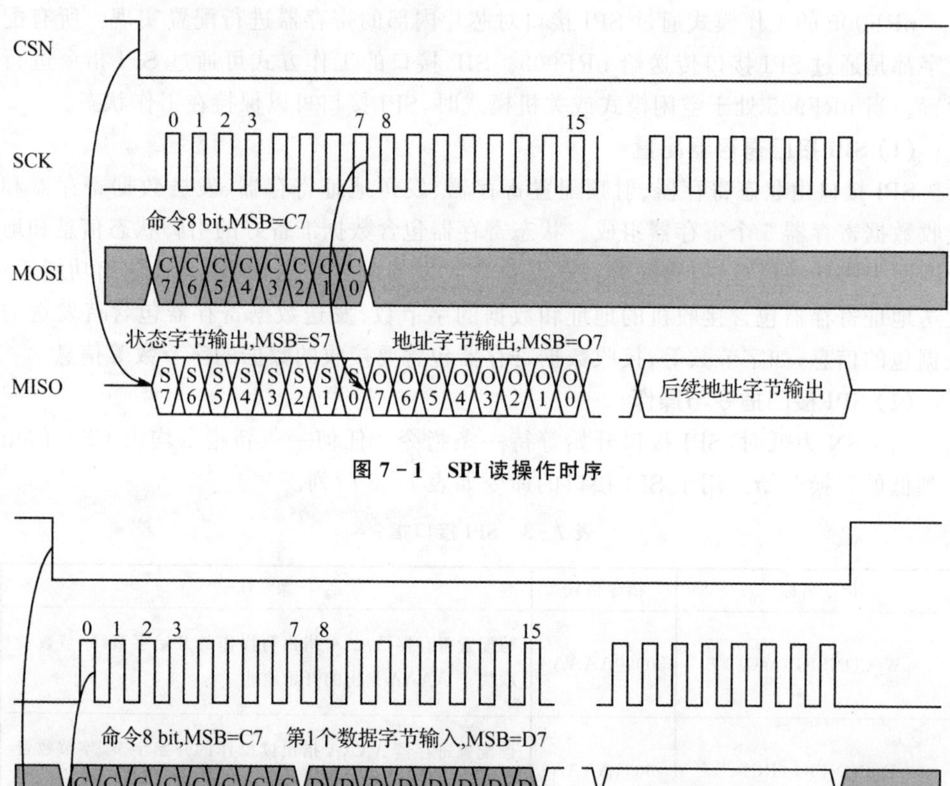

图 7-1 SPI 读操作时序

图 7-2 SPI 写操作时序

4. nRF905 应用电路

nRF905 的应用电路如图 7-3 所示,PCB 图如图 7-4 所示。nRF905 外围元件均采用贴片封装形式。

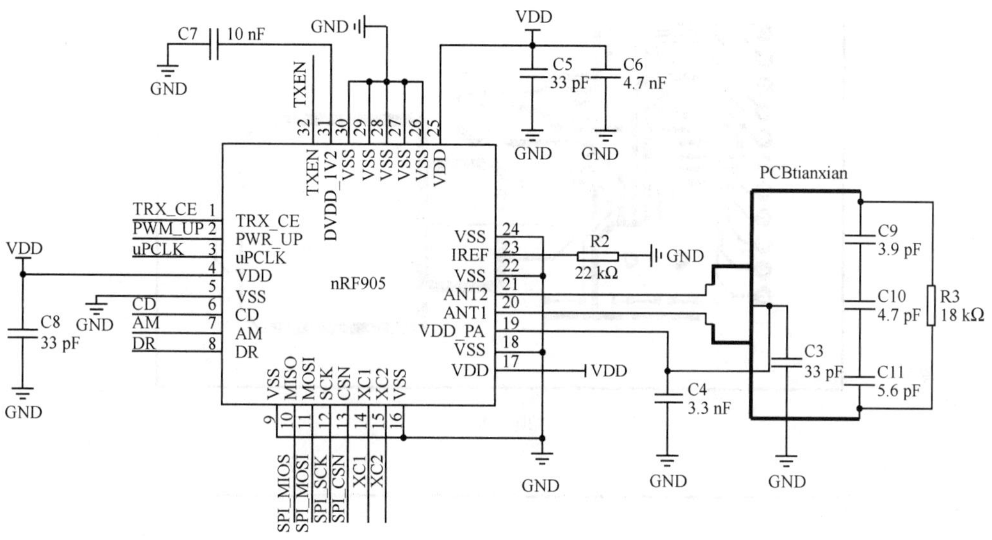

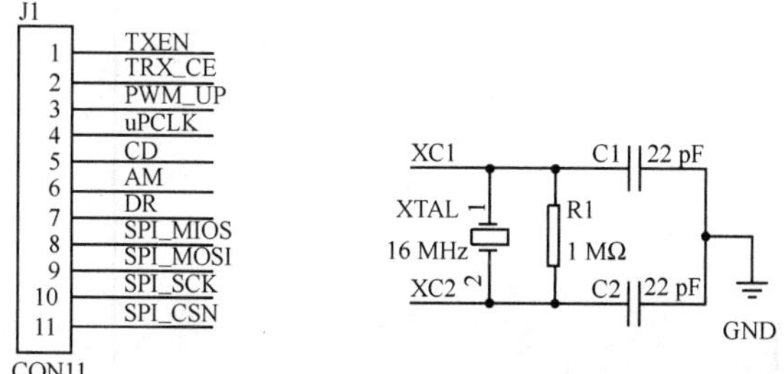

图 7-3　nRF905 应用电路原理图

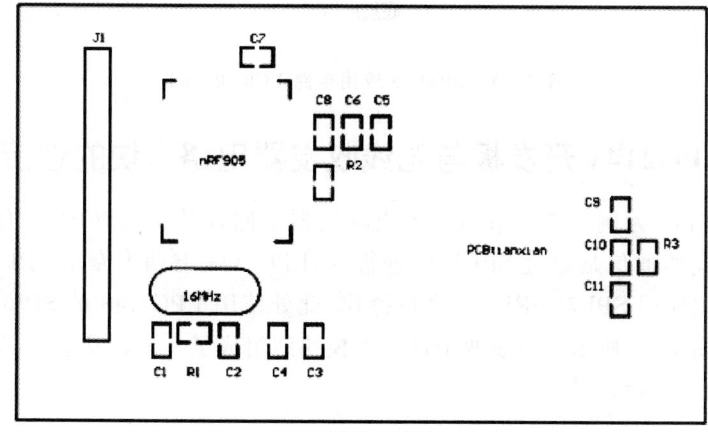

(a) 元器件布局图

图 7-4　nRF905 应用电路 PCB 图

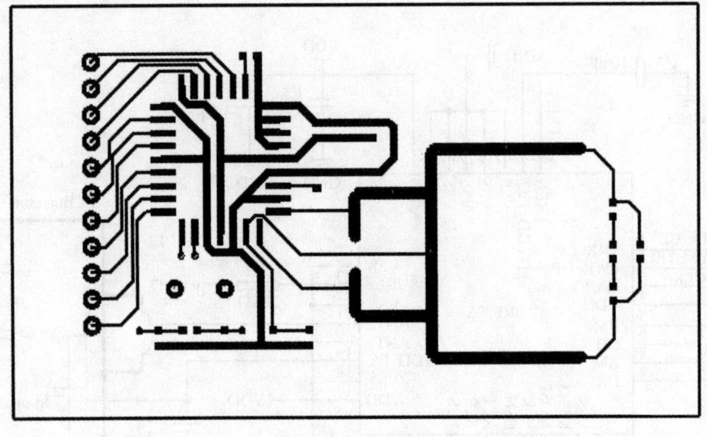

(b) 顶层PCB图

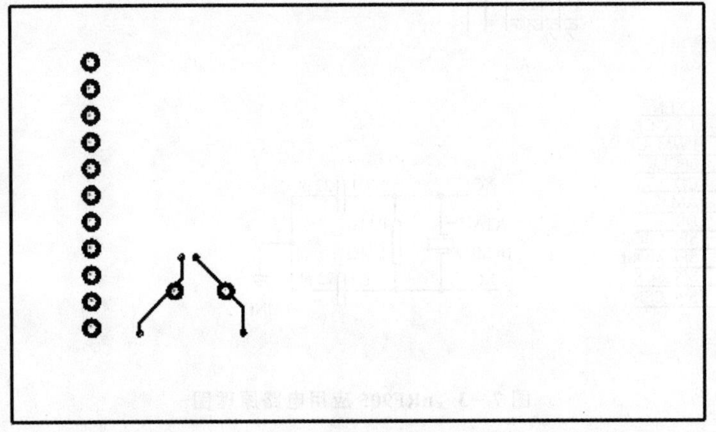

(c) 底层PCB图

图 7－4　nRF905 应用电路 PCB 图(续)

7.1.2　LPC214x 开发板与无线收发器电路模块的连接

以 LPC2148 为例说明 nRF905 无线收发器电路模块与 LPC214x 开发板的连接。两者之间的通信是通过 SPI 接口进行的，LPC2148 有两个专用 SPI 总线接口，用户也可使用模拟 SPI 对 nRF905 进行操作，此处使用 LPC2148 的 SPI0，二者的连接示意图如图 7－5 所示。搭建两个这样的模块即组成了一个收发系统，收发系统示意图如图 7－6 所示。

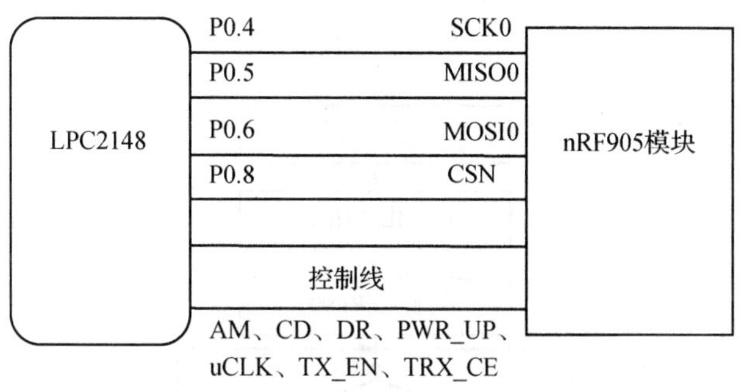

图 7-5　nRF905 与 LPC2148 连接示意图

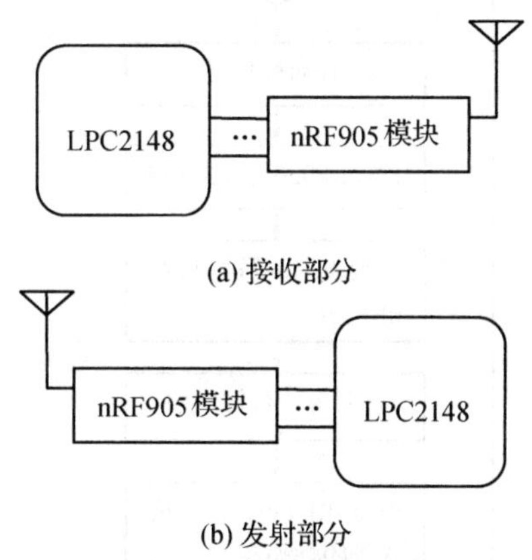

图 7-6　收发系统示意图

7.1.3　无线收发器电路模块的编程示例

根据上述的硬件设计,下面将分别给出发射和接收的编程流程图和示例。

1. 发射部分

使用定时器 0 定时 1 s,每隔 1 s 蜂鸣器开启且发送一次数据,发送完一次数据后蜂鸣器停止鸣叫。发射部分程序流程如图 7-7 所示。

启动 ADS 1.2,使用 ARM Executable Image for lpc2148 工程模板创建一个工程。使用过程中,要将 nRF905.c 和 nRF905.h 包含入工程。nRF905.c 中包含了 nRF905 的引脚初始化和配置函数,nRF905.h 中包含了 nRF905 的引脚定义、地址定义等。发射部分的程序代码如程序清单 7.1～程序清单 7.4 所示。程序清单 7.1 中 system.h 头文件主要用于对所有函数的声明。

程序清单 7.1　nRF905.c 示例

```
#include "config.h"
#include "nRF905.h"
```

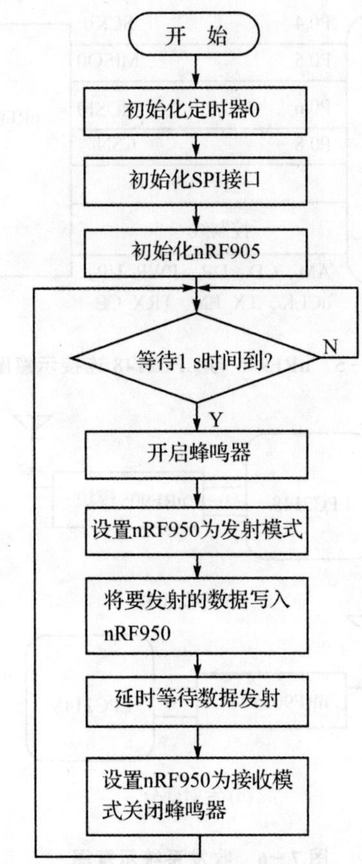

图 7－7　发射部分程序流程图

```
#include "system.h"
#define  uchar  unsigned char
#define  uint   unsigned int
uchar  TxAddress[4] = {TX_ADDR_Byte0,TX_ADDR_Byte1,TX_ADDR_Byte2,TX_ADDR_Byte3};
RFConfig RxTxConf =
{
   10,
   RFConfig_Byte0,  RFConfig_Byte1,  RFConfig_Byte2,  RFConfig_Byte3,  RFConfig_Byte4,
   RFConfig_Byte5,  RFConfig_Byte6,  RFConfig_Byte7,  RFConfig_Byte8,  RFConfig_Byte9
};
/*************************************************************************
*  函数名称:Delay()
*  函数功能:软件延时。
*  入口参数:n   延时时长
*  出口参数:无
*************************************************************************/
static void Delay(uchar n)
```

```
{
    uint i;
    while(n--)
    for(i=0;i<80;i++);
}
/****************************************************************
*  函数名称:nRF905Init()
*  函数功能:初始化 nRF905 控制 I/O 口
*  入口参数:无
*  出口参数:无
****************************************************************/
void nRF905Init(void)
{
    IO1SET = DR;                    // 设置 DR 为输入
    IO1SET = AM;                    // 设置 AM 为输入
    IO1SET = PWR_UP;                // nRF905 上电
    Delay(4);
    IO1CLR = TRX_CE;                // 设置 nRF905 为待机模式
    IO1CLR = TX_EN;
}
/****************************************************************
*  函数名称:WriteTxAddress ()
*  函数功能:写入目标地址
*  入口参数:无
*  出口参数:无
****************************************************************/
void WriteTxAddress(void)
{
    unsigned char i;
    unsigned char len = (RxTxConf.buf[2]>>4);
    IO0CLR = CSN;                   // 选通 nRF905
    MSPI_Senddata(WTA);             // 写地址命令
    for (i=0;i<len;i++)
    {
        MSPI_Senddata(TxAddress[i]);
    }
    IO0SET = CSN;                   // 使能端关闭
}
/****************************************************************
*  函数名称:Config905()
*  函数功能:配置 nRF905 寄存器
*  入口参数:无
*  出口参数:无
****************************************************************/
void Config905(void)
{
    uchar i;
```

```
    IO0CLR = CSN;                        // 选通 nRF905
    MSPI_Senddata(WC);                   // 写放配置命令
    for (i = 0;i<RxTxConf.n;i ++ )       // 写放配置字
    {
        MSPI_Senddata(RxTxConf.buf[i]);
    }
    IO0SET = CSN;                        // 使能端关闭
    WriteTxAddress();
}
/***********************************************************************
 * 函数名称:SetTxMode ()
 * 函数功能:设置为发送模式
 * 入口参数:无
 * 出口参数:无
***********************************************************************/
void SetTxMode(void)
{
    IO1CLR = TRX_CE;
    IO1SET = TX_EN;
    Delay(1);                            // 延时 650 μs 以上
}
/***********************************************************************
 * 函数名称:SetRxMode()
 * 函数功能:设置为接收模式
 * 入口参数:无
 * 出口参数:无
***********************************************************************/
void SetRxMode(void)
{
    IO1CLR = TX_EN;
    IO1SET = TRX_CE;                     // 延时 650 μs 以上
    Delay(10);
}
/***********************************************************************
 * 名称:TxPacket()
 * 功能:发送数据包。
 * 入口参数:TxBuf  发送的数据
 * 出口参数:无
***********************************************************************/
void TxPacket(unsigned char * TxBuf)
{
    uchar i;
    IO0CLR = CSN;
    MSPI_Senddata(WTP);                  // 写命令
    for (i = 0;i<RxTxConf.buf[4];i ++ )
    {
```

```
        MSPI_Senddata(TxBuf[i]);        // 写入 32 字节数据
    }
    IO0SET = CSN;
    IO1SET = TRX_CE;                    // 设置 TRX_CE 为高,开启 nRF905 发送数据
    Delay(1);

    //若发送较多数据,应用 while (DR! = 1);等待
    IO1CLR = TRX_CE;                    // 设置 TRX_CE 为低
}


/*****************************************************************
*  函数名称:RxPacket()
*  函数功能:读取数据包
*  入口参数:TxBuf    读取的数据存储区
*  出口参数:0 表示没有接收到数据;1 表示成功接收到数据
*****************************************************************/
unsigned char RxPacket(unsigned char * RxBuf)
{
    uchar i = 1;
    if ( (DR & IO1PIN) ! = 0)                       // DR 高电平,接收到数据
    {
        IO1CLR = TRX_CE;                            // 设置 nRF905 为待机模式
        IO0CLR = CSN;
        MSPI_Senddata(RRP);                         // 读命令
        for (i = 0 ;i < RxTxConf.buf[3] ;i ++ )
        {
            RxBuf[i] = MSPI_Receivedata();          // 读取数据
        }
        IO0SET = CSN;
        while( (DR  & IO1PIN) ! = 0) ;              // 若是高电平则一直等待
        IO1SET = TRX_CE;
        return 1;
    }
    else
    {
        return 0;
    }
}
```

程序清单 7.2　nRF905.h 示例

```
#ifndef NRF905_H
#define NRF905_H
//<SPI 命令>
#define WC              0x00
#define RC              0x10
```

```
#define WTP            0x20
#define RTP            0x21
#define WTA            0x22
#define RTA            0x23
#define RRP            0x24
//----------------------------------------------------------------
//<RF-Configuration-Register 配置信息>
#define CH_NO              76//0x01   // freq = 422.4 + 76/10 = 430 MHz
#define HFREQ_PLL          0x0        // 433 MHz/868 MHz/915 MHz
#define PA_PWR             0x3        // 最大输出功率
#define RX_RED_PWR         0x0        // 接收功率
#define AUTO_RETRAN        0x0        // 自动重新发送
//<地址宽度与数据宽度>

#define RX_AWF             0x4        // Rx本地接收地址宽度
#define TX_AWF             0x4        // Tx本地发送地址宽度
#define RX_PW              0x1c       // Rx本地接收有效数据宽度
#define TX_PW              0x1c       // Tx本地发送有效数据宽度
#define UP_CLK_FREQ        0x0        // 输出时钟频率
#define UP_CLK_EN          0x0        // 输出时钟使能
#define XOF                0x3        // 根据外部晶体的标称频率设置晶体振荡器频率
#define CRC_EN             0x1        // CRC校验允许
#define CRC_MODE           0x1        // CRC模式,0b0表示8位CRC校验位;0b1表示16位CRC校验位
//<本机地址>
#define RX_ADDR_Byte3         0xcc    // 本机地址最高字节
#define RX_ADDR_Byte2         0xcc
#define RX_ADDR_Byte1         0xcc
#define RX_ADDR_Byte0         0xcc    // 本机地址最低字节
#define TX_ADDR_Byte3         0xcc
#define TX_ADDR_Byte2         0xcc
#define TX_ADDR_Byte1         0xcc
#define TX_ADDR_Byte0         0xcc
//----------------------------------------------------------------
//<将设置信息组合成每个字节的数据信息,此区域无需修改>
#define RFConfig_Byte0        (CH_NO & 0xff)
#define RFConfig_Byte1        (AUTO_RETRAN<<5 | RX_RED_PWR<<4 | PA_PWR<<2 |
                              HFREQ_PLL<<1 | CH_NO>>8)
#define RFConfig_Byte2        (TX_AWF<<4 | RX_AWF)
#define RFConfig_Byte3        RX_PW
#define RFConfig_Byte4        TX_PW
#define RFConfig_Byte5        RX_ADDR_Byte0
#define RFConfig_Byte6        RX_ADDR_Byte1
#define RFConfig_Byte7        RX_ADDR_Byte2
#define RFConfig_Byte8        RX_ADDR_Byte3
#define RFConfig_Byte9        (CRC_MODE<<7 | CRC_EN<<6 | XOF<<3 | UP_CLK_EN<<2 | UP_CLK_FREQ)
//----------------------------------------------------------------
```

```
typedef struct RFConfig
{
    unsigned char n;
    unsigned char buf[10];
} RFConfig;
/****************************************************************
void nRF905Init(void);                          // 初始化 I/O 口引脚
void Config905(void);                           // 配置 nRF905 模式
void SetTxMode(void);                           // 设置 nRF905 为发送模式
void SetRxMode(void);                           // 设置 nRF905 为接收模式
void TxPacket(unsigned char * TxBuf);           // 通过 nRF905 发送数据
unsigned char RxPacket(unsigned char * RxBuf);  // 通过 nRF905 接收数据

extern unsigned char TxAddress[4];
extern RFConfig RxTxConf;

#endif
```

程序清单 7.3　主函数示例

```
#include "config.h"
#include "nRF905.h"
#include "system.h"

#define    beep        1<<7;           // 定义蜂鸣器引脚
uint8 s_flag = 0;                      // 定义秒标志
/****************************************************************
*  函数名称:mDelaymS()
*  函数功能:延时 ms  毫秒
*  入口参数:ms   延时参数,值越大,延时越久
*  出口参数:无
****************************************************************/
void    mDelaymS( uint32 ms )
{
    uint32     i;
    while ( ms -- ) for ( i = 25000; i != 0; i -- );
}

void __irq IRQ_Time0(void)
{
     s_flag = 1;
}
/****************************************************************
 * 函数名称:Time0Init()
 * 函数功能:初始化定时器,定时时间为 1 s
 * 入口参数:无
 * 出口参数:无
****************************************************************/
void Time0Init(void)
{
    T0PR = 99;                          // 设定定时器分频为 100 分频
```

```
    T0MCR = 0x03;                       // 设置匹配通道 0 匹配中断并复位 T0TC
    T0MR0 = 3 * 120000;                 // 设置预分频寄存器值,定时 1 s
    T0TCR = 0x03;                       // 设置定时器计数器和预分频计数器启动并复位
    T0TCR = 0x01;                       // 开启定时器工作
    /* 设置定时器 0 向量中断 IRQ */
   VICIntSelect = 0x00000000;           // 设置所有中断分配为 IRQ 中断
   VICVectCntl0 = 0x24;                 // 给定时器 0 中断分配向量中断 0,开启 IRQ 中断使能
   VICVectAddr0 = (uint32)IRQ_Time0;    // 设置中断服务程序地址
   VICIntEnable = 1<<4;                 // 使能定时器 0 中断
}
/******************************************************************************
* 函数名称:MSPI_Ini()
* 函数功能:初始化 SPI 接口,设置 LPC2148 为主机模式
* 入口参数:无
* 出口参数:无
******************************************************************************/
void MSPI_Ini(void)
{
      // 设置 P0.4、P0.5、P0.6、P0.7 为 SIP 接口
      PINSEL0 = (PINSEL0 & 0xFFFF00FF) | 0x5500 ;
      S0SPCCR = 0x52;                   // 设置 SPI 时钟分频,即 SCK 的频率为 Fpclk/S0SPCCR
      S0SPCR   = 0x28;                  // 设置 SPI 接口模式,MSTR = 1,COPL = 1,CPHA = 1,LSBF = 1
}
/******************************************************************************
* 函数名称:MSPI_Senddata()
* 函数功能:向 SPI 总线发送数据
* 入口参数:data   待发送的数据
* 出口参数:无
******************************************************************************/
uint8 MSPI_Senddata(uint8 data)
{
   IO0CLR = CSN;                        // 选通 nRF905
   S0SPDR = data;                       // 将要发送的数据写入 S0SPDR 数据寄存器
   while (S0SPSR&0x80 == 0);            // 等待数据发送完毕
   IO0SET = CSN;
   return (0) ;
}
/******************************************************************************
* 函数名称:MSPI_Receivedata()
* 函数功能:从 SPI 总线接收数据
* 入口参数:无
* 出口参数:接收到的数据
******************************************************************************/
uint8 MSPI_Receivedata(void)
{
   uint8 Rxdata;
```

```
    IO0CLR = CSN;                          // 选通 nRF905
    Rxdata = S0SPDR;                       // 读取 S0SPDR 的数据
    while (S0SPSR&0x80 = = 0);             // 等待数据接收完毕
    IO0SET = CSN;
    return (Rxdata) ;
}
/*********************************************************************
* 函数名称:nRF905_IO_Init()
* 函数功能:nRF905 控制 I/O 口初始化
* 入口参数:无
* 出口参数:无
*********************************************************************/
void nRF905_IO_Init(void)
{
    PINSEL0 = PINSEL0 & 0xfffcffff;             // 设置 P0.8 为 GPIO
    IO0DIR   = CSN ;                            // 设置 nRF905 片选引脚
    IO1DIR & = ~(DR | CD | AM);                 // 设置为输入
    IO1DIR | = (PWR_UP | TRX_CE | TX_EN );      // 设置为输出
}
/*********************************************************************
* 函数名称:nRF_Init()
* 函数功能:nRF905 配置初始化
* 入口参数:无
* 出口参数:无
*********************************************************************/
void nRF_Init(void)
{
    nRF905_IO_Init();
    nRF905Init();
    Config905();                                // 配置 nRF905
}
/*********************************************************************
* 名称:main ()
* 功能:nRF905 发送部分主函数,使用硬件 SPI 接口输出 TxBuf 数组的数据给 nRF905,nRF905 再将
*         数据发射出去
*********************************************************************/
int  main(void)
{
    uint8  TxBuf[8] = {0xaa,0xcc,0x45,0x24,0x55,0xb6,0x78,0xdf};  // 要发送的数据
    IO0DIR | = beep;
    MSPI_Ini();
    nRF_Init();
    IO0SET = beep;                              // 关蜂鸣器
    while(1)
    {
        if(s_flag = = 1)                        // 1 s 时间到
```

```
    {
        s_flag = 0;
        IO0CLR = beep;                          // 开启蜂鸣器
        SetTxMode();                            // 设置 nRF905 为发送模式
        TxPacket(TxBuf);                        // 发送 TxBuf 中存储的数据
        mDelaymS(5);                            // 此处需延时 mDelaymS(1)以上,才能将数据发送出去
        IO0SET = beep;                          // 关闭蜂鸣器
        SetRxMode();                            // 设置 nRF905 为接收模式
    }
  }
  return(0);
}
```

2. 接收部分

接收部分的程序流程图与发射部分类似,如图 7-8 所示。

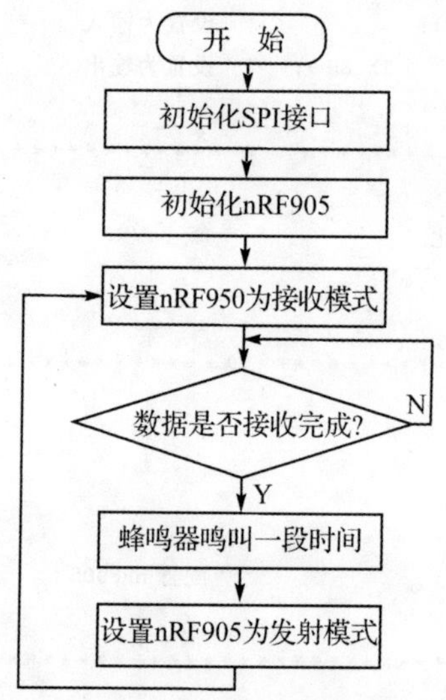

图 7-8　接收部分程序流程图

与发射部分一样建立好工程后,需要将 nRF905.c 和 nRF905.h 文件包含进去。程序清单 7.4 只给出主函数程序。

程序清单 7.4　接收部分主函数

```
#include "config.h"
#include "nRF905.h"
#include "system.h"
#define   beep   1≪7;
/**************************************************************************
*  函数名称:mDelaymS()
*  函数功能:延时 ms 毫秒
```

```
* 入口参数:ms  延时参数,值越大,延时越久
* 出口参数:无
****************************************************************/
void    mDelaymS( uint32 ms )
{
    uint32     i;
    while ( ms -- ) for ( i = 25000; i != 0; i -- );
}
/****************************************************************
* 函数名称:MSPI_Ini()
* 函数功能:初始化 SPI 接口,设置 LPC2148 为主机模式
* 入口参数:无
* 出口参数:无
****************************************************************/
void MSPI_Ini(void)

{
    PINSEL0 = (PINSEL0 & 0xFFFF00FF) | 0x5500 ;     // 设置 P0.4、P0.5、P0.6、P0.7 为 SPI 接口
    S0SPCCR = 0x52;          // 设置 SPI 时钟分频,即 SCK 的频率为 Fpclk/S0SPCCR
    S0SPCR   = 0x28;         // 设置 SPI 接口模式,MSTR = 1,COPL = 1,CPHA = 1,LSBF = 1
}
/****************************************************************
* 函数名称:MSPI_Senddata()
* 函数功能:向 SPI 总线发送数据
* 入口参数:data  待发送的数据
* 出口参数:无
****************************************************************/
uint8 MSPI_Senddata(uint8 data)
{
    IO0CLR = CSN;                          // 选通 nRF905
    S0SPDR = data;                         // 将要发送的数据写入 S0SPDR 数据寄存器
    while (S0SPSR&0x80 == 0);              // 等待数据发送完毕
    IO0SET = CSN;
    return (0) ;
}
/****************************************************************
* 函数名称:MSPI_Receivedata()
* 函数功能:从 SPI 总线接收数据
* 入口参数:无
* 出口参数:接收到的数据
****************************************************************/
uint8 MSPI_Receivedata(void)
{
    uint8 Rxdata;
    IO0CLR = CSN;                          // 选通 nRF905
    Rxdata = S0SPDR;                       // 读取 S0SPDR 的数据
    while (S0SPSR&0x80 == 0);              // 等待数据接收完毕
    IO0SET = CSN;
```

```
    return (Rxdata);
}
/*****************************************************************
* 函数名称:nRF905_IO_Init()
* 函数功能:nRF905 控制 I/O 口初始化
* 入口参数:无
* 出口参数:无
*****************************************************************/
void nRF905_IO_Init(void)
{

    PINSEL0 = PINSEL0 & 0xfffcffff;          // 设置 P0.8 为 GPIO
    IO0DIR   = CSN ;                         // 设置 nRF905 片选引脚
    IO1DIR &= ~(DR | CD | AM);               // 设置为输入
    IO1DIR |= (PWR_UP | TRX_CE | TX_EN );    // 设置为输出
}
/*****************************************************************
* 函数名称:nRF_Init()
* 函数功能:nRF905 配置初始化
* 入口参数:无
* 出口参数:无
*****************************************************************/
void nRF_Init(void)
{
    nRF905_IO_Init();
    nRF905Init();
    Config905();                             // 设置 nRF905
}
/*****************************************************************
* 名称:main ()
* 功能:nRF905 接收部分主函数,接收由发射部分发来的数据,存入 RxBuf 数组中
*****************************************************************/
int  main(void)
{
    uint8  RxBuf[100] = {0};                 // 存储接收到的数据
    IO0DIR |= beep;
    MSPI_Ini();
    nRF_Init();
    IO0SET = beep;                           // 关蜂鸣器
    while(1)
    {
        SetRxMode();                         // 设置 nRF905 为接收模式
        while (RxPacket(RxBuf) == 0);        // 等待有数据被接收
        IO0CLR = beep;                       // 开启蜂鸣器
        mDelaymS(100);
        IO0SET = beep;                       // 关闭蜂鸣器
        SetTxMode();                         // 设置 nRF905 为发送模式
    }
```

```
    return(0);
}
```

7.2　CAN 总线应用

7.2.1　CAN 总线简介

CAN(Controller Area Network,控制器局域网)是德国 Bosch 公司于 1983 年为汽车应用而开发的,它是一种现场总线(FieldBus),能有效支持分布式控制和实时控制的串行通信网络。1993 年 11 月,ISO 正式颁布了控制器局域网 CAN 国际标准(ISO 11898)。

一个理想的由 CAN 总线构成的单一网络中可以挂接任意多个节点,但在实际应用中节点数目受网络硬件的电气特性所限制。例如:当使用 Philips P82C250 作为 CAN 收发器时,同一网络中允许挂接 110 个节点。CAN 可提供 1 Mbit/s 的数据传输速率。CAN 总线是一种多主方式的串行通信总线。基本设计规范要求有高的位速率,高抗电磁干扰性,并可以检测出产生的任何错误。当信号传输距离达到 10 km 时 CAN 总线仍可提供高达 50 kbit/s 的数据传输速率。CAN 总线具有很高的实时性能,已经在汽车工业、航空工业、工业控制、安全防护等领域中得到了广泛应用。

CAN 总线的通信介质可采用双绞线、同轴电缆和光导纤维,最常用的是双绞线。通信距离与波特率有关,最大通信距离可达 10 km,最大通信波特率可达 1 Mbit/s。CAN 总线仲裁采用 11 位标识和非破坏性位仲裁总线结构机制,可以确定数据块的优先级,保证在网络节点冲突时最高优先级节点不需要冲突等待。CAN 总线采用了多主竞争式总线结构,具有多主站运行和分散仲裁的串行总线以及广播通信的特点。CAN 总线上任意节点可在任意时刻主动向网络上其他节点发送信息而不分主次,因此可在各节点之间实现自由通信。

CAN 总线信号使用差分电压传送,两条信号线被称为 CAN_H 和 CAN_L,静态时均为 2.5 V 左右,此时状态表示为逻辑 1,也可以叫做“隐性”。采用 CAN_H 比 CAN_L 高表示逻辑 0,称为“显性”,通常电压值为 CAN_H=3.5 V,CAN_L=1.5 V。当“显性”位和“隐性”位同时发送的时候,最后总线数值将为“显性”。

CAN 总线的一个位时间可以分成 4 个部分:同步段、传播时间段、相位缓冲段 1 和相位缓冲段 2。每段的时间份额数目都是可以通过 CAN 总线控制器编程控制,而时间份额的大小 t_q 由系统时钟 t_{sys} 和波特率预分频值 BRP 决定:$t_q=BRP/t_{sys}$。图 7-9说明了 CAN 总线的一个位时间的各个组成部分。

- 同步段:用于同步总线上的各个节点,在此段内期望有一个跳变沿出现(其长度固定)。如果跳变沿出现在同步段之外,那么沿与同步段之间的长度叫做沿相位误差。采样点位于相位缓冲段 1 的末尾和相位缓冲段 2 的开始处。

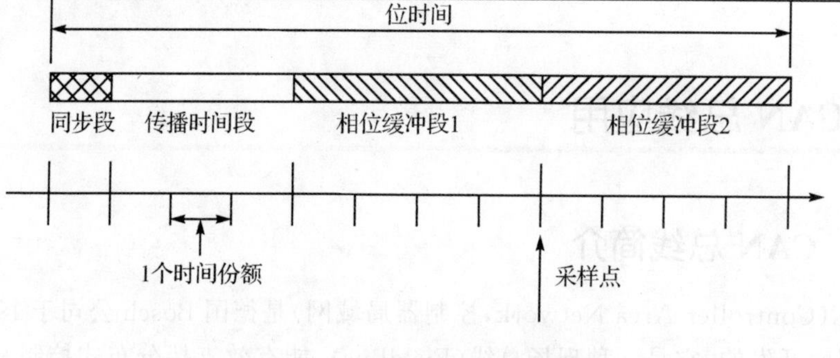

图 7-9 CAN 总线的一个位时间

- 传播时间段：用于补偿总线上信号传播时间和电子控制设备内部的延迟时间。因此，要实现与位流发送节点的同步，接收节点必须移相。CAN 总线非破坏性仲裁规定，发送位流的总线节点必须能够收到同步于位流的 CAN 总线节点发送的显性位。
- 相位缓冲段 1：重同步时可以暂时延长。
- 相位缓冲段 2：重同步时可以暂时缩短。
- 同步跳转宽度：长度小于相位缓冲段。

同步段、传播时间段、相位缓冲段 1 和相位缓冲段 2 的设定和 CAN 总线的同步、仲裁等信息有关。其主要思想是要求各个节点在一定误差范围内保持同步。必须考虑各个节点时钟（振荡器）的误差和总线的长度带来的延迟（通常每米延迟为 5.5 ns）。正确设置 CAN 总线各个时间段，是保证 CAN 总线良好工作的关键。

按照 CAN 2.0B 协议规定，CAN 总线的帧数据有如图 7-10 所示的两种格式：标准格式和扩展格式。作为一个通用的嵌入式 CAN 节点，应该支持这两种格式。

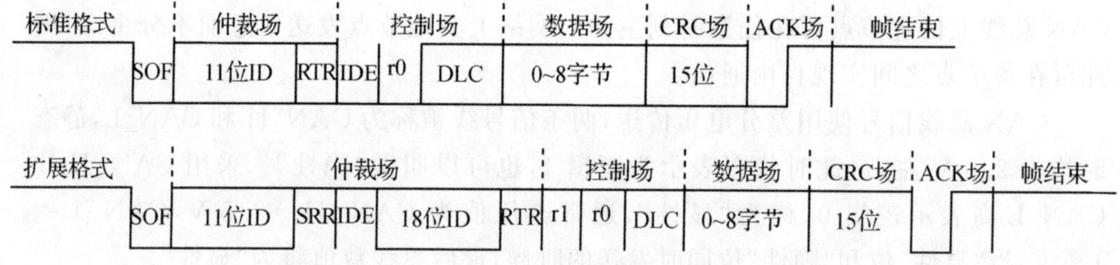

图 7-10 CAN 总线数据帧格式

7.2.2 在嵌入式处理器上扩展 CAN 总线接口

一些面向工业控制的嵌入式处理器本身就集成了一个或者多个 CAN 总线控制器。例如，韩国现代公司的 hms30c7202（ARM720T 内核）带有两个 CAN 总线控制器；Philips 公司的 LPC2194 和 LPC2294（ARM7TDMI 内核）带有 4 个 CAN 总线控制器。CAN 总线控制器主要是完成时序逻辑转换等工作，要在电气特性上满足

CAN总线标准，还需要一个CAN总线的物理层芯片，用它来实现TTL电平到CAN总线电平特性的转换，即CAN收发器。

实际上，多数嵌入式处理器都不带CAN总线控制器，如LPC214x系列。通常的解决方案是在嵌入式处理器的外部总线上扩展CAN总线接口芯片，例如，Philips公司的SJA1000 CAN总线接口芯片，Microchip公司的MCP251x系列(MCP2510和MCP2515)CAN总线接口芯片，这两种芯片都支持CAN 2.0B标准。

SJA1000的总线采用的是地址线和数据线复用的方式，多数嵌入式处理器采用SJA1000扩展CAN总线时，控制较为复杂。详细的描述请登录http://www.semiconductors.philips.com，查阅"Stand-alone CAN controller SJA1000"。

MCP2510是由Microchip公司生产的CAN协议控制器，完全支持CAN总线V2.0A/B技术规范。0～8字节的有效数据长度，支持远程帧；最大1 Mbit/s的可编程波特率；两个支持过滤器(Filter，Mask)的接收缓冲区，三个发送缓冲区；支持回环(Loop Back)模式，便于测试；SPI高速串行总线，最大5 MHz；3～5.5 V供电。MCP2510支持CAN 2.0B技术规范中所定义的标准数据帧、扩展数据帧以及远程帧(标准和扩展)，详细描述请登录http://www.microchip.com，查阅MCP2510数据手册。

7.2.3　CAN总线网络结构

一个基于LPC2148的CAN总线网络示意图如图7-11所示，本设计采用CAN

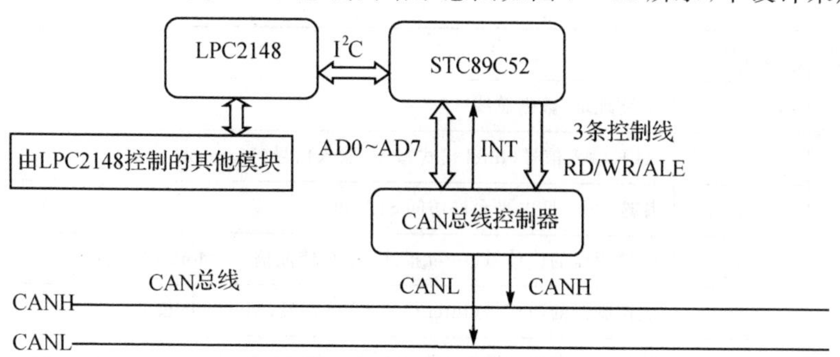

(a) 单个网络节点内部通信方框图

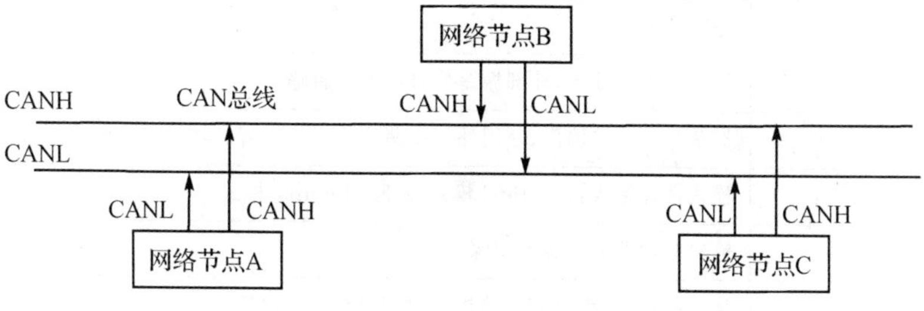

(b) 网络节点之间通信方框图

图7-11　基于LPC2148的CAN总线网络示意图

总线来完成3个网络节点之间的通信,单个网络节点采用LPC2148作为主控制器,主控制器将要发往总线上的数据先采用I^2C协议的方式传给本节点的辅助控制器STC89C52,再由辅助控制器STC89C52控制CAN总线控制器将数据发往总线。单个网络节点内部通信方框图如图7-11(a)所示。网络节点之间的通信方框图如图7-11(b)所示。

7.2.4 CAN总线模块设计

1. CAN总线控制器SJA1000简介

本设计的CAN总线控制器采用Philips公司的符合CAN 2.0B协议的SJA1000,它是一个可以应用于汽车和一般工业环境的独立CAN总线控制器。SJA1000具有完成CAN通信协议所要求的全部特性。经过简单总线连接的SJA1000可完成CAN总线的物理层和数据链路层的所有功能。其硬件与软件设计可兼容基本CAN模式(BasicCAN)和新增加的增强CAN模式(PeliCAN)CAN 2.0B协议。

2. SJA1000的封装与引脚端功能

SJA1000采用28引脚的DIP封装或SO封装,引脚端功能如表7-4所列。

表7-4 SJA1000引脚功能

引脚符号	引脚号	说 明
AD7~AD0	2,1,28~23	多路地址/数据总线
ALE/AS	3	ALE输入信号(Intel模式),AS输入信号(Motorola模式)
$\overline{CS}$	4	片选输入,低电平允许访问SJA1000
$\overline{RD}$/E	5	微控制器的读信号(Intel模式)或E使能信号(Motorola模式)
$\overline{WR}$	6	微控制器的写信号(Intel模式)或读信号(Motorola模式)
CLKOUT	7	SJA1000产生的提供给微控制器的时钟输出信号,此信号由内部振荡器经可编程分频器得到,时钟控制寄存器的时钟关闭位可禁止该引脚
VSS1	8	逻辑电路地
XTAL1	9	振荡放大器输入,外部振荡信号由此引脚输入
XTAL2	10	振荡放大电路输出,使用外部振荡信号时此引脚必须开路输出
MODE	11	模式选择输入:1为Intel模式,2为Motorola模式
VDD3	12	输出驱动器的5 V电压源
TX0	13	从CAN输出驱动器0输出到物理总线的输出端

续表 7-4

引脚符号	引脚号	说　明
TX1	14	从 CAN 输出驱动器 1 输出到物理总线的输出端
VSS3	15	输出驱动器地
$\overline{\text{INT}}$	16	中断输出，用于中断微控制器
$\overline{\text{RST}}$	17	复位输入端，用于重新启动 CAN 接口，低电平有效
VDD2	18	输入比较器的 5 V 电压源
RX0，RX1	19，20	从物理的 CAN 总线输入到 SJA1000 的输入比较器；控制电平将会唤醒 SJA1000 的睡眠模式
VSS2	21	输入比较器的接地端
VDD1	22	逻辑电路的 5 V 电压源

3. CAN 总线模块电路

图 7-12 为 CAN 总线模块电路。

图 7-13 为 CAN 总线模块 PCB 图。

CAN 总线模块电路(见图 7-12)和 PCB 图(见图 7-13)，主要由系统辅助控制器 STC89C52 电路、CAN 控制器 SJA1000 电路、CAN 总线收发器 PCA82C250 电路和高速光电耦合器 6N137 组成。

由于 LPC2148 采用 SJA1000 扩展 CAN 总线较为复杂，CAN 总线模块利用了一个辅助控制器 STC89C52，辅助控制器 STC89C52 作为主控制器和 CAN 控制器的数据转发器。

(1) 辅助控制器 STC89C52 和 CAN 控制器电路

系统辅助控制器 STC89C52 负责 SJA1000 的初始化，通过控制 SJA1000 实现数据的接收和发送等通信任务。SJA1000 的 6 个控制信号引脚和 8 条数据/地址复用引脚通过 J1 排插引出，方便与 MCU 的连接。将 SJA1000 的片选信号引脚接低电平，复位引脚接高电平，以保证 SJA1000 能够正常工作；其余的 4 条控制信号引脚 RD、WR、ALE、INT 分别与 STC89C52 对应的引脚相连，INT 引脚接 STC89C52 的 INT1，8 条数据/地址复用引脚 AD0～AD7 接到 STC89C52 的 P0 口；SJA1000 选择内部时钟，因此在 9 引脚和 10 引脚之间接 16 MHz 的晶振；11 引脚 MODE 接高电平，以选择 SJA1000 的工作模式为 Intel 模式；TX1 引脚悬空，RX1 引脚的电位必须维持在 VCC 的值在 0.5 V 上，否则将不能形成 CAN 协议所要求的电平逻辑。

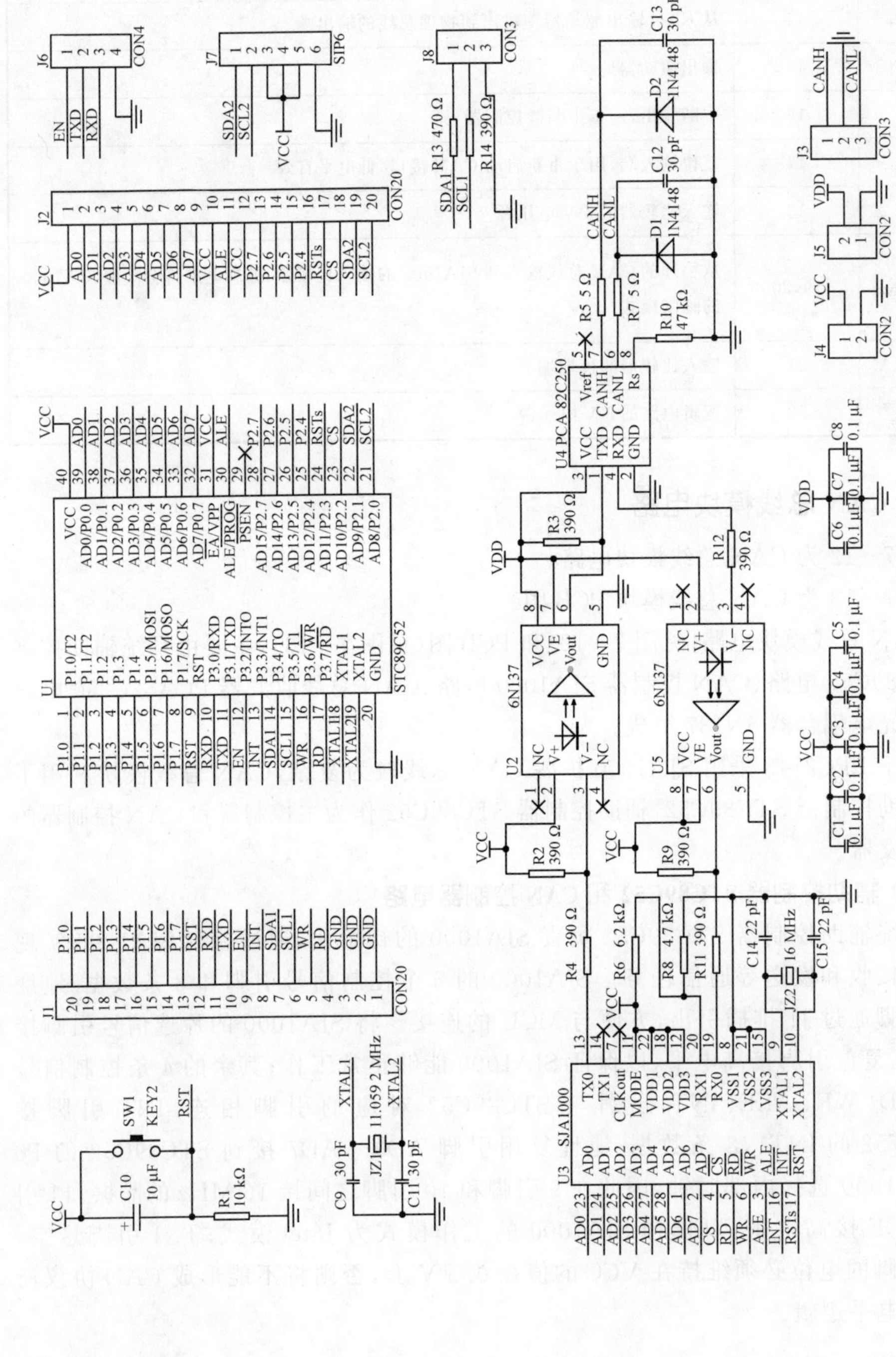

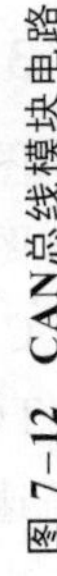

图 7-12　CAN总线模块电路

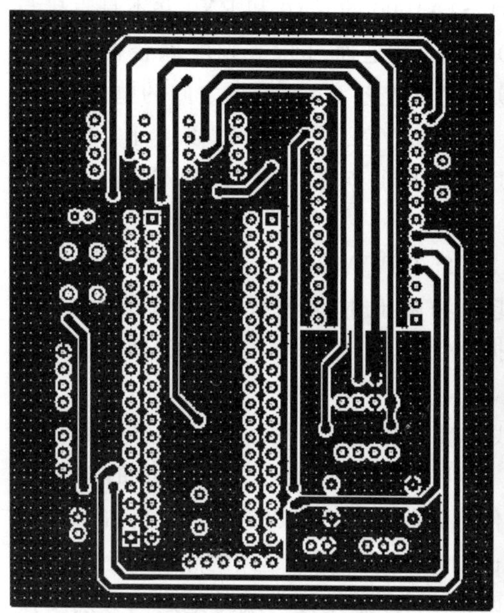

(a) 顶层PCB图

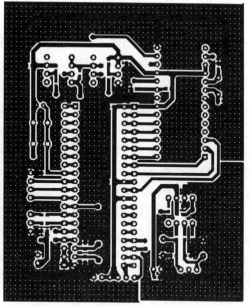

(b) 底层PCB图

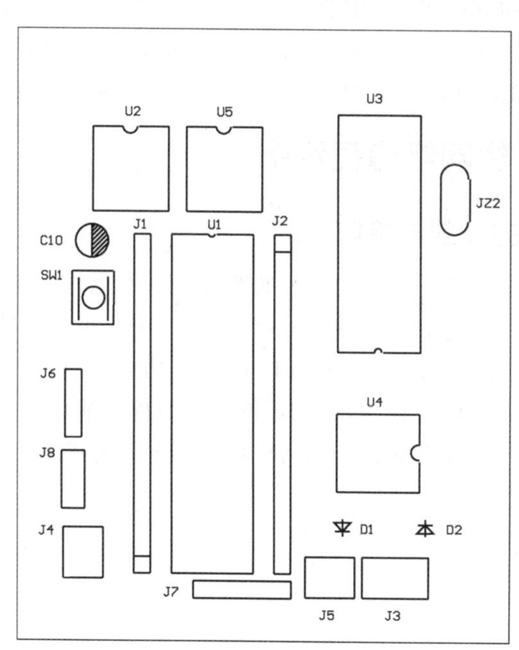

(c) 顶层元器件布局图

图 7-13 CAN 总线模块 PCB 图

(2) 高速光电耦合器电路

为了增强 CAN 总线节点的抗干扰能力,SJA1000 的 TX0 和 RX0 并不是直接与 CAN 收发器 PCA82C250 的 TXD 和 RXD 相连,而是通过高速光电耦合器 6N137 后,与 PCA82C250 相连,这样就能很好地实现总线上各 CAN 节点之间的电气隔离。电路中将光耦部分电路所采用的两个电源 VCC 和 VDD 完全隔离,以保证实现电气隔离,使用高速光耦虽增加了电路的复杂度,却提高了节点的稳定性和安全性。

(3) CAN 收发器电路

CAN 收发器电路选择专用芯片 PCA82C250,PCA82C250 的传输速度可高达 1 Mbit/s,具有抗瞬间干扰保护总线的能力和降低射频干扰的斜率控制能力。PCA82C250 将 SJA1000 中 TXFIFO 的数据发往总线,也将所接收到的总线上的数据写入 SJA1000 的 RXFIFO。

在本设计中,PCA80C250 与 CAN 总线的接口部分也采用了一定的安全和抗干扰措施,如在 PCA82C250 的 CANH 和 CANL 引脚各自通过 1 个 5 Ω 的电阻与 CAN 总线相连,电阻可以起到一定的限流作用,保护 PCA82C250 免受过流的冲击。CANH 和 CANL 与地之间并联了两个 30 pF 的小电容,可以起到滤除总线上的高频干扰和一定的防电磁辐射的能力。另外在两根 CAN 总线接入端与地之间分别反接了一个保护二极管,当 CAN 总线有较高的负电压时,通过二极管的短路可起到过压保护作用。在 PCA80C250 的 Rs 引脚接一个斜率电阻,以选择 PCA80C250 的工作模式为斜率方式,同时可降低 CAN 总线的向外辐射,电阻大小可根据总线的通信速度适当调整,一般在 16～140 kΩ 之间,此处连接的电阻为 47 kΩ。

7.2.5　CAN 总线网络编程示例

1. 控制 CAN 的发射与接收

使用 STC89C52 作为控制器控制 CAN 的发射与接收,软件流程图如图 7-14 所示。

开启 Keil μVision2 编译软件,建立一个工程。在使用过程中,需要在工程中添加 sja_bcanfunc.c 和 sja_bcanconf.h,可以在周立功公司网站(http://www.zlgmcu.com)上下载。根据软件流程图,给出示例程序如程序清单 7.5 所示。

程序清单 7.5　CAN 收发程序示例程序

```
#include <reg52.H>
#include <intrins.h>
#include  "sja_bcanFunc.C"

#define  uchar unsigned char
#define  uint unsigned int

bit s;                                          //配置 sja 标志
bit flag_send = 0;                              //发送命令标志
uchar data    send_data[10],rcv_data[10];       //发送和接收数组
```

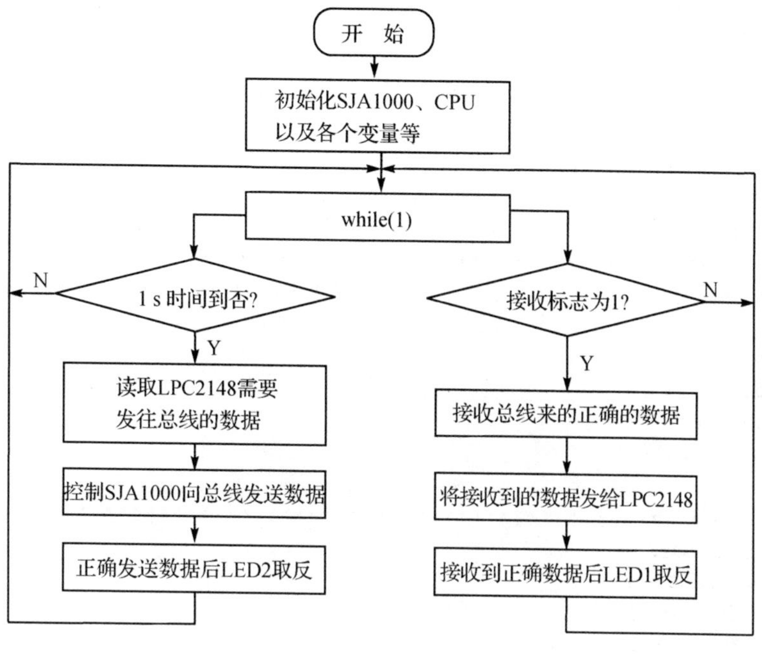

图 7－14　软件流程图

```
uchar bdata flag_init;                      // 保存中断寄存器值
/ * bdata 不是数据类型,而是指存放的内存空间 * /
uint count_k;                               // 延时计数用
sbit rcv_flag = flag_init^0;                // 接收中断标志
sbit err_flag = flag_init^2;                // 错误中断标志
char flag_sec  = 0;
int receive_flag = 0;
unsigned char times_50ms  =  0;
void Sja_1000_Init(void);

sbit LED0  =  P1^3;
sbit LED1  =  P1^4;

sbit CS  =  P1^0;
/ * * * * * * * * * * * * * * * * * * * * * * * * * * * * * * * * * * * * * * * * * * * * * * * * * * * * * * * * * * *
 * 函数名称:Time0_Init
 * 函数功能:定时器 0 初始化
 * 入口参数:无
 * 出口参数:无
* * * * * * * * * * * * * * * * * * * * * * * * * * * * * * * * * * * * * * * * * * * * * * * * * * * * * * * * * * * * /
void    Time0_Init(void)
{
    EA = 1;                                 // 开启中断总开关
    ET0 = 1;                                // 开启定时器 0 中断开关
    TMOD = 0x21;                            // 选择定时器工作方式二
    TH0 = (65536 - 50000)/256;              // 定时器 0 定时 50 ms
```

```
    TL0 = (65536 - 50000) % 256;
    TR0 = 1;                                         // 开启定时器
}
/*********************************************************************
* 函数名称:Init_Cpu
* 函数功能:MCU 初始化
* 入口参数:无
* 出口参数:无
*********************************************************************/
void Init_Cpu(void)                                  // 单片机初始化,开放外部中断 0
{
     PX1 = 1;
     EX1 = 1;
     EA = 1;
     Time0_Init();
}
/*********************************************************************
* 函数名称:delay
* 函数功能:延时
* 入口参数:延时时间
* 出口参数:无
*********************************************************************/
void delay(uint z)
{
    uchar x,y;
    for(x = z;x>0;x-- )
      for(y = 250;y>0;y-- );
}
/*********************************************************************
* 函数名称:ex0_int
* 函数功能:外部中断 0 中断服务函数
* 入口参数:无
* 出口参数:无
*********************************************************************/
void ex0_int(void) interrupt 2 using 1               // 外部中断 0
{
     SJA_BCANAdr = REG_INTERRUPT;                    // 指针指向中断寄存器
     flag_init = * SJA_BCANAdr;                      // 保持中断寄存器值
    if(rcv_flag )                                    // 若 SJA1000 收到数据
     {
         rcv_flag = 0;                               // 清零接收位
         BCAN_DATA_RECEIVE(rcv_data);                // 接收数据
         BCAN_CMD_PRG(RRB_CMD);                      // 释放接收缓冲区
         receive_flag = 1;
     }
if (err_flag)                                        // 错误中断
    {
```

```
        err_flag = 0;                           // 错误标志位清零
        Sja_1000_Init();                        // 初始化 SJA
    }
}
/********************************************************************
* 函数名称:Time0_interrupt
* 函数功能:定时器 0 中断服务函数
* 入口参数:无
* 出口参数:无
********************************************************************/
void Time0_interrupt(void) interrupt 1    using 1
{
    TH0 = (65536 - 50000)/256;                  // 定时器 0 定时 50 ms
    TL0 = (65536 - 50000) % 256;
    times_50ms ++ ;
    if(times_50ms == 20)                        // 1 s
    {
        times_50ms = 0;
        flag_sec = 1;                           // 秒标志
    }
}
/********************************************************************
* 函数名称:Sja_1000_Init
* 函数功能:Sja1000 初始化
* 入口参数:无
* 出口参数:无
********************************************************************/
void Sja_1000_Init(void)
{
    s = BCAN_CREATE_COMMUNATION();              // SJA 自测
    s = BCAN_ENTER_RETMODEL();                  // 进入复位
    s = BCAN_SET_BANDRATE(0x04);                // 设置波特率 100 kbit/s
    s = BCAN_SET_OBJECT(0x88,0x00);             // 设置地址 ID:44
    s = BCAN_SET_OUTCLK(0xaa,0x48);             // 设置输出方式,禁止 COLOCKOUT 输出
    s = BCAN_QUIT_RETMODEL();                   // 退出复位模式
    SJA_BCANAdr = REG_CONTROL;                  // 地址指针指向控制寄存器
    *SJA_BCANAdr| = 0x1e;                       // 开放错误、接收、发送中断
}
/********************************************************************
*  函数名称:main()
*  函数功能: 1 s 到 MCU 控制 SJA1000 发送数据,并接收总线数据
* 入口参数:无
* 出口参数:无
********************************************************************/
void main()
{
    LED0 = 1;                                   // 熄灭 LED 灯
```

```
    LED1 = 1;
    CS = 0;                                     // 选通 SJA1000
    s = 0;                                      // 配置 SJA1000 出现错误时,重新初始化
    do
    {
       Sja_1000_Init();
    }while(s! = 0);
    Init_Cpu();                                 // 初始化 MCU
    flag_init = 0x00;                           // 保存中断寄存器值清零
    send_data[0] = 0x66;                        // 向地址为 0x66 发送数据
    send_data[1] = 0x08;
    send_data[2] = 0x53;                        // send_data[2]~ send_data[5]为所发送的数据
    send_data[3] = 0x02;
    send_data[4] = 0x00;
    send_data[5] = 0x44;
    while(1)
    {
        if(flag_sec == 1)                       // 1 s 时间到
        {
            flag_sec = 0;                       // 清零秒标志
            BCAN_DATA_WRITE(send_data);         // 发送数据
            while(BCAN_CMD_PRG(TR_CMD));        // 置位发送请求位,返回 1 为不成功
            LED0 = ~LED0;                       // 发送成功则取反 LED0
        }
        if(receive_flag)                        // 接收标志为 1
        {
            receive_flag = 0;                   // 接收标志位清零
            LED1 = ~LED1;                       // 取反 LED1
        }
    }
}
```


2. LPC2148 与 STC89C52 之间的 I²C 通信编程示例

启动 ADS 1.2,使用 ARM Executable Image for lpc2148 工程模板创建一个工程。使用过程中,需要将I²C软件包文件 I2CINT.c 和 I2CINT.h 包含进工程,并且在 config.h 中将 I2CINT.h 包括进去,且还需要 Startup.s 修改系统模式,堆栈设置为 0x5f,开启中断。I²C 软件包可以在周立功公司网站(http://www.zlgmcu.com)上下载。软件流程如图 7-15 所示,程序如程序清单 7.6 所示。

程序清单 7.6 LPC2148 与 STC89C52RC 之间的 I²C 通信示例程序

```
/*************************************************************************
*  文件名称:I2C1_Receive.c
*  功    能:将 ARM 的 I2C1 设置为从机模式,接收 STC89C52RC 的数据,并将数据转移以备使用
*  控 制 器:LPC214X
*  编译环境:ADS 1.2
```

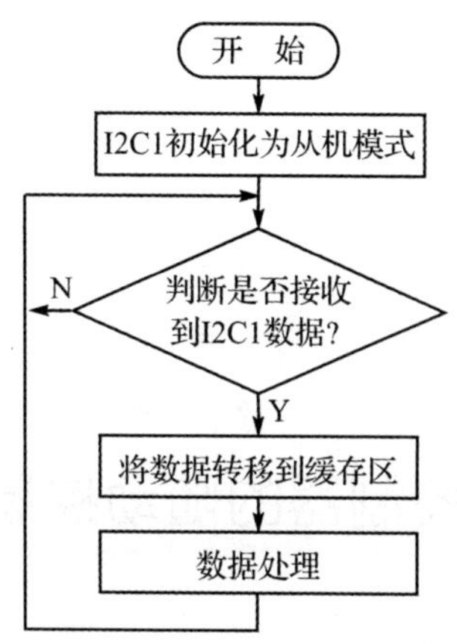

图7-15 LPC2148与STC89C52RC之间的I²C通信流程图

```
* 说    明:LPC2148 的端口 P0.11、P0.14 分别与 STC89C52RC 的模拟 I²C 端口 SCL 和 SDA 连接
*****************************************************************************/
#include "config.h"
extern uint8 eeprom_I2C1[256];          // I2C1 接收数据缓存区,256 字节
extern uint8 I2C1_Rcv_flag;             // I2C1 接收数据标志
uint8   DataBuf[256];                   // 系统数据缓存,用于转移 I2C1 接收到的数据
/*****************************************************************************
* 名称:main()
* 功能:LPC2148 I²C 从模式接收程序主函数
*****************************************************************************/
int main(void)
{
    uint32 i;
    PINSEL0  =  0;
    PINSEL1  =  0;

    IO0DIR  =  0;
    IO1DIR  =  0;
    I2C1_SlavInit(0xb0);                // 将 I2C1 设为从机模式,地址 0xb0
    for(;;)
    {
        if(I2C1_Rcv_flag)               // 若 I²C 接收数据标志被置 1
        {
            mDelaymS(20);               // 延时等待数据接收完毕
            I2C1_Rcv_flag  =  0;        // 清零标志位
            for(i = 0;i<256;i++)        // 将数据转移,以便下次接收数据
                DataBuf[i]  =  eeprom_I2C1[i];
        }
    }
    return 0;
}
```

第 8 章

系统应用

8.1 基于 ARM 微控制器的随动控制系统

8.1.1 设计要求

设计一套随动控制系统，由手动和随动（自动）两部分构成。手动部分和随动部分具有相同的结构，都是有两节可转动的臂和两个转轴构成。两个动臂长度均为 8 cm（臂端点到轴心之间的距离），它们的不同之处在于手动部分在转轴 1、转轴 2 处加装角度传感器，而在随动部分中是加装电机。整个系统在水平平面上运行。

制作两块相同的平板，尺寸均大于 20 cm×20 cm，在表面铺上坐标纸，将通过转轴连接在一起的两臂安装在平板上，如图 8-1 所示。注意转轴 1 的轴心（即原点）应与坐标纸的格子交叉，转轴 1 的轴心固定在平板上。手动部分和随动部分的节点 C 处都各安装一支画笔。

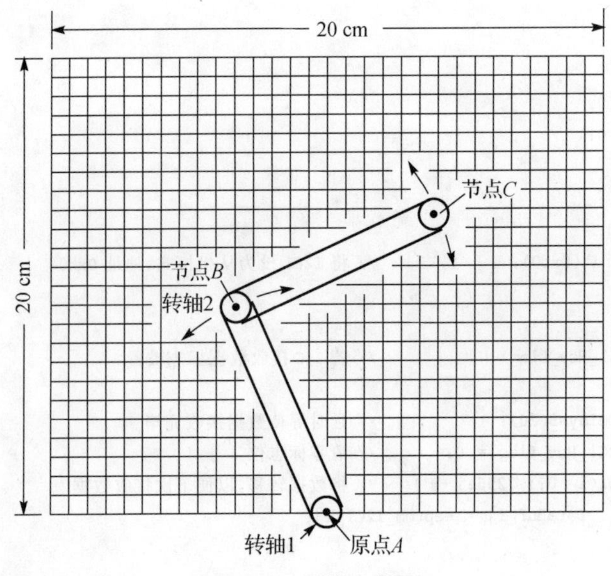

图 8-1 系统示意图

8.1.2　总体方案设计

按照系统任务要求，搭建手动和随动控制平台。为达到系统设计要求，需要采用以下模块：控制器模块、电机驱动模块、角度传感器模块、原点检测模块、人机界面、语音模块。系统方框图如图8-2所示。控制器模块作为整个系统的核心，起着指挥整个系统运行的作用；使用电机驱动模块驱动电机运转，从而带动随动部分的两臂；使用角度传感器模块检测手动部分两臂的旋转角度；使用原点检测臂的原点定位检测；采用人机界面对整个系统进行操作信息的输入和结果显示；使用语音模块进行操作提示以及结果播报。

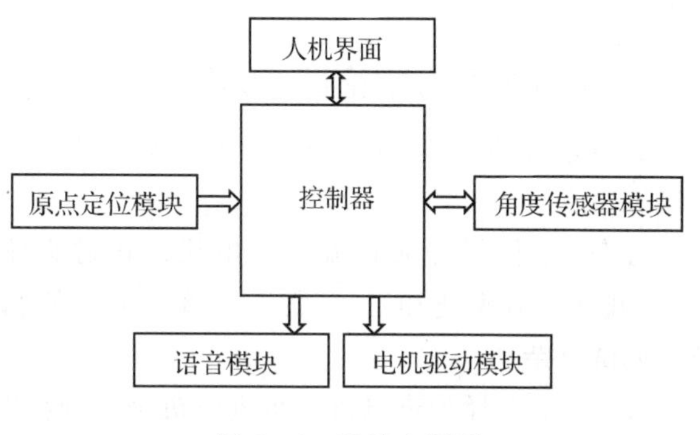

图8-2　系统方框图

8.1.3　系统各模块方案论证与选择

1. 控制器方案的论证与选择

方案一：采用现在比较通用的51系列单片机STC89C52RC。51系列单片机的发展已经经历了比较长的时间，应用比较广泛，各种技术都比较成熟。5 V供电的STC89C52RC共40个引脚，其中有32个I/O口，具有8 KB Flash和512 B的RAM。支持两种软件可选择的节电模式。空闲模式下，CPU停止工作，允许RAM、定时器/计数器、串口、中断继续工作，典型功耗为0.5 μA；掉电保护方式下，RAM内容被保存，振荡器被冻结，单片机一切工作停止，直到下一个中断或硬件复位为止，典型功耗为2 mA；而在正常工作模式下典型功耗为4～7 mA。

方案二：采用ARM7TDMI-S微控制器LPC2148。32位的LPC2148共64个引脚，工作电压为3.3 V，具有45个可承受5 V电压的I/O口，内嵌512 KB的高速Flash存储器和32 KB RAM，内置了宽范围的串行通信接口，包括全速USB 2.0 Device、2个UART异步串行口、SPI和SSP串行外设接口、I²C总线接口、2个32位的定时器以及1个看门狗定时器，低功耗的实时时钟RTC，9个边沿或电平触发的外部中断引脚，2个8通道10位的A/D和1个10位的D/A转换器以及6路输出的PWM单元。而LPC2148提供的掉电和空闲两种低功耗工作模式更能起到减小功耗的作用。另外，LPC2148支持实时仿真和跟踪，可方便地实现软硬件调试。

方案选择：对比方案一和方案二可知，LPC2148具有比51系列单片机STC89C52RC更强的功能和更充足的资源，且因ARM7采用的是3级流水线的工作模式,故处理速度比单片机快。另外,LPC2148的工作电压为3.3 V,在空闲模式下,处理器、存储器和相关控制器以及内部总线都不再消耗功率,同时,在掉电模式下,芯片的功耗降低到几乎为零。故本系统控制器模块选择ARM7TDMI-S微控制器LPC2148。

2. 电机类型方案论证与选择

分析题目,要求通过控制电机来精确控制臂的运动,因此只能选择精度较高的电机。

方案一:使用普通直流伺服电机。普通直流伺服电机控制简单,利用双极性PWM即可实现调速和正、反转。但该类电机的制动能力较差,需要闭环控制速度,安装调试复杂。

方案二:选择步进电机。步进电机是一种将电脉冲信号转换成相应的角位移或线位移的机电组件,它的转角及转速分别取决于脉冲信号的数量和频率。这一线性关系的存在,加上步进电机只有周期性的误差等特点,使得它可以达到很高的控制精度,且控制难度较小。

方案选择:步进电机调试比伺服电机简单容易,故本系统采用第二种方案。考虑基于永磁式步进电机步进角过大,而反应式步进电机噪声和振动又很大,故本系统选择步进角为1.8°的二相混合式步进电机。节点B的电机只用于控制带画笔的臂运动,而节点A处的电机需带动整个节点B运动,故节点B处的电机需要的力矩相对较小,所以节点B处的电机选择20系列,型号为20BYG,而节点A处的电机选择39系列,型号为STH-39D213-15。

3. 电机驱动模块的方案论证与选择

方案一:使用步进电机专用驱动器。步进电机专用驱动器可为步进电机提供稳定的工作电压和工作电流。但是其较重,且价格比较昂贵。

方案二:使用L297+L298驱动芯片驱动电路。L298N可以驱动直流电机和两个二相电机,也可以驱动一个四相电机,可直接通过电源来调节输出电压。最大输入电流DC 2 A,最高输入电压为DC 50 V,最大输出功率25 W。L297的心脏部件——译码器能将控制器的控制信号译成所需的相序,再将产生的四相A、B、C、D输入到L298进行功率放大调整,最后输出四相给步进电机控制其步进,其设计方框图如图8-3所示。

方案三:使用TA8435H驱动芯片驱动电路。该芯片是东芝公司推出的一款单片正弦细分二相步进电机驱动专用芯片,可以驱动二相步进电机,电路简单,工作可靠。工作电压为10~40 V,平均输出电流可达1.5 A,峰值可达2.5 A,具有整步、半步、1/4细分、1/8细分运行方式可供选择,采用脉宽调制式斩波驱动方式,具有正/反转控制功能,带有复位和使能引脚,可选择使用单时钟输入或双时钟输入。

方案选择:本系统只需要通过控制电机的步进来控制臂的运动,负载并不大,故

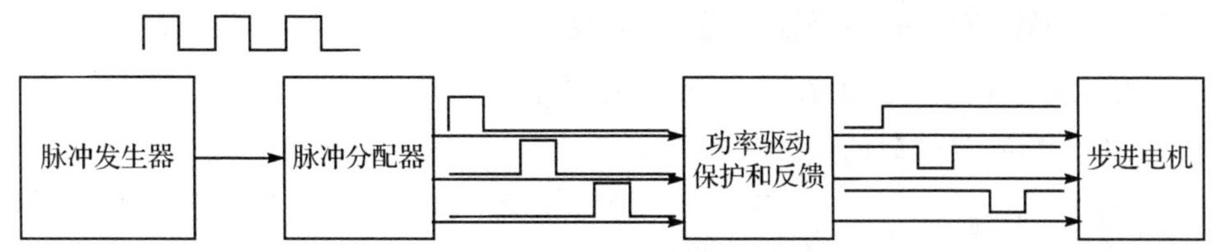

图 8-3　选用 L297+L298N 作为驱动电路的设计方框图

对驱动器没有过高的要求，所以淘汰方案一。方案二和方案三电路设计都较简单，且都易实现对步进电机的驱动，但若使用方案二驱动电路，步进电机的最小步进角是在半步模式下的 0.9°，而方案三驱动电路在 1/8 细分模式下可使步进电机的最小步进角为 0.225°，因此方案三比方案二驱动电机更精确。故本模块采用方案三。

4. 角度传感器模块的方案论证与选择

角度传感器用于手动部分检测两臂的旋转角度，属于平面角度检测范围。方案选择如下。

方案一：采用 UZZ9001 和 KMZ41 组成磁阻式角度传感器。使用矩形磁铁进行配合测量磁铁与 KMZ41 芯片之间的角度，测量范围为 180°。UZZ9001 将 KMZ41 输出的正余弦角度信号转换为数字信号，并通过 SPI 串行输出。

方案二：使用 WDD35D-4 角度传感器。此传感器是一个高精度、标称阻值为 5 kΩ的电位器，独立线性度达到 0.1%，机械转角可达 360°，理论电旋转角为 345°，故 WDD5D-4 测量角度的最大偏差为 0.345°。将此传感器与转动臂相结合，电位器的阻值随着臂的转动而被改变，即将臂的旋转角度值转换成了电位器的阻值，因此可利用 A/D 采样的方法，通过计算处理，得出臂的旋转角度值。其应用示意简图如图 8-4所示。

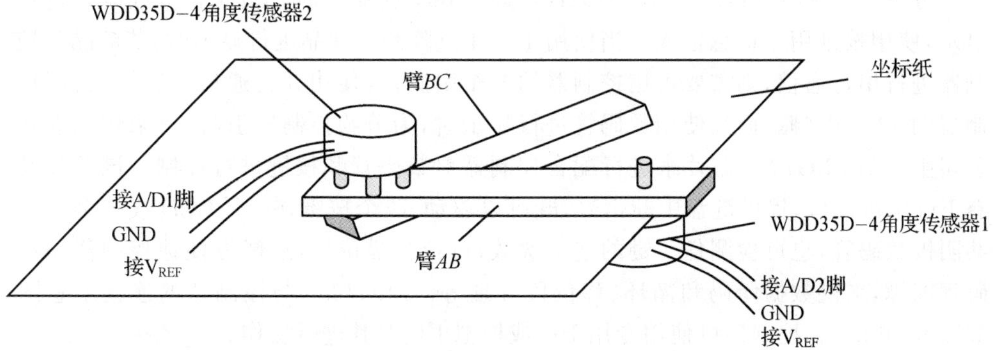

图 8-4　WDD35D-4 角度传感器应用示意简图

方案选择：虽然方案一输出为数字信号，但 KMZ41 调试比较困难，且需外加磁铁，增加了硬件的搭载复杂度。故本部分选择方案二。

5. 原点定位模块的方案论证与选择

为使系统的运动看起来有条理，系统设计随动部分在完成了一项任务后，臂 A 和臂 B 自动复位，复位后的示意图如图 8-5 所示。这就要求由原点定位模块来检测 A 和 B 两臂是否复位。对于原点复位模块的方案选择有以下两种。

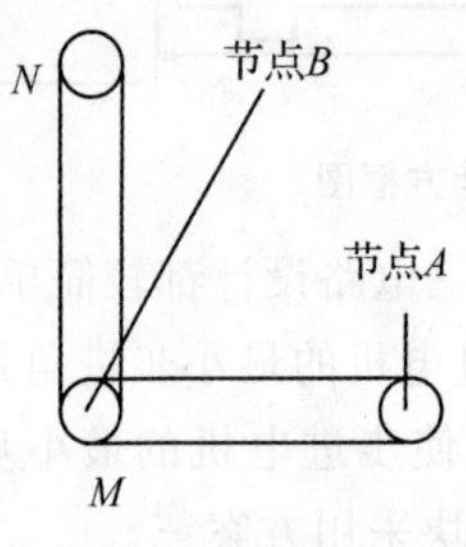

图 8-5 A、B 两臂复位后示意图

方案一：使用触碰开关。将 2 个触碰开关分别固定安装在上 N 点和 M 点，当随动部分的两臂到达复位点并触碰到开关时，触碰开关连接控制器的 I/O 口电平将发生变化，控制器通过检测电平判断两臂是否到达初始位置。

方案二：使用霍尔传感器。将 2 个霍尔传感器分别安装在平板底部的 N 点和 M 点，2 个磁钢分别安装在两臂复位时对应磁钢的位置，当两臂到达设定的复位位置时，霍尔传感器受磁钢的作用，会输出一个脉冲信号，因此控制器可检测此脉冲信号来判断两臂是否正确复位。

方案选择：对于控制器而言，方案一和方案二没有差异，但是对于硬件搭建来说，霍尔传感器比触碰开关体积小，安装起来相对简单，且若将触碰开关安装在 N 点，对臂的运动也会有一定的阻碍作用。因此，选择方案二。

6. 人机界面模块的方案论证与选择

为方便对系统的每一项要求进行测试，采用人为控制测试启动的方法，且题目发挥部分要求设定圆心坐标，因此需要使用人机界面来完成系统与测试者之间的信息交流。同时人机界面还可用于显示测试结果。

方案一：使用 FYD12864-0402B 普通 12864 点阵图形液晶或 8 位数码管进行显示，使用按键用于信息输入。当使用 12864 点阵图形液晶进行显示时，若液晶与控制器进行串行通信，则需要占用控制器的 5 个引脚；若使用并行通信，则需要占用控制器的 13 个引脚。而当使用数码管进行显示时，为节约控制器引脚，可采用周立功公司生产的 ZLG7290 芯片来进行配合控制器对数码管和按键进行控制。该芯片具有 I^2C 串行接口，提供键盘中断信号，既可以驱动 64 个按键，64 个 LED 或 8 个 8 位共阴极数码管，也可检测任一键的连击次数，可控扫描位数，还能方便地控制任一数码管闪烁，提供数据译码和循环、移位段寻址等。ZLG7290 与控制器的连接示意图如图 8-6 所示，控制器可使用专用 I^2C 或模拟 I^2C 对其进行操作。

方案二：使用迪文触摸屏人机交互模组(HMI)。该模组型号为 DMT32240S035_01WT，其分辨率为 320×240，工作温度范围为 -20～+70 ℃；工作电压范围为 5～28 V，功耗为 1 W；12 V 时，背光最亮和背光熄灭时的工作电流分别为 90 mA 和 50 mA。该模块共有 33 MB 字库空间，可存放 60 个字库，支持多语言、多字体、字体大小可变的文本显示,还支持用户自行设计的字库;96 MB 的图片存储空间,最多可存储 384 幅全屏图片,支持 USB 高速图片下载更新,图形功能完善;用户最大串口访问存储器空间为 32 MB,与图片存储器空间重叠;具有触摸屏漂移处理技术,同时还内嵌拼音输入法、数据排序等简单算法处理。使用异步、全双工串口 RS232 与控制器通信,只需占用控制器 2 个引脚,其与控制器之间的连接示意图如图 8-7 所示。

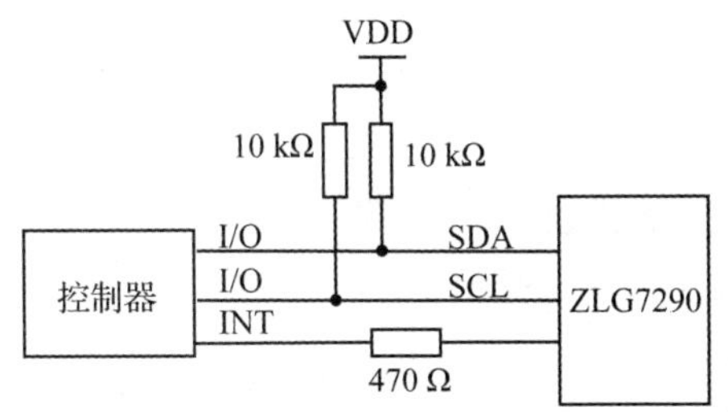

图 8-6　ZLG7290 与 LPC2148 进行 I^2C 通信示意图

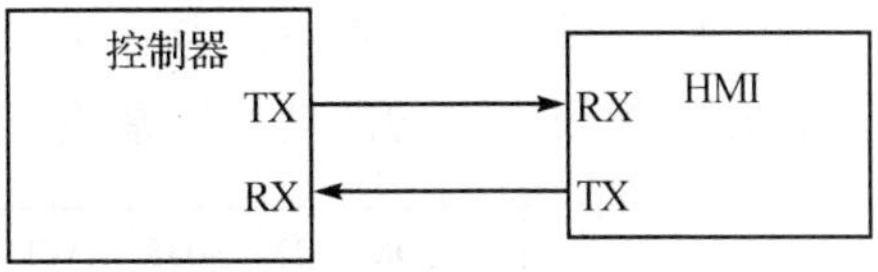

控制器
TX
RX
RX
TX
HMI

图 8-7　迪文人机交互模组与控制器通信示意图

方案选择:方案一使用液晶作为显示比使用数码管作为显示占用的控制器引脚多,而使用数码管显示又受到不能显示汉字的限制,但无论方案一使用哪种方法,需要占用的控制器引脚都比方案二多,又因迪文人机交互模组(HMI)带触摸功能,可省去另外加按键的电路,可使画面更加生动有趣,增强可观赏性。故此人机交互界面模块选择方案二。

7. 语音模块方案的选择与论证

语音模块为自由发挥项,可用于适时地播报一些操作提示或数据。如当臂完成一项任务时,进行语音提示,设定好圆心坐标后,对坐标进行播报等。

方案一:选择专用语音录放芯片,如 ISD1400/ISD1110 系列单片语音录放电路,采用的是在 E^2PROM 中直接模拟存储技术(DAS),省去了数字存储器、数据转换及备用电源等外围电路,具有零功率存储信息、无需编辑开发机、高保真语音录放等特点,使用方便。

方案二:选择带 SD 卡的 MP3 语音播放模块。通过录音方式将需要播报的内容以 MP3 格式存放在 SD 卡中,可使用标准模式、按键模式、并口模式、串口模式读取,操作简单。

方案选择:选用专用语音录放芯片可省去先录音再存放在 SD 卡中这一步骤,但采用 SD 卡存储器的 MP3 模块,具有音质好,存储容量大等优点,2G 的 SD 卡可以存储播放数小时的语音文件,且通过串口控制,通信方便,占用 I/O 口少。所以本部分选用方案二。

8. 系统组成

经过上述方案分析和论证,决定了系统最终方案如下:

① 控制器选择：ARM7TDMI-S微控制器LPC2148。

② 步进电机选择：节点A使用39系列混合式步进电机STH-39D213-15，节点B使用20系列混合式步进电机20BYG。

③ 步进电机驱动模块选择：TA8435H驱动电路。

④ 角度传感器选择：5 kΩ精密电位器WDD35D-4。

⑤ 原点定位模块选择：霍尔传感器、磁钢。

⑥ 人机界面模块选择：迪文DMT32240S035_01WT触摸屏人机交互模组(HMI)。

⑦ 语音模块选择：带SD卡的MP3语音播放模块。

由以上模块组成的随动控制系统方框图如图8-8所示。

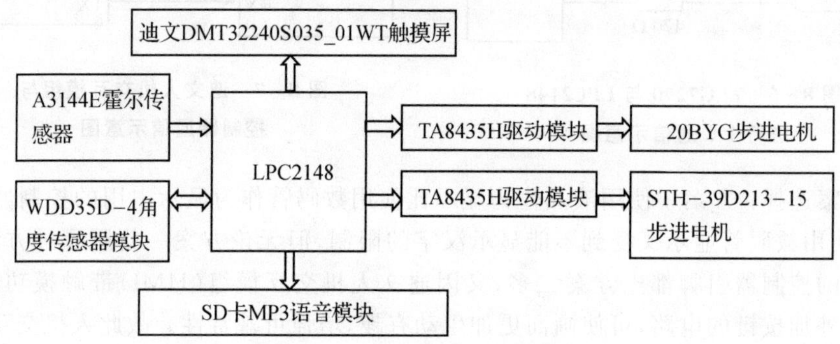

图8-8 随动控制系统方框图

8.1.4 理论分析及计算

系统原点坐标定义如图8-9所示，图中A点坐标为(100,0)，单位为mm。

1. 步进电机的模式

为提高精度，系统采用1/8细分激励方式控制电机，步进角为1.8°/8=0.225°，如图8-9所示，步进电机走一步，AB臂从B点运动到B'点，则α=0.225°。

2. *AB*臂和*BC*臂的角度差值计算

因系统设计每完成一次任务两臂都要复位，(复位示意图如图8-5所示)，这样，要控制系统到达任意一点，只需分别计算出AB臂、BC臂以及其复位时的角度差值便可，为方便计算，根据任意给定点的坐标将计算分为两种情况。

(1) 设定点的横坐标大于100 mm，示意图如图8-10所示

假设要使随动系统到达坐标点(x,y)，其中$x>100$ mm，即C点的坐标为(x,y)，则只需控制AB臂从复位点旋转α，使BC臂与AB臂之间的夹角变为β即可。

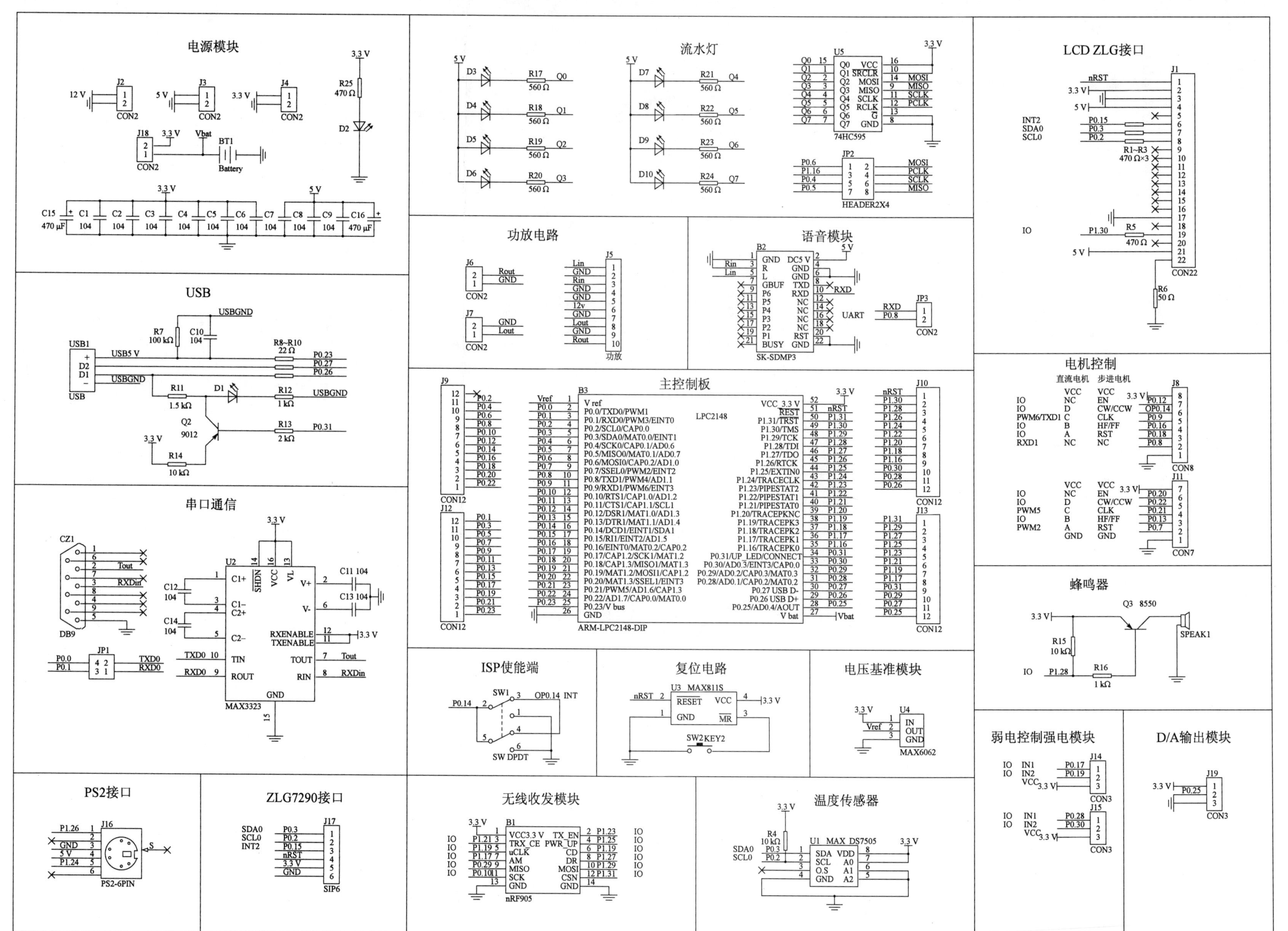

图 8-13 LPC2148主控制板电路图

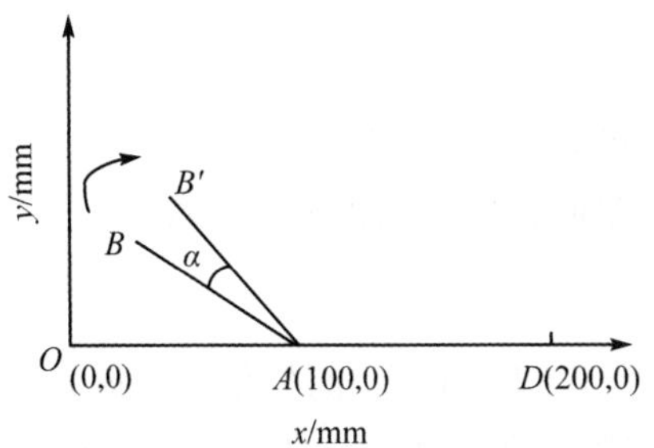

图 8-9　AB 臂从 B 点到运动到 B′点示意图

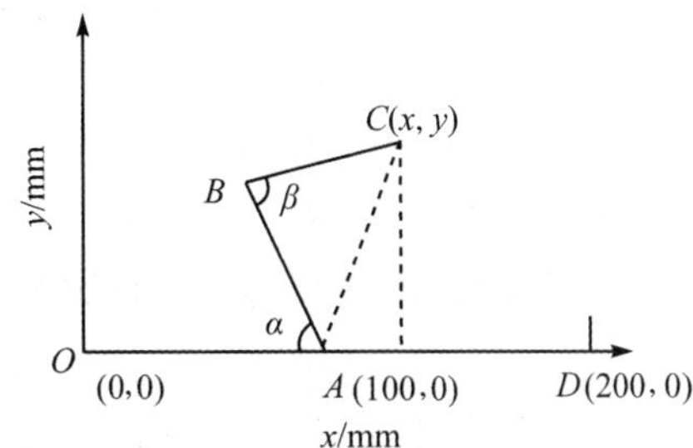

图 8-10　AB 臂和 BC 臂运动计算示意图 1

又已知 AB 臂和 BC 臂长为 80 mm，A 点坐标为(100,0)mm，由两点间距离公式(8.1.1)可计算出任意两点间的距离 d，其中(x_1,y_1)、(x_2,y_2)为任意两点的坐标。

$$d=\sqrt{(x_1-x_2)^2+(y_1-y_2)^2} \tag{8.1.1}$$

故由已知的 A 和 C 点坐标计算出 AC 间的距离如下：

$$AC=\sqrt{(x-100)^2+(y-0)^2} \tag{8.1.2}$$

知道 AC 的长度后，又由已知的 BC 臂长 80 mm 可得出 β 的计算如下：

$$\beta=2\times\arcsin\frac{AC}{2\times 90^\circ} \tag{8.1.3}$$

因此可算出$\angle BAC=(360^\circ-\beta)/2$，便得出$\angle CAD$ 的值如下：

$$\angle CAD=\arcsin\frac{y}{AC} \tag{8.1.4}$$

从而得出了 $\alpha=180^\circ-\angle CAD-\angle BAC$。

(2) 设定点的横坐标小于 100 mm，示意图如图 8-11 所示

情况二与情况一原理相同，只是将 α 的值变为了$\angle CAO-\angle CAB$。

有了以上两种情况的分析计算，要求随动系统从复位点运动到任意给定的坐标(x,y)，首先应判断 x 与 100 的大小关系，确定是两种情况中的哪一种，从而计算出 α 和 β 的值。接着便可将 α 和 β 转换成电机的步进步数，因电机采用的是半速模式，步进角为 0.225°，AB 臂要旋转 α，BC 臂要旋转$(90^\circ-\beta)$，只需控制节点 A 的步进电机步进 $\alpha/0.225$，节点 B 的步进电机步进$(90^\circ-\beta)/0.225^\circ$便可。若 α 和 $90^\circ-\beta$ 都为0.225的整数倍，则在电机不失步的情况下，能精确到达 C 点；若两者中有一个不为 0.225 的整数倍，则在电机不失步的情况下，将会有误差，AB 臂误差大小为 $\alpha-N\times0.225$，其中 N 为α 除以 0.225 的整数部分，BC 臂误差大小为$(90^\circ-\beta)-K\times 0.225$，其中 K 为 $90^\circ-\beta$ 除以 0.225 的整

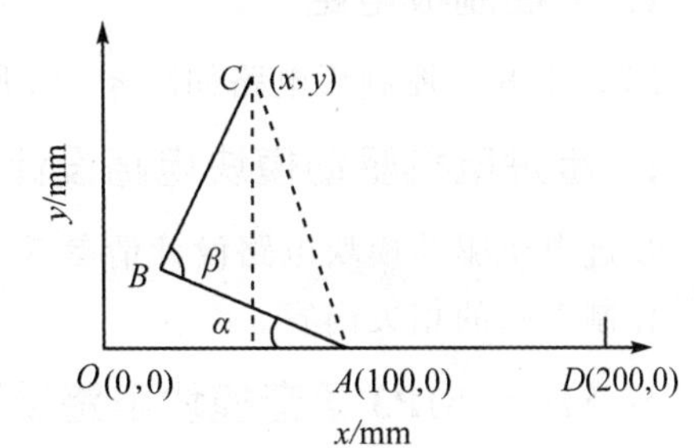

图 8-11　AB 臂和 BC 臂运动计算示意图 2

数部分。

若要使系统从任意一点(M_1,N_1)到任意另一点(M_2,N_2),则只需按照同样的方法,计算出到(M_1,N_1)需要的 α_1 和 β_1 值,和到(M_2,N_2)需要的 α_2 和 β_2 值,所不同的是,还需求出 α_1 和 α_2,β_1 和 β_2 之间的差值,再将差值转换为电机的步进数。

3. 利用角度传感器进行角度测量

由前面角度传感器方案论证可知角度传感器 WDD35D-4 测量角度的最大偏差为0.345°。系统使用控制器 LPC2148 自带采样频率为 1 MHz 的 10 位 ADC 来对 WDD35D-4 的阻值进行采样,从而计算出手动系统臂的旋转角度,示意图如图 8-12所示。

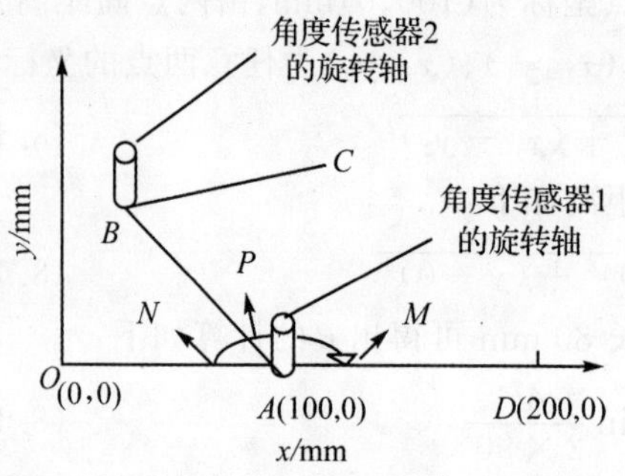

图 8-12 手动部分角度传感器计算示意图

臂 AB 和臂 BC 旋转角度计算方法一样,以节点 A 处的角度传感器为例,说明计算方法。节点 A 处的角度传感器最大只需要旋转 180°,以箭头方向表示角度传感器 1 的旋转方向,当角度传感器 1 的旋转轴标号在起始位置 N 时,假设 A/D1 采用得到的电压为 V_N,再旋转角度传感器 1 的旋转轴,使其标号到达 M 位置,再使用 A/D1 采样,得到电压值 V_M,则由 V_N 和 V_M 可计算出角度传感器每旋转 1°所对应的电压值 V 为

$$V=\frac{|V_N-V_M|}{180°} \tag{8.1.5}$$

因此,假设当臂运动到 K 点时,角度传感器的旋转轴标号到达 P 位置,而此时 A/D1 采样到的电压值为 V',则可计算出 AB 轴的旋转角度为 V'/V。

8.1.5 系统主要单元电路设计

1. 主控制板电路

LPC2148 主控制板电路如图 8-13 所示,PCB 和元器件布局图如图 8-14 所示。

2. 步进电机驱动模块电路设计

步进电机驱动模块电路设计请参考 4.3.4 小节的基于 TA8435H 的步进电机驱动与控制电路的相关内容。

3. SD 卡 MP3 语音模块电路设计

SD 卡 MP3 语音模块电路设计请参考 6.2 节的 SK-SDMP3 语音模块及其应用的相关内容。

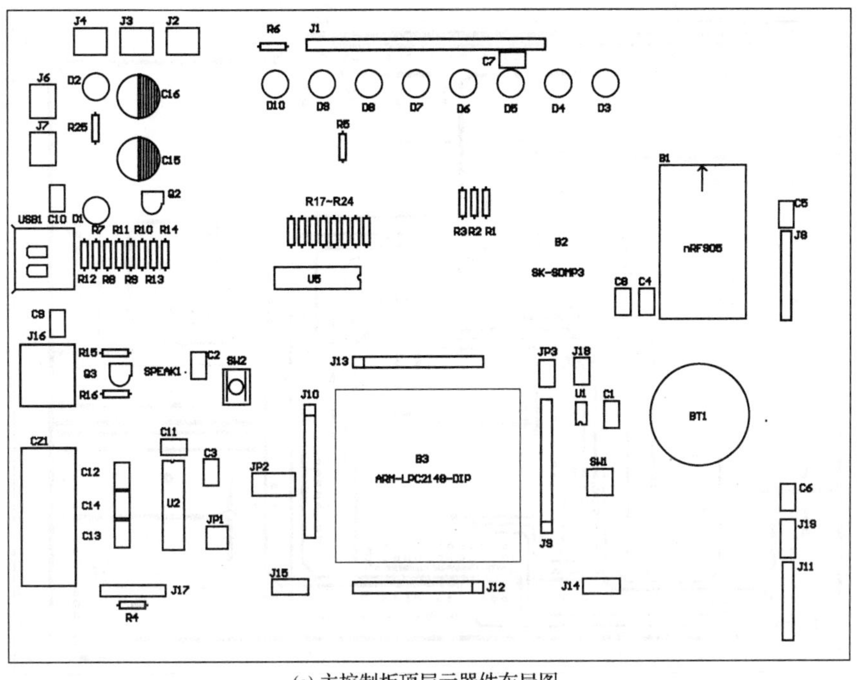

(a) 主控制板顶层元器件布局图

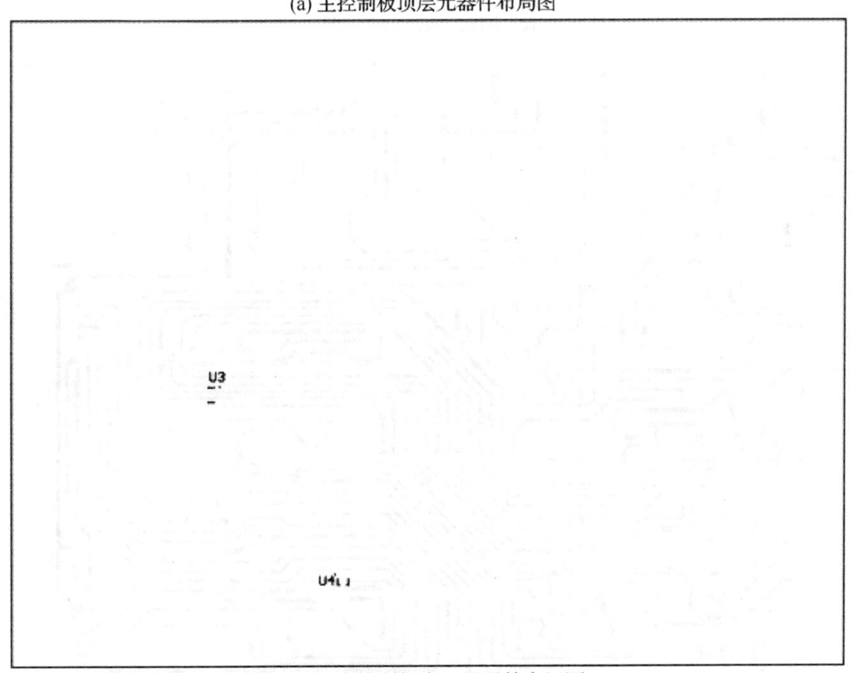

(b) 主控制板底层元器件布局图

图 8－14　主控制板 PCB 和元器件布局图

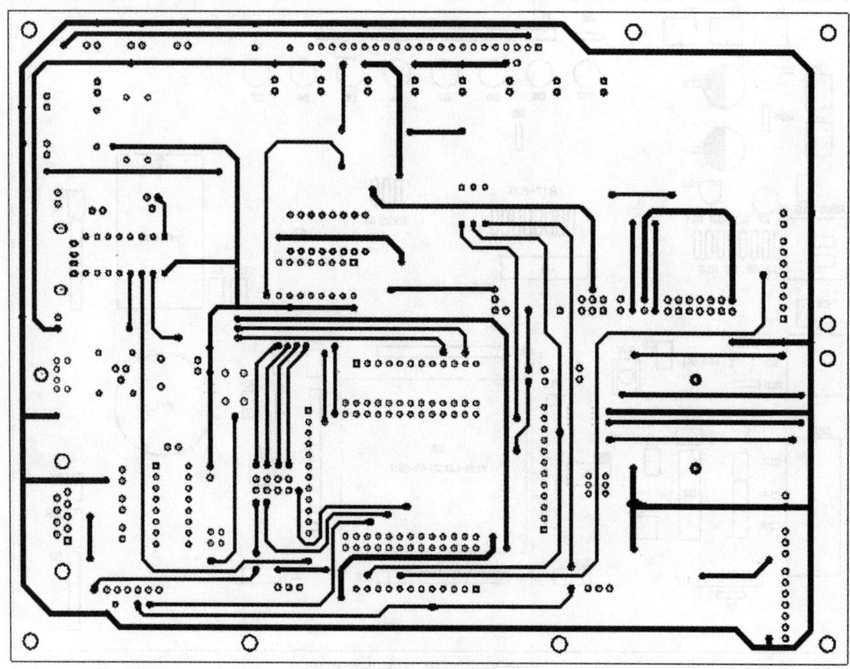

(c) 主控制板顶层PCB图

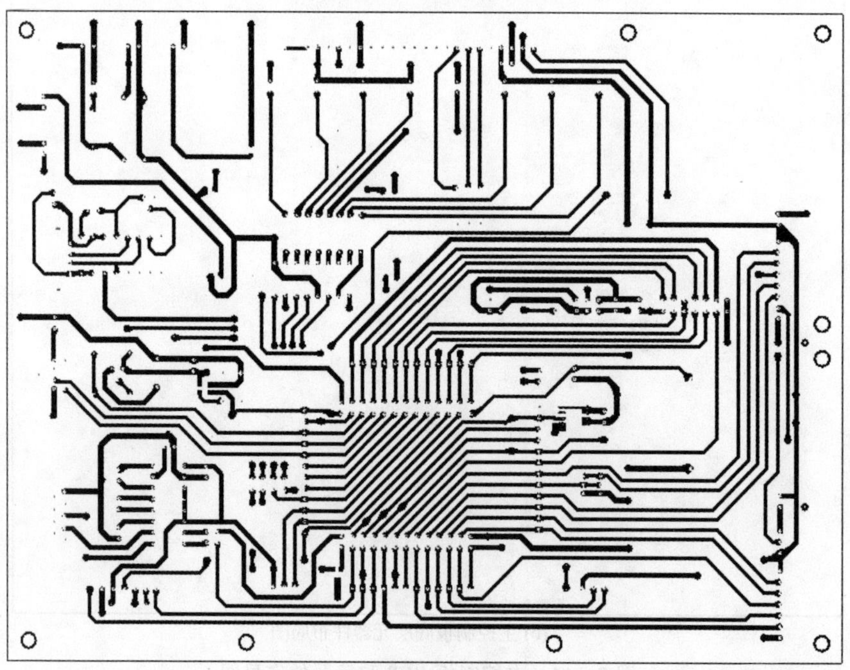

(d) 主控制板底层PCB图

图 8-14　主控制板 PCB 和元器件布局图(续)

8.1.6　系统软件设计

1. 系统软件开发环境介绍

本系统使用 C 语言在 ARM Developer Suite——ADS 1.2 集成开发环境中编译。这个编译器通过了 Plum Hall C Validation Suite 为 ANSI C 的一致性测试，用于将 ANSI C 编写的程序编译成 32 位 ARM 指令代码。ADS 的 CodeWarrior 集成开发环境(IDE)为管理和开发项目提供了简单多样化的图形用户界面。用户可以使用 ADS 的 CodeWarrior IDE 为 ARM 和 Thumb 处理器开发用 C 语言，C++语言，或 ARM 汇编语言的程序代码，界面友好。ADS 1.2的集成开发环境如图 8-15 所示。

图 8-15　ADS 1.2 的集成开发环境

2. 系统控制总流程图

系统控制总流程图如图 8-16 所示。系统上电后，触摸屏液晶显示开机画面，单击屏幕进入随动控制系统主菜单，此时，语音提示选择执行基本要求还是执行发挥部分。若选择执行基本要求，则显示切换到基本要求子菜单，子菜单内可通过屏幕键盘

图 8-16　系统控制的总流程图

输入 AB 臂的角度设定值,也可选择执行基本要求 2 和基本要求 3,同样使用语音提示选择执行基本要求的哪一项,当通过触摸屏选择任务后,系统便进行该任务的执行指令。若选择的是发挥部分,则显示切换到发挥部分子菜单,子菜单内同样也是通过屏幕键盘输入圆心的设定值和半径值,或执行发挥部分 3,根据语音提示选择了执行任务后,系统开始执行该任务指令。

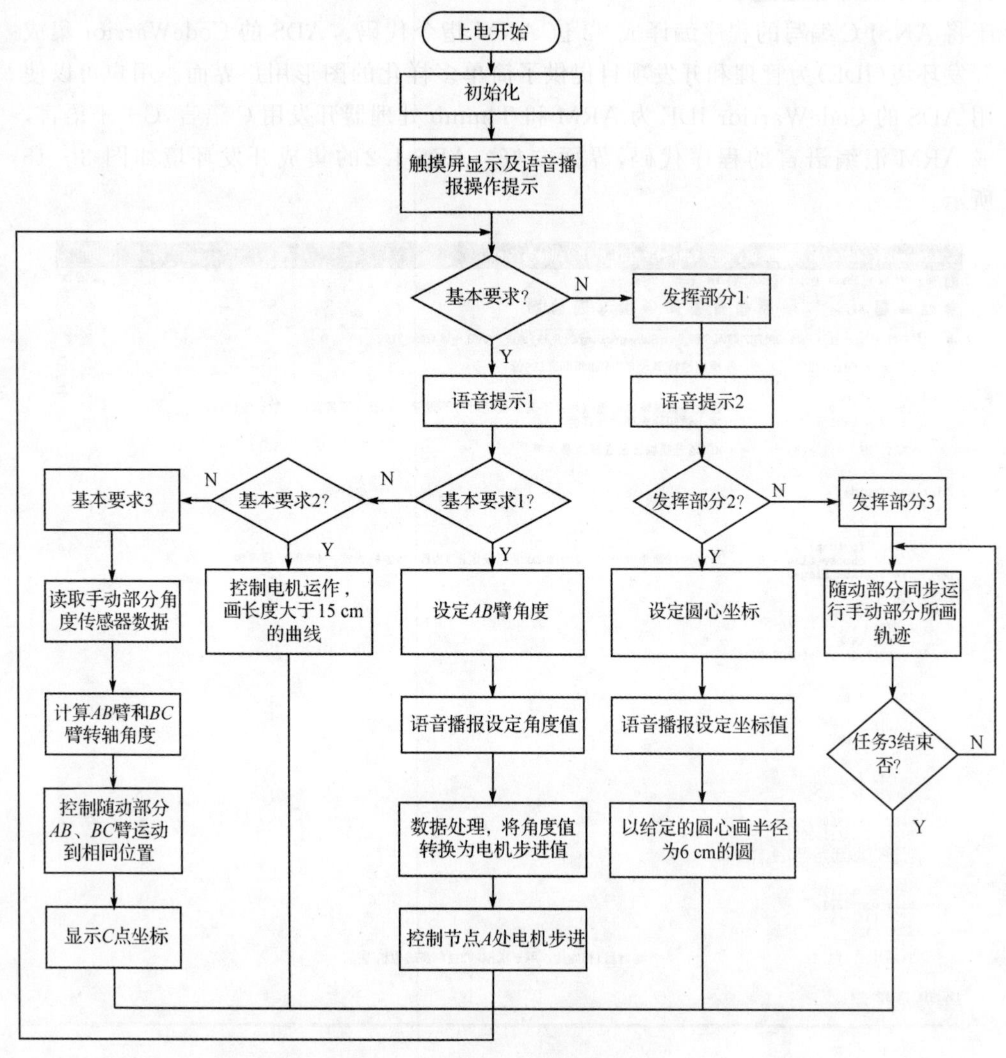

图 8-16　系统控制的总流程图

3. 系统主要子程序流程图

步进电机的运动是整个系统的重要部分,其运动程序的流程图如图 8-17 所示,当需要控制电机步进时,只需设定 PWM 值和电机需要运转的步数,然后调用此子函数即可;角度数据的读取是整个系统的另一个重要部分,其程序流程图如图 8-18 所示,控制器 LPC2148 以1 kHz的速度采用角度传感器的输出电压值,然后储存待用;随动系统连接任意两点的子程序流程图如图 8-19 所示,该程序中包括对随动部分的几何计算,可准确地控制 2 臂到达指定位置;当系统设定好两点的坐标后画圆,子程序流程图如图 8-20 所示;触摸屏负责人机交互功能,通过很好的交互界面能够方便地调用系统各项功能,触摸屏子程序流程图如图 8-21 所示。

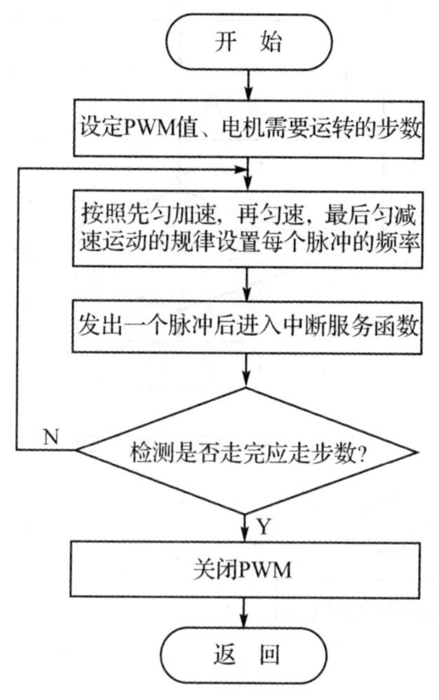

图 8-17　步进电机步进子程序流程图

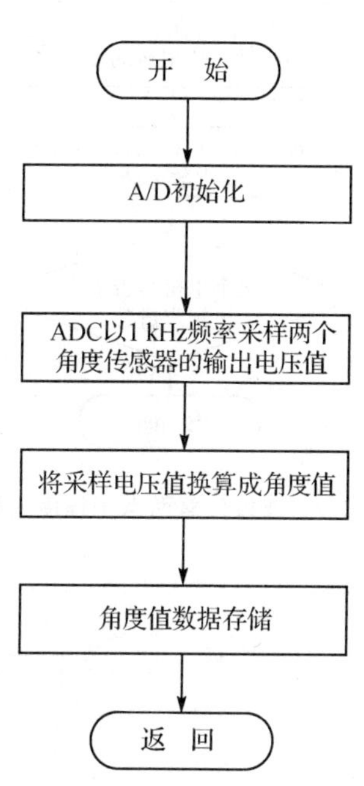

图 8-18　角度测量子程序流程图

4. 系统程序清单

(1) 系统主函数清单

```
#include "config.h"
extern uint8    rcv_buf[100],num;                          // 定义存放串口接收数据的数组
extern uint8    rcv_flag;                                  // 定义表示串口收到数据的标志
extern  uint32 act_Angle_integer = 0,act_Angle_decimal = 0;  // 放设定的角度数据
extern  uint32 act_centralA_integer = 0,act_centralA_decimal = 0;  // 设定圆心坐标 a
extern  uint32 act_centralB_integer = 0,act_centralB_decimal = 0;  // 设定圆心坐标 b
extern  uint32 act_radius_integer = 0,act_radius_decimal = 0;      // 给定圆的半径值
extern  uint8    commen_flag = 0;
// commen_flag = 2 基本要求 2   commen_flag = 3 基本要求 3   commen_flag = 4 发挥部分 3
uint8       number[5] = {0};                                // 存放输入数据
uint16      x,integer,decimal;
```

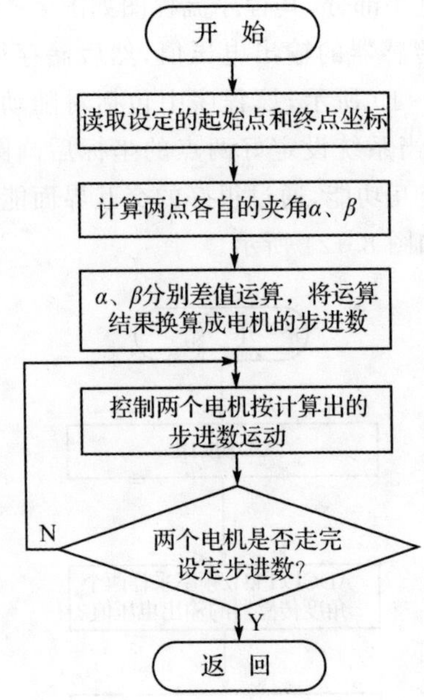

图 8-19 连接任意两点子程序流程图

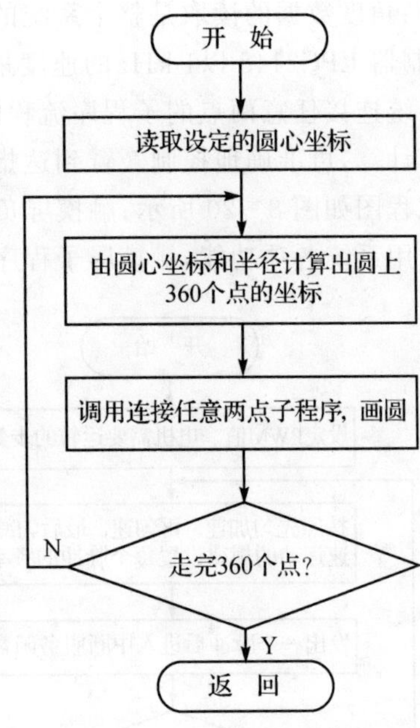

图 8-20 画圆子程序流程图

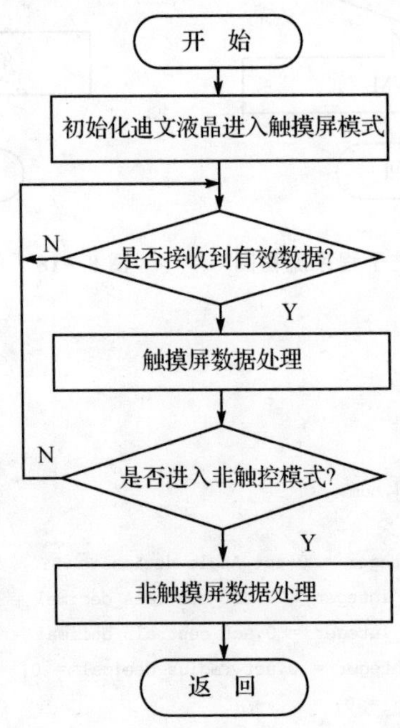

图 8-21 触摸屏子程序流程图

```
uint8       dot_num = 0;
char        dot = 0;
uint16      words_style = 0;
uint16      x_max = 0;
uint16      Angle_x = 122;
uint16      central_x_a = 116;
uint16      central_x_b = 138;
uint16      radius_x = 117;
uint16      y_begin = 0;
/**********************************************************************
 * 名称:main()
 * 功能:随动系统主函数,通过读取触摸屏数据控制系统选择执行各种功能
 **********************************************************************/
int main(void)
{
    uint32 i;
    fp64   fp_temp;
    uint8  choose_flag = 0;
    uint8  TASK_Flag = 0;
    i = 0;
    PINSEL0 = 0;
    PINSEL1 = 0;
    PINSEL2 &= ~(0x00000006);
    IO0DIR = 0;
    IO1DIR = 0;
    IO1DIR |= BEEP;
    IO1SET = BEEP;
    IO0DIR |= MOTO_IO_1;
    IO0DIR |= MOTO_IO_2;
    UART1_Ini(9600);
    UART0_Init(115200);
    Dianji_init();
    Dianji_init2();
    Half_Speed();
    Half_Speed2();
    STOP();
    STOP2();
    Time1Init();                              // 定时器 1 初始化,用于模拟 PWM 控制步进电机
    PWM_X_Wait(200,1);
    PWM_X_Wait2(200,1);
    IO0SET = EN;
    IO0SET = EN2;
    AD0_1_P0_28();
    AD0_3_P0_30();
    RTCIni();                                 // RTC 初始化,用于系统时钟计时
    Time0Init();
    mDelaymS(100);
```

```
MOTO_RST();
NOW_X = 15;                                    // 初始化起点坐标
NOW_Y = 80;
NOW_Angle1 = 0.0;
NOW_Angle2 = 90.0;
set_mode(0x30);                                // 进入触控模式
Display(123456);
UART1_MP3_voice(8);                            // 语音模块音量调节
mDelaymS(50);
UART1_MP3display_Folder_song(0,12);            // 语音播报欢迎进入随动系统
while(1)
{
    if(TASK_Flag == 6)
    {
        Hight_TASK();
    }
    if(TASK_Flag == 7)
    {
        if(Time0_Flag)
        {
            Time0_Flag = 0;
            Hight_TASK3();
        }
    }
   /* 若按下触屏模式的按键 */
    if(rcv_flag)                              // 收到彩屏数据
    {
        mDelaymS(5);
        rcv_flag = 0;
        num = 0;
        if((rcv_buf[0] == 0xaa)&(rcv_buf[1] == 0x78))
        {
            if(rcv_buf[3] == 0x74)      // 若进入基本要求子菜单中的AB臂角度值设定
            {
                choose_flag = 1;
                x = Angle_x;
                words_style = 0x53;
                y_begin = 76;
                x_max = 169;
            }
            if(rcv_buf[3] == 0x12)      // 若按下基本要求 2
            {
                commen_flag = 2;
            }
            if(rcv_buf[3] == 0x13)      // 若按下基本要求 3
```

```
    {
        commen_flag = 3;
    }
    if(rcv_buf[3] == 0x76)         // 若按下"给定圆的圆心"
    {
        choose_flag = 2;
        x = central_x_a;
        words_style = 0x54;
        y_begin = 73;
        x_max = 128;
    }
    if(rcv_buf[3] == 0x75)         // 若按下"给定圆的半径"
    {
        choose_flag = 4;
        x = radius_x;
        words_style = 0x54;
        y_begin = 112;
        x_max = 156;
    }
    if(rcv_buf[3] == 0x30)         // 按下进入发挥部分
    {
       commen_flag = 4;
    }
   if(rcv_buf[3] == 0x66)          // 按下基本功能 2 初始化
   {
       TASK2_RST();                // 基本功能 2,画曲线初始化位置
   }
   if(rcv_buf[3] == 0x16)
   {
       MOTO_RST();                 // 2 个转轴初始化位置
       TASK_Flag = 0;
   }
   if(rcv_buf[3] == 0x17)
   {
       MOTO_RST();                 // 2 个转轴初始化位置
       TASK_Flag = 0;
   }
   if(rcv_buf[3] == 0x14)
   {
       TASK_Flag = 0;
   }
   else if(choose_flag == 1)       //设定 AB 臂角度
   {
        Dis_number0(Angle_x);
        if(rcv_buf[3] == 0x15)  // 按下确定键
        {
            x = Angle_x;
```

```
            act_Angle_integer = integer;
            act_Angle_decimal = decimal;
            integer = 0;
            decimal = 0;
            choose_flag = 0;
            /* 开始执行AB角度设定,基本功能1 */
            fp_temp = (fp64)act_Angle_integer +
            (fp64)act_Angle_decimal/100;
            UART1_MP3display_Folder_song(0,9);
            Set_AB(fp_temp,1,100);
        }
    }
    else if(choose_flag == 2)                                    // 设定圆心坐标a
    {
        Dis_number0(central_x_a);
        if(rcv_buf[3] == 0x15)                                   // 按下确定键
        {
            x = central_x_a;
            act_centralA_integer = integer;
            act_centralA_decimal = decimal;
            integer = 0;
            decimal = 0;
            choose_flag = 3;                                     // 设定圆心坐标b标志
            x = central_x_b;
            words_style = 0x54;
            y_begin = 73;
            x_max = 150;
            /* 若纵坐标已经设定完成,输入圆心坐标完毕,开始画圆 */
            if((act_centralB_integer! = 0) & (act_radius_integer == 0))
            {
                UART1_MP3display_Folder_song(0,1);       // 语音播报开始画圆
                Circle_1080(act_centralA_integer,act_centralB_integer,30);
                act_centralA_integer = 0;
                act_centralB_integer = 0;
                act_radius_integer = 0;
            }
        }
    }
    else if( choose_flag == 3)                                   //设定圆心坐标b
    {
        Dis_number0(central_x_b);
        if(rcv_buf[3] == 0x15)                                   //按下确定键
        {
            x = central_x_b;
            act_centralB_integer = integer;
```

```
                act_centralB_decimal = decimal;
                integer = 0;
                decimal = 0;
                choose_flag = 0;
                /* 若纵坐标已经设定完成,输入圆心坐标完毕,开始画圆 */
               if((act_centralB_integer! = 0) & (act_radius_integer == 0))
               {
               UART1_MP3display_Folder_song(0,1);       // 语音播报开始画圆
               Circle_1080(act_centralA_integer,act_centralB_integer,30);
               act_centralA_integer = 0;
               act_centralB_integer = 0;
               act_radius_integer = 0;
               }
            }
        }
    else if(choose_flag == 4)
    {
        Dis_number0(radius_x);
        if(rcv_buf[3] == 0x15)                          // 按下确定键
        {
            x = radius_x;
            act_radius_integer = integer;
            act_radius_decimal = decimal;
            integer = 0;
            decimal = 0;
            choose_flag = 0;
            if( (act_centralA_integer! = 0) &
            (act_centralB_integer! = 0) )                // 若已经设好圆心坐标
            {
                /* 指定圆心、半径画圆 */
                UART1_MP3display_Folder_song(0,1);       // 语音播报开始画圆
                Circle_1080(act_centralA_integer,
                act_centralB_integer,act_radius_integer);
                act_centralA_integer = 0;
                act_centralB_integer = 0;
                act_radius_integer = 0;
            }
        }
    }
}
if(commen_flag == 2)                         // 基本任务 2
{
    commen_flag = 0;
    // 开始执行画曲线
    UART1_MP3display_Folder_song(0,14); // 语音播报选择基本要求 2
    TASK2();
```

```
            }
            if(commen_flag == 3)                           // 基本任务 3
            {
                commen_flag = 0;
                // 实现手动部分单个点的跟踪
                UART1_MP3display_Folder_song(0,15); // 语音播报选择基本要求 3
                TASK3();
            }

            if(commen_flag == 4)                           //发挥部分 3
            {
                commen_flag = 0;
                TASK_Flag = 7;
                UART1_MP3display_Folder_song(0,2);  // 语音播报实时跟踪手动系统
            }
        }
    }
    return 0;
}
```

(2) 随动部分程序清单

```
/*********************************************************************
* 文件名：Follow.c
* 功能:随动系统各计算部分算法函数,包括计算各臂的夹角、距离、坐标等
*********************************************************************/
#include "config.h"
#include <math.h>

#define  AB_0       237          // AB 轴 0°时 A/D 值
#define  AB_90      509          // AB 轴 90°时 A/D 值
#define  AB_180     779          // AB 轴 180°时 A/D 值
#define  BC_90    794          // BC 轴 90°时 A/D 值
#define  BC_180   522          // BC 轴 270°时 A/D 值
#define  BC_270   254          // BC 轴 270°时 A/D 值
#define  BC_MIN   995          // BC 轴最小角度时 A/D 值
#define  BC_MAX   58           // BC 轴最大角度时 A/D 值
#define  MAX_X   250           // 定义 X 坐标范围小于 250 mm
#define  MAX_Y   250           // 定义 Y 坐标范围小于 250 mm

#define  L       80            // 转轴长度为 80 mm
#define  OA      100           // OA 长度为 80 mm

#define  STAR_Angle_AB   2     // AB 轴起始角度
#define  STAR_Angle_BC   2     // BC 轴起始角度
#define  AB_STAR_Dat    239
#define  Step_Angle 0.225      // 电机的步进角度
```

```
uint32 Read_num = 0;
fp64    AB_Per_Angle = 180.0/(fp64)(AB_180 - AB_0);
fp64    BC_Per_Angle = 180.0/(fp64)(BC_90 - BC_270);
uint32 B_X = 0;                     // B 点横坐标
uint32 B_Y = 0;                     // B 点纵坐标
uint32 C_X = 0;                     // C 点横坐标
uint32 C_Y = 0;                     // C 点纵坐标
fp64    Angle1 = 0;                 // 手动 A 点转轴转过的角度
fp64    Angle2 = 0;                 // 手动 B 点转轴转过的角度
fp64    Angle3 = 0;                 // 中间计算需要的角度,BC 轴与坐标纸 X 轴的夹角
uint32 AD,BD,BE,CE;
uint32 NOW_X,NOW_Y;                 // 随动 C 点当前时刻的坐标
fp64    NOW_Angle1,NOW_Angle2;      // 随动部分 2 转轴的当前时刻角度
fp64 Next_Angle1;
fp64 Next_Angle2;
/**************************************************************************
* 名称:Average()
* 功能:求平均值
* 入口参数:uint32 *S            输入的数据指针变量存储地址
*          uint32 no            输入数据个数
* 出口参数:Sum                  输出平均值
**************************************************************************/
uint32 Average(uint32 *S,uint32 no)
{
    uint32 i;
    uint32 Sum = 0;
    for(i = 0;i<no;i++)
    {
        Sum += S[i];
    }
    Sum = Sum/no;
    return Sum;
}
/**************************************************************************
* 名称:ZhongJian()
* 功能:求中间值
* 入口参数:uint32 *S            输入的数据指针变量存储地址
*          uint32 no            输入数据个数
* 出口参数:Sum                  输出中间值
**************************************************************************/
uint32 ZhongJian(uint32 *S,uint32 no)
{
    uint32 a,b,temp = 0;
    uint32 Sum = 0;
    for(a = 0;a<no;a++)
```

```
    {
        for(b = 0;b<a;b++)
        {
            if(S[a] > S[b+1])
            {
                temp = S[a];
                S[a] = S[b+1];
                S[b+1] = temp;
            }
        }
    }
    no = no/2;
    Sum = S[no];
    return Sum;
}
/*******************************************************************
* 名称:Angle_Radian()
* 功能:把角度转化为弧度
* 入口参数:Angle           输入角度值
* 出口参数:Rad             输出弧度值
*******************************************************************/
fp64 Angle_Radian(fp64 Angle)
{
    fp64 Rad;
    Rad = Angle/180.0 * PI;
    return Rad;
}
/*******************************************************************
* 名称:Radian_Angle()
* 功能:把弧度转化为角度
* 入口参数:Rad             输入弧度值
* 出口参数:Angle           输出角度值
*******************************************************************/
fp64 Radian_Angle(fp64 Rad)
{
    fp64 Angle;
    Angle = Rad/PI * 180.0;
    return Angle;
}
/*******************************************************************
* 名称:Angle_Sinusoidal()
* 功能:由角度计算正弦值
* 入口参数:Angle           输入角度值
* 出口参数:b               输出正弦值
*******************************************************************/
fp64 Angle_Sinusoidal(fp64 Angle)
{
```

```
    fp64 a,b;
    a = Angle_Radian(Angle);
    b = sin(a);
    return b;
}
/***********************************************************************
* 名称:Angle_Cosine()
* 功能:由角度计算余弦值
* 入口参数:Angle           输入角度值
* 出口参数:b               输出余弦值
***********************************************************************/
fp64 Angle_Cosine(fp64 Angle)
{
    fp64 a,b;
    a = Angle_Radian(Angle);
    b = cos(a);
    return b;
}

/***********************************************************************
* 名称:Angle_Location()
* 功能:角度计算当前B点和C点坐标,计算的结果存在B_X、B_Y、C_X、C_Y中,
*       手动部分和随动部分都可以用此函数
* 入口参数:Angle_A     A点转轴角度
*          Angle_B     B点转轴角度
* 出口参数:无
* 说    明:以B点为原点坐标,重新定义X1,Y1坐标平面,X1,Y1分别平行于X,Y判断BC臂所在
*          象限时,因B点是动点,所以需要比较Angle1和Angle2的大小再确定BC所在象限
***********************************************************************/
void Angle_Location(fp64 Angle_A,fp64 Angle_B)
{
    /* AB轴在左边,分为四种情况 */
    /* AB轴在左边,BC轴在第四象限 */
    if((Angle_A<90.0) & (Angle_B<Angle_A))
    {
        Angle3 = Angle_A - Angle_B;
        AD = L*Angle_Cosine(Angle_A);
        BD = L*Angle_Sinusoidal(Angle_A);
        CE = L*Angle_Sinusoidal(Angle3);
        BE = L*Angle_Cosine(Angle3);

        B_X = OA - AD;             //B点坐标
        B_Y = BD;
        C_X = B_X + BE;            //C点坐标
        C_Y = BD - CE;
    }
```

```
/* AB轴在左边,BC轴在第一象限 */
if((Angle_A<90.0) & (Angle_A<Angle_B) & (Angle_B<(180.0-Angle_A)) )
{
    Angle3 = Angle_B - Angle_A;
    AD = L*Angle_Cosine(Angle_A);
    BD = L*Angle_Sinusoidal(Angle_A);

    CE = L*Angle_Sinusoidal(Angle3);
    BE = L*Angle_Cosine(Angle3);

    B_X = OA-AD;                //B点坐标
    B_Y = BD;

    C_X = B_X + BE;             //C点坐标
    C_Y = BD + CE;
}
/* AB轴在左边,BC轴在第二象限 */
if( (Angle_A<90.0) & (Angle_B> (180-(90-Angle_A)) ) & ( Angle_B < (270.0-(90.0-Angle_A))) )
{
    Angle3 = 270.0-(90.0-Angle_A) - Angle_B;
    AD = L*Angle_Cosine(Angle_A);
    BD = L*Angle_Sinusoidal(Angle_A);

    CE = L*Angle_Sinusoidal(Angle3);
    BE = L*Angle_Cosine(Angle3);

    B_X = OA-AD;                //B点坐标
    B_Y = BD;

    C_X = B_X - BE;             //C点坐标
    C_Y = BD + CE;
}
/* AB轴在左边,BC轴在第三象限 */
if((Angle_A<90.0) & (Angle_B > (270.0-(90.0 - Angle_A)) )  )
{
    Angle3 = 360.0-Angle_B-(90.0-Angle_A);
    AD = L*Angle_Cosine(Angle_A);
    BD = L*Angle_Sinusoidal(Angle_A);

    CE = L*Angle_Sinusoidal(Angle3);
    BE = L*Angle_Cosine(Angle3);

    B_X = OA-AD;                //B点坐标
    B_Y = BD;

    C_X = B_X - BE;             //C点坐标
    C_Y = BD - CE;
}
/* AB轴在右边时,分为四种情况 */
/* AB轴在右边时,BC轴在第四象限 */
if(Angle_A>90.0)
{
```

```
    Angle_A = 180.0 - Angle_A;
    if( (Angle_A + Angle_B) < 180.0   )
    {
        Angle3 = Angle_B - (90.0 - Angle_A);
        AD = L * Angle_Cosine(Angle_A);
        BD = L * Angle_Sinusoidal(Angle_A);
        CE = L * Angle_Sinusoidal(Angle3);
        BE = L * Angle_Cosine(Angle3);
        B_X = OA + AD;         // B 点坐标
        B_Y = BD;
        C_X = B_X + CE;        // C 点坐标
        C_Y = BD - BE;
    }
    else Angle_A = 180.0 - Angle_A;
}
/* AB 轴在右边时,BC 轴在第一象限 */
if(Angle_A>90.0)
{
    Angle_A = 180.0 - Angle_A;
    if( ((Angle_A + Angle_B)>180.0) & (Angle_B<(180 + (90.0 - Angle_A)) ) )
    {
        Angle3 = Angle_B - (180.0 - Angle_A);
        AD = L * Angle_Cosine(Angle_A);
        BD = L * Angle_Sinusoidal(Angle_A);
        CE = L * Angle_Sinusoidal(Angle3);
        BE = L * Angle_Cosine(Angle3);
        B_X = OA + AD;         // B 点坐标
        B_Y = BD;
        C_X = B_X + BE;        // C 点坐标
        C_Y = BD + CE;
    }
    else Angle_A = 180.0 - Angle_A;
}
/* AB 轴在右边时,BC 轴在第二象限 */
if(Angle_A>90.0)
{
    Angle_A = 180.0 - Angle_A;
    if( (Angle_B>(180.0 + (90.0 - Angle_A))) & (Angle_B<(270 + (90.0 - Angle_A)) ) )
    {
        Angle3 = 360.0 - Angle_B - Angle_A;
        AD = L * Angle_Cosine(Angle_A);
        BD = L * Angle_Sinusoidal(Angle_A);
        CE = L * Angle_Sinusoidal(Angle3);
        BE = L * Angle_Cosine(Angle3);
```

```
            B_X = OA + AD;          // B点坐标
            B_Y = BD;

            C_X = B_X - BE;         // C点坐标
            C_Y = BD + CE;
        }
        else Angle_A = 180.0 - Angle_A;
    }
    /* AB轴在右边时,BC轴在第三象限 */
    if(Angle_A>90.0)
    {
        Angle_A = 180.0 - Angle_A;
        if( Angle_B>(270.0 + (90.0 - Angle_A)) )
        {
            Angle3 = 90.0 - (360.0 - Angle_B) - (90.0 - Angle_A);

            AD = L * Angle_Cosine(Angle_A);
            BD = L * Angle_Sinusoidal(Angle_A);

            CE = L * Angle_Sinusoidal(Angle3);
            BE = L * Angle_Cosine(Angle3);

            B_X = OA + AD;          // B点坐标
            B_Y = BD;

            C_X = B_X - BE;         // C点坐标
            C_Y = BD - CE;
        }
        else Angle_A = 180.0 - Angle_A;
    }
}
/*************************************************************************
* 名称:Location_Angle()
* 功能:由C点坐标计算转轴角度,结果存在Next_Angle1、Next_Angle2全局变量中
* 入口参数:X        C点横坐标值
*          Y        C点纵坐标值
* 出口参数:无
*************************************************************************/
void Location_Angle(uint32 X,uint32 Y)
{
    uint32 AC,CG,AG,BG;
    fp64  fp_tempa,Angle_temp;
    uint32 a,b;
    if(X> = OA)
    {
        a = (X - OA) * (X - OA);
    }
    else
    {
        a = (OA - X) * (OA - X);
```

```
    }

    b = Y * Y;
    AC = sqrt(a + b);                                  // 求解等腰三角形底边长度
    CG = AC/2;
    BG = sqrt(L * L - CG * CG);
    AG = CG;
    fp_tempa = asin((fp64)CG/(fp64)L);                 // 求反正弦
    Next_Angle2 = 2 * Radian_Angle(fp_tempa);          // 得到 B 点转轴角度

    fp_tempa = asin((fp64)Y/(fp64)AC);                 // 求反正弦
    Angle_temp = Radian_Angle(fp_tempa);               // 得到 A 点转轴角度
    if(X>= OA)
    {
        Next_Angle1 = 180.0 - (180.0 - Next_Angle2)/2.0 - Angle_temp;
    }
    else
    {
        Next_Angle1 = Angle_temp - (180.0 - Next_Angle2)/2.0;
    }
}
/**************************************************************
* 名称:AB_Left()
* 功能:控制 AB 臂向左边运转
* 入口参数:Angle          AB 臂需运动的角度
*          Order          是否等待运转完毕,0 表示不等待,1 为等待
*          frequency      电机运转频率
* 出口参数:无
**************************************************************/
void AB_Left(fp64 Angle,uint8 Order,uint32 frequency)
{
    uint32 Steup;
    Steup = Angle/Step_Angle;                          // 计算电机步数
    F_Z2();                                            // 使电机左转
    if(Order)
    {
        PWM_X_Wait2(frequency,Steup);
    }
    else
    {
        PWM_X_Y2(frequency,Steup);
    }

    NOW_Angle1 = NOW_Angle1 - Step_Angle * (fp64)Steup;
}
/**************************************************************
```

```
 * 名称:AB_Right()
 * 功能:控制AB臂向右边运转
 * 入口参数:Angle        AB臂需运动的角度
 *          Order        是否等待运转完毕,0表示不等待,1为等待
 *          frequency    电机运转频率
 * 出口参数:无
*************************************************************************/
void AB_Right(fp64 Angle,uint8 Order,uint32 frequency)
{
    uint32 Steup;
    Steup = Angle/Step_Angle;                              // 计算电机步数
    Z_Z2();
    if(Order)
    {
        PWM_X_Wait2(frequency,Steup);
    }
    else
    {
        PWM_X_Y2(frequency,Steup);
    }

    NOW_Angle1 = NOW_Angle1 + Step_Angle * (fp64)Steup;
}
/*************************************************************************
 * 名称:BC_Left()
 * 功能:控制AB臂向左边运转
 * 入口参数:Angle        BC臂需运动的角度
 *          Order        是否等待运转完毕,0表示不等待,1为等待
 *          frequency    电机运转频率
 * 出口参数:无
*************************************************************************/
void BC_Left(fp64 Angle,uint8 Order,uint32 frequency)
{
    uint32 Steup;
    Steup = Angle/Step_Angle;                              // 计算电机步数
    Z_Z1();
    if(Order)
    {
        PWM_X_Wait1(frequency,Steup);
    }
    else
    {
        PWM_X_Y1(frequency,Steup);
    }

    NOW_Angle2 = NOW_Angle2 + Step_Angle * (fp64)Steup;
}
```

```
/*****************************************************************
* 名称:BC_Right()
* 功能:控制 AB 臂向右边运转
* 入口参数:Angle       BC 臂需运动的角度
*           Order       是否等待运转完毕,0 表示不等待,1 为等待
*           frequency   电机运转频率
* 出口参数:无
*****************************************************************/
void BC_Right(fp64 Angle,uint8 Order,uint32 frequency)
{
    uint32 Steup;
    Steup = Angle/Step_Angle;                          // 计算电机步数
    F_Z1();
    if(Order)
    {
        PWM_X_Wait1(frequency,Steup);
    }
    else
    {
        PWM_X_Y1(frequency,Steup);
    }

    NOW_Angle2 = NOW_Angle2 - Step_Angle * (fp64)Steup;
}
/*****************************************************************
* 名称:From_GoTo_Point()
* 功能:从起点(NOW_X,NOW_Y) 到达任意点(X,Y)
* 入口参数:X           目标点的横坐标值
*           Y           目标点的纵坐标值
*           frequency   电机运转频率
* 出口参数:无
*****************************************************************/
void From_GoTo_Point(uint32 X,uint32 Y,uint32 frequency)
{
    fp64 Difference_Angle1,Difference_Angle2,temp_Angle1,temp_Angle2;
    /* 由坐标计算前一点的 2 个转轴角度,结果存在 Next_Angle1、Next_Angle2 中 */
    Location_Angle(X,Y);
    temp_Angle1 = Next_Angle1;
    temp_Angle2 = Next_Angle2;
    /* AB 转轴控制 */
    if(temp_Angle1>NOW_Angle1)
    {
        Difference_Angle1 = temp_Angle1 - NOW_Angle1;
        AB_Right(Difference_Angle1,1,frequency);     // 执行 AB 转动动作
    }
```

```
    else
    {
        Difference_Angle1 = NOW_Angle1 - temp_Angle1;
         AB_Left(Difference_Angle1,1,frequency);     // 执行 AB 转动动作
    }
    /* BC 转轴控制 */
    if(temp_Angle2>NOW_Angle2)
    {
        Difference_Angle2 = temp_Angle2 - NOW_Angle2;
        BC_Left(Difference_Angle2,1,frequency);      // 执行 BC 转动动作
    }
    else
    {
        Difference_Angle2 = NOW_Angle2 - temp_Angle2;
        BC_Right(Difference_Angle2,1,frequency);     // 执行 BC 转动动作
    }
    NOW_X = X;
    NOW_Y = Y;
}
/*****************************************************************
* 名称:Read_AB_ADC()
* 功能:读取 AB 臂 A 点角度传感器的 A/D 值,用于计算手动部分 AB 臂角度
* 入口参数:无
* 出口参数:返回的 A/D 值
* 说明:调用 A/D 读取数据函数
*****************************************************************/
uint32 Read_AB_ADC(void)
{
    return AD0_1_P0_28_Read();
}
/*****************************************************************
* 名称:Read_BC_ADC()
* 功能:读取 BC 臂 B 点角度传感器的 A/D 值,用于计算手动部分 BC 臂相对于 AB 臂的角度
* 入口参数:无
* 出口参数:返回的 A/D 值
* 说明:调用 A/D 读取数据函数
*****************************************************************/
uint32 Read_BC_ADC(void)
{
    return AD0_3_P0_30_Read();
}
/*****************************************************************
* 名称:Read_Angle()
* 功能:通过角度传感器数据计算手动部分转轴夹角,结果存在全局变量 Angle1,Angle2 中
* 入口参数:无
* 出口参数:无
*****************************************************************/
```

```
void Read_Angle(void)
{
    uint32 a;
    uint32 temp;

    temp = Read_AB_ADC();
    if(temp<AB_0)  Beep_mS(50);                    // 若角度超出范围,则报警
    if(temp>AB_180)  Beep_mS(50);

    if(temp < AB_180)
    {
        a = AB_180 - temp;
        Angle1 = (fp64)a * AB_Per_Angle;           // 读取 AB 臂 A 点角度传感器的 A/D 值
    }
    else Angle1 = 0.0;

    temp = Read_BC_ADC();                          // 读取 BC 臂 B 点角度传感器的 A/D 值
    if(temp>BC_MIN)  Beep_mS(50);
    if(temp<BC_MAX)  Beep_mS(50);

    temp = 1064 - temp;
    Angle2 =  (fp64)temp * BC_Per_Angle;

    Read_num ++ ;
}
/*********************************************************************
* 名称:Calculate_Angle()
* 功能:由角度传感器数据计算手动部分 A、B 点转轴角度
* 入口参数:Sensor_A     A 点角度传感器数据
*            Sensor_B     B 点角度传感器数据
* 出口参数:无
*********************************************************************/
void Calculate_Angle(uint16 Sensor_A,uint16 Sensor_B)
{
    uint32 a;
    uint32 temp;

    temp = Sensor_A;
    if(temp<AB_0)  Beep_mS(100);
    if(temp>AB_180)  Beep_mS(100);

    if(temp < AB_180)
    {
        a = AB_180 - temp;
        Angle1 = (fp64)a * AB_Per_Angle;
    }
    else Angle1 = 0.0;

    temp = Sensor_B;
    if(temp<BC_MIN)  Beep_mS(100);
    if(temp>BC_MAX)  Beep_mS(100);

    temp = 1064 - temp;
```

```
    Angle2 =  (fp64)temp * BC_Per_Angle;
}
/*********************************************************************
 * 名称:MOTO_RST()
 * 功能:电机运动到设定原点
 * 入口参数:无
 * 出口参数:无
*********************************************************************/
void MOTO_RST(void)
{
    if(NOW_Angle1>0)
    {
        Set_AB(30,1,100);
    }
    if(NOW_Angle2>0)
    {
        Set_BC(60,1,100);
    }
    /* B点复位 */
    while((IO0PIN&B_RST))                    // 等待 B 点霍尔传感器端口电平变化
    {
        AB_Left(Step_Angle,1,100);           // AB 臂向左转
    }
    AB_Right(1.8,1,100);                     // 因霍尔传感器存在误差,软件校准原点

    /* C点复位 */
    while((IO0PIN&C_RST))                    // 等待 C 点霍尔传感器端口电平变化
    {
        BC_Left(Step_Angle,1,100);           // BC 臂向左转
    }
    BC_Right(Step_Angle,1,100);              // 因霍尔传感器存在误差,软件校准原点

    NOW_X = 15;                              // 初始化起点坐标
    NOW_Y = 80;
    NOW_Angle1 = 0.0;
    NOW_Angle2 = 90.0;
}
/*********************************************************************
 * 文件名: Follow_TASK.c
 * 功能:随动系统各任务函数
*********************************************************************/
#include "config.h"
#include <math.h>

extern volatile  uint8   Time0_Flag;         // 定时器 0 中断标志

extern volatile  uint32 PWM_Num1;
extern volatile  uint32 PWM_Num2;
```

```
extern fp64 NOW_Angle1;
extern fp64 NOW_Angle2;
extern uint32 NOW_X,NOW_Y;                    // 随动C点当前时刻的坐标

extern fp64 Angle1;                           // 手动A点转轴转过的角度
extern fp64 Angle2;                           // 手动B点转轴转过的角度
extern fp64 Angle3;                           // 中间计算需要的角度,BC轴与坐标纸X轴的夹角

uint16 Angle_Follow_AB[6000];
uint16 Angle_Follow_BC[6000];

uint32 Angle_Follow_AB_Next_num = 0;          // 下一个数据存储位置
uint32 Angle_Follow_BC_Next_num = 0;

uint32 Angle_FollowNow_num = 0;               // 下一个数据执行位置
extern uint32 C_X;                            // C点横坐标
extern uint32 C_Y;                            // C点纵坐标
/*********************** 基本任务1 ********************************
* 基本任务1:任意设定AB角度
* (1)随动部分能够通过键盘或其他方式任意设定AB臂的角度(与坐标纸横线间的夹角)。精度:误差≤5°
*****************************************************************/

/*****************************************************************
* 名称:Set_AB()
* 功能:AB臂旋转到预定位置
* 入口参数:Angle        AB臂需旋转角度
* 出口参数:无
*****************************************************************/
void Set_AB(fp64 Angle,uint8 Order,uint32 frequency)
{
    fp64 Difference_Angle1;
    // AB转轴角度控制
    if(Angle>NOW_Angle1)
    {
        Difference_Angle1 = Angle - NOW_Angle1;
        if(Order)
        {
            AB_Right(Difference_Angle1,1,frequency);          // 执行AB转动动作
        }
        else
        {
            AB_Right(Difference_Angle1,0,frequency);          // 执行AB转动动作
        }
    }
    else
    {
        Difference_Angle1 = NOW_Angle1 - Angle;
        if(Order)
        {
```

```
        AB_Left(Difference_Angle1,1,frequency);                // 执行 AB 转动动作
    }
    else
    {
        AB_Left(Difference_Angle1,0,frequency);                // 执行 AB 转动动作
    }
  }
}
/**********************************************************************
* 名称:Set_BC()
* 功能:BC 臂旋转到预定位置
* 入口参数:Angle          BC 臂需旋转角度
* 出口参数:无
**********************************************************************/
void Set_BC(fp64 Angle,uint8 Order,uint32 frequency)
{
  fp64 Difference_Angle;
  // AB 转轴控制
  if(Angle>NOW_Angle2)
  {
    Difference_Angle = Angle - NOW_Angle2;
    if(Order)
    {
        BC_Left(Difference_Angle,1,frequency);                 // 执行 BC 转动动作
    }
    else
    {
        BC_Left(Difference_Angle,0,frequency);                 // 执行 BC 转动动作
    }
  }
  else
  {
    Difference_Angle = NOW_Angle2 - Angle;
    if(Order)
    {
        BC_Right(Difference_Angle,1,frequency);                // 执行 BC 转动动作
    }
    else
    {
        BC_Right(Difference_Angle,0,frequency);                // 执行 BC 转动动作
    }
  }
}
/************************    基本任务 2     ****************************
* 基本任务 2:在坐标纸上画一条长度大于 200 mm 的线,30 s 内完成
*(2)随动部分的节点 C 自动在坐标纸上画出一条长度不小于 20 cm 的任意曲线,
*    曲线不超过 30 cm×30 cm 的坐标纸的范围。限 30 s 内完成
```

```
***************************************************************************/
/***************************************************************************
* 名称:TASK2_RST()
* 功能:C 点画一条长度大于 20 cm 的曲线,第一步设置起始位置
* 入口参数:无
* 出口参数:无
***************************************************************************/
void TASK2_RST(void)
{
    Set_AB(60,1,100);
    Set_BC(180,1,100);                    // 设置起始状态
}
/***************************************************************************
* 名称:TASK2()
* 功能:C 点画一条长度大于 20 cm 的曲线,第二步运动到预定坐标
* 入口参数:无
* 出口参数:无
***************************************************************************/
void TASK2(void)
{
    TASK2_RST();

    Set_AB(120,1,100);                    // 执行画线
    Angle_Location(NOW_Angle1,NOW_Angle2); // 由角度计算手动部分 C 点坐标
    Disp_Location(C_X,C_Y);               // 液晶显示 C 点坐标
}
/*************************** 基本任务 3 ******************************
*(3)用手将手动部分的 C 点移动到某一坐标,随动部分的 C 点能移动到相应位置,
    手动部分与随动部分的 C 点坐标误差≤(1 cm,1 cm),限 60 s 内完成
***************************************************************************/
/***************************************************************************
* 名称:TASK3()
* 功能:读取手动系统的角度值,随动系统运动到相应坐标
* 入口参数:无
* 出口参数:无
***************************************************************************/
void TASK3(void)
{

    Read_Angle();                         // 读取角度传感器
    Set_AB(Angle1,0,100);                 // 设置 AB 轴角度
    Set_BC(Angle2,0,200);                 // 设置 BC 轴角度
    Angle_Location(Angle1,Angle2);        // 由角度计算随动部分 C 点坐标
    Disp_Location(C_X,C_Y);               // 液晶显示 C 点坐标
    while(PWM_Num1>0);                    // 等待电机转完
```

```
    while(PWM_Num2>0);
}
void Hight_TASK3(void)
{
    Read_Angle();                                 //读取角度传感器
    Set_AB(Angle1,0,100);                         //设置AB轴角度
    Set_BC(Angle2,0,200);                         //设置BC轴角度

    while(PWM_Num1>0);                            //等待电机转完
    while(PWM_Num2>0);
}
```

8.2 音频信号分析仪

8.2.1 赛题要求

赛题(2007年A题本科组)要求设计、制作一个可分析音频信号频率成分,并可测量正弦信号失真度的仪器。

1. 基本要求(50分)

① 输入阻抗:50 Ω。

② 输入信号电压范围(峰-峰值):0.1～5 V。

③ 输入信号包含的频率成分范围:0.2～10 kHz。

④ 频率分辨力:100 Hz(可正确测量被测信号中,频差不小于100 Hz的频率分量的功率值)。

⑤ 检测输入信号的总功率和各频率分量的频率和功率,检测出的各频率分量的功率之和不小于总功率值的95%;各频率分量功率测量的相对误差的绝对值小于10%,总功率测量的相对误差的绝对值小于5%。

⑥ 分析时间:5 s。应以5 s周期刷新分析数据,信号各频率分量应按功率大小依次存储并可回放显示,同时实时显示信号总功率和至少前两个频率分量的频率值和功率值,并设暂停键保持显示的数据。

2. 发挥部分(50分)

① 扩大输入信号动态范围,提高灵敏度。(10分)

② 输入信号包含的频率成分范围:20 Hz～10 kHz。(10分)

③ 增加频率分辨力20 Hz挡。(10分)

④ 判断输入信号的周期性,并测量其周期。(10分)

⑤ 测量被测正弦信号的失真度。(5分)

⑥ 其他。(5分)

3. 说 明

① 电源可用成品,必须自备,亦可自制。

② 设计报告正文中应包括系统总体框图、核心电路原理图、主要流程图、主要的测试结果。完整的电路原理图、重要的源程序和完整的测试结果用附件给出。

4. 评分标准

音频信号分析仪评分标准如表 8-1 所列。

表 8-1 音频信号分析仪评分标准

<table>
<tr><td rowspan="7">设计报告</td><td>项 目</td><td>主要内容</td><td>分 数</td></tr>
<tr><td>系统方案</td><td>比较与选择;方案描述</td><td>5</td></tr>
<tr><td>理论分析与计算</td><td>放大器设计;功率谱测量方法;周期性判断方法</td><td>15</td></tr>
<tr><td>电路与程序设计</td><td>电路设计;程序设计</td><td>10</td></tr>
<tr><td>测试方案与测试结果</td><td>测试方案及测试条件;测试结果完整性;测试结果分析</td><td>12</td></tr>
<tr><td>设计报告结构及规范性</td><td>摘要;设计报告正文的结构;图表的规范性</td><td>8</td></tr>
<tr><td colspan="2">总分</td><td>50</td></tr>
<tr><td>基本要求</td><td colspan="2">实际制作完成情况</td><td>50</td></tr>
<tr><td rowspan="8">发挥部分</td><td colspan="2">主要内容</td><td>分 数</td></tr>
<tr><td colspan="2">完成第①项</td><td>10</td></tr>
<tr><td colspan="2">完成第②项</td><td>10</td></tr>
<tr><td colspan="2">完成第③项</td><td>10</td></tr>
<tr><td colspan="2">完成第④项</td><td>10</td></tr>
<tr><td colspan="2">完成第⑤项</td><td>5</td></tr>
<tr><td colspan="2">其他</td><td>5</td></tr>
<tr><td colspan="2">总分</td><td>50</td></tr>
</table>

8.2.2 基于单片机和 FPGA 的设计方案

根据赛题要求需要设计制作一个可分析音频信号频率特性的频谱分析仪和可测量音频信号失真度的失真度仪。

音频信号分析仪的主要功能是能够对信号进行频谱分析,从而得到信号的功率谱、失真度和周期性等参数。对信号进行频谱分析可以采用扫频超外差法、傅里叶分析法等方法。

扫频超外差法采用扫频振荡器作为本机振荡器,输入信号与扫频本机振荡器信号进行混频,通过中频放大器电路进行放大并滤波,滤波器为窄带形式,按超外差方式选择所需频率分量,形成频谱图。扫频超外差法的扫描范围大,但对硬件电路有较高要求，而且只适合于测量稳态信号的频谱。

傅里叶分析法也称为数字分析法，即在一个特定时间周期内对信号进行采样，做傅里叶变换以获得频率和幅度等信息。实现傅里叶分析法的方案简单，但通常受到ADC转换速度以及MCU傅里叶变换算法的限制，测量频率范围较窄。本赛题只要求测量分析20 Hz～10 kHz音频信号的频率成分，故可以采用此方案。

失真度表征一个信号偏离纯正弦信号的程度。根据失真度的定义：失真度定义为信号中全部谐波分量的能量与基波能量之比的平方根值，如果负载与信号频率无关，则信号的失真度也可以定义为全部谐波电压的有效值与基波电压的有效值之比，并以百分数表示，即

$$C = \sqrt{\frac{P - P_1}{P_1}} = \sqrt{\frac{U_2^2 + U_3^2 + \cdots + U_n^2}{U_1}} \qquad (8.2.1)$$

式中，C为失真度；P为信号总功率；P_1为基波信号的功率；U_1为基波电压的有效值；$U_2 \sim U_n$为谐波电压有效值。

对输入信号采样后进行离散傅里叶变换(DFT)，求出各次谐波的幅值或者功率谱，按式(8.2.1)即可计算出信号的失真度。

1. 基于FPGA DFT算法的逻辑结构的音频信号分析仪方案

一个基于FPGA DFT算法的逻辑结构的音频信号分析仪方框图如图8-22所示。

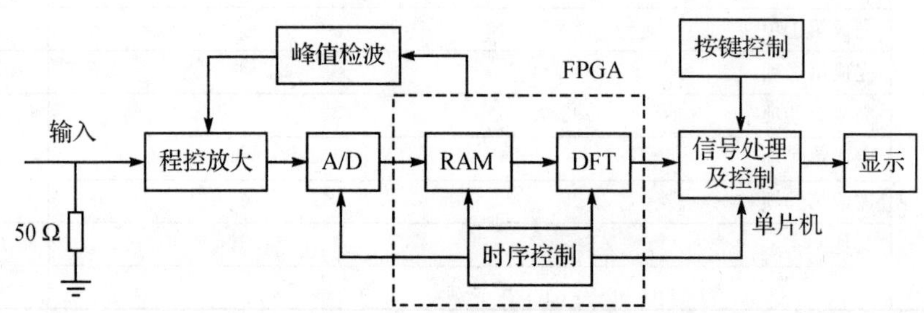

图8-22 基于FPGA的音频信号分析仪方框图

输入信号通过由运算放大器组成的前级调理电路调理到ADC的输入范围，然后进行高速A/D采样，利用在FPGA中实现DFT算法的逻辑结构进行DFT分析，得到信号频谱。对得到的幅度谱求模取平方可以得到功率谱，再将功率谱信息送到单片机中进一步分析，获得各频率成分的功率及失真度等，单片机将处理结果送入液晶显示器或示波器上显示。

完成DFT算法的逻辑结构如图8-23所示，将旋转因子W_n^{nk}的值存储在FPGA内部的一块ROM中形成查找表，避免计算旋转因子W_n^{nk}耗用大量资源及带来误差。为保证处理精度，使用40位的累加器和40位的乘法器，仅在最后求模取平方之后截取高16位结果输出，可以避免运算中间的截断误差。

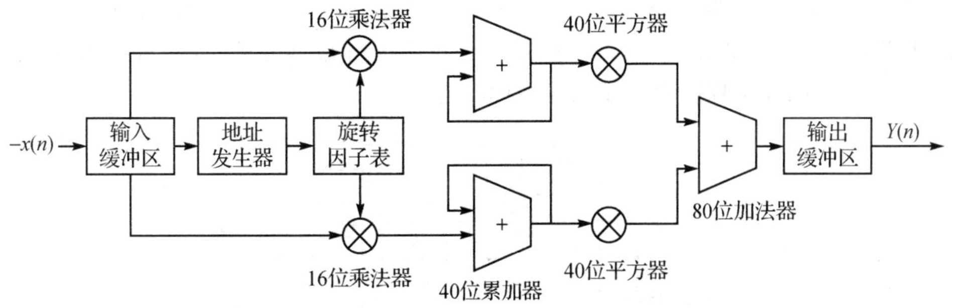

图 8-23　完成 DFT 算法的逻辑结构

2. 基于 DSP 快速傅里叶变换(FFT)的数字音频信号分析仪方案

一个基于 DSP FFT 的数字音频信号分析仪方案如图 8-24 所示，语音信号经过由运算放大器组成的前端跟随器和抗混叠低通滤波器滤波后，由高性能 A/D 完成被测信号的采样，在 FPGA 的内部实现一个 FIFO，缓存 A/D 采样的信号。单片机用来控制 LCD 液晶显示和键盘；DSP 实现数据计算和处理，进行 DFT 变换，并将处理过的数据返回到 FPGA。

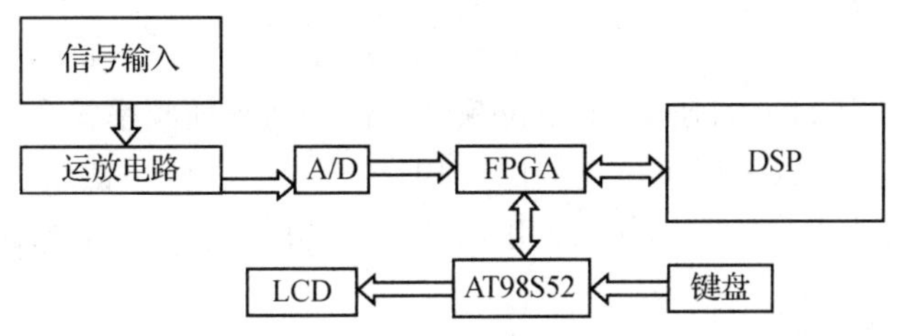

图 8-24　基于 DSP FFT 的数字音频信号分析仪方框图

这个方案采用数字方法直接由 ADC 对输入信号取样，经过 FPGA 的 FIFO 等待，送到 DSP 进行快速傅里叶变换(FFT)处理和运算。然后分析频谱，进而通过运算得到相应的频谱和功率值，由单片机控制的 LCD 来显示相应数值。为获得高分辨率，ADC 的取样率最少等于输入信号最高频率的两倍；FFT 运算时间与取样点数成对数关系，要实现高频率、高分辨率和高速运算时，需要选用与其相应的高速数字信号处理器(DSP)芯片。采用 DSP 进行信号分析，硬件电路简单，主要依靠软件运算提高分辨率，是一个比较成熟的方法。DSP 芯片可以选择 TMS320VC33。

8.2.3　基于 LPC214x ARM 微控制器的设计方案实例

1. 系统设计方案

一个采用 LPC214x ARM 微控制器的设计方案实例如图 8-25 所示，LPC214x ARM 微控制器直接控制程控放大器的放大倍数，并通过控制芯片内部的 DAC 的输出来控制可程控放大器的精度调节。程控放大器放大后的信号，经过偏置电路输入给 LPC214x 芯片内部的 ADC，通过 A/D 转换后进行数据处理，并将处理结果通过 LCD 显示器显示，参数设置与输入通过键盘电路进行。

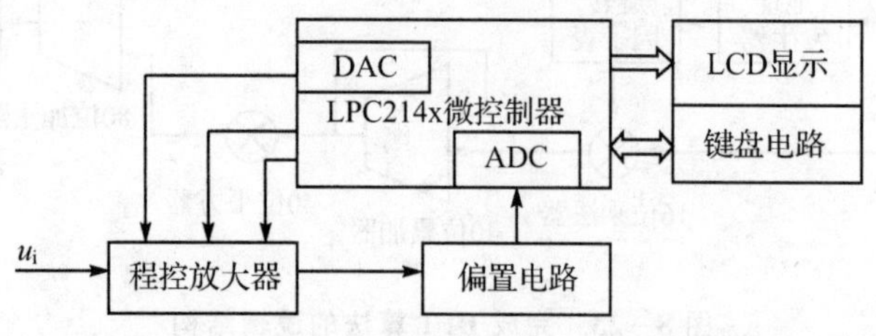

图 8-25 采用 LPC214x 的音频信号分析仪方框图

2. 理论分析与计算

本设计采用 32 位的 ARM 微控制器，关键部分是 FFT(Fast Fourier Transform)，通过使用定点运算及 FFT 算法计算功率谱。

(1) 功率谱测量方法

功率谱的定义为

$$P(u,v) = |F(u,v)| \tag{8.2.2}$$

其值表示空间频率(u,v)的强度，即单位频带内信号功率随频率的变化情况。

功率谱的计算可以采用布拉克-杜开(Blackman - Tukey)法、模拟滤波器法，以及库立-杜开(Cooley - Tukey)法，即用 FFT 计算功率谱。本系统中采用的是通过 FFT 计算功率谱。

控制器通过 ADC 对程控放大器的输出信号进行实时采样，对采集到的数据进行 FFT 变换，得到的数据即是该信号的各次谐波的幅值，根据公式

$$P = U^2/R \tag{8.2.3}$$

即可算出信号在各次谐波处的功率，本系统中设计的输入电阻是 50 Ω。

(2) 周期性判断方法

对于一个周期函数有

$$g(t) = g(t+nT) \tag{8.2.4}$$

式(8.2.3)对于任意整数 n 都成立，T 即为函数的周期。

函数重复的最小周期称为基波周期 T_0。基波频率 f_0 就是每秒循环的次数，即对于一个周期信号来说，它的傅里叶变换得到的谱线是稳定的，且任意两次的谱线是相似的。

在本系统中，通过 ADC 对信号进行采样，将采样的数据进行两次快速傅里叶分析，如果发现两次采样的数据傅里叶分析的结果是稳定且相似的，则可认为该信号是周期性信号，它的周期就是该信号的基波周期 T_0。

3. 主要单元电路设计

(1) 程控放大器电路

集成的程控运算放大器是程控放大器的首选，如 PGA101、PGA103、PGA202/203 和 PGA206 等芯片。

在本设计中程控放大器电路选择数字可编程增益的仪表放大器 PGA202，偏置电流为 50 pA，非线性为 0.012%，CMRR 为 80 dB，采用 DIP－14 封装，电源电压为 ±6～±15 V，可控增益为 1，10，100，1 000。

说明：PGA203 除可控增益为 1，2，4，8 外，其余与 PGA202 完全相同。

数字可编程增益的仪表放大器 PGA202 的内部结构图如图 8－26 所示，改变引脚端 A_0 和 A_1 的输入电平，可以改变放大器输入级的增益，如表 8－2 所列。改变引脚端 4 的 V_{REF} 也可以调节输出级的增益，调节电路如图 8－27 所示，电阻 R1 和 R2 与输出级的增益关系如表 8－3 所列。改变引脚端 Filter A 和 Filter B 的外部连接电容器的容量，可以改变放大器的输出截止频率，连接电路如图 8－28 所示，电容器的容量与截止频率的关系如表 8－4 所列。

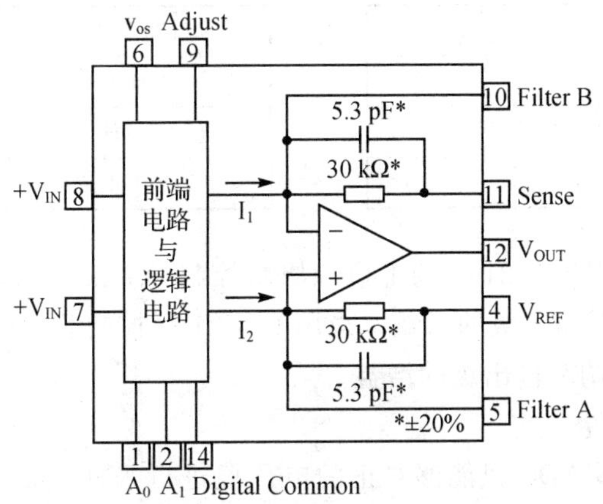

图 8－26　PGA202 的内部结构图

表 8－2　引脚端 A_0 和 A_1 的输入电平与放大器的增益关系

输入电平		PGA202		PGA203	
A_1	A_0	增　益	误　差	增　益	误　差
0	0	1	0.05%	1	0.05%
0	1	10	0.05%	2	0.05%
1	0	100	0.05%	4	0.05%
1	1	1 000	0.05%	8	0.05%

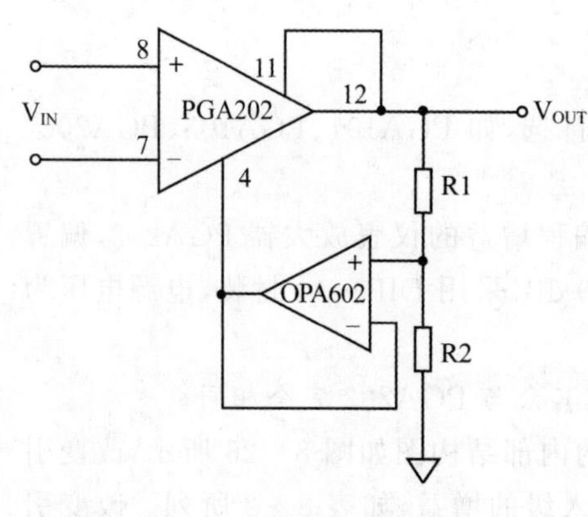

图 8-27 输出级的增益调节电路

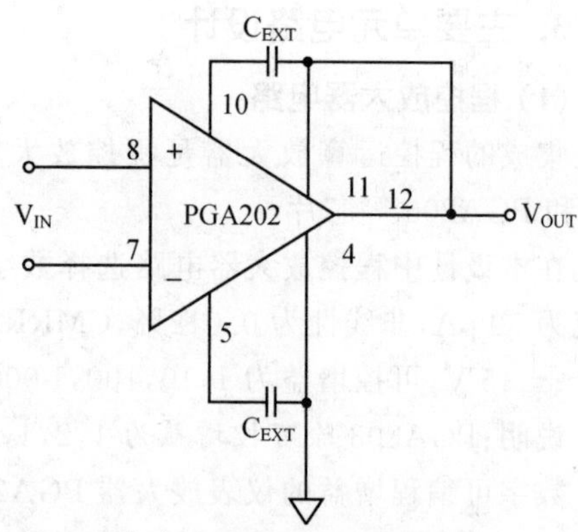

图 8-28 引脚端 Filter A 和 Filter B 的外部连接电容器

表 8-3 输出级的增益与 R1 和 R2 的关系

输出增益	R1/kΩ	R2/kΩ
2	5	5
5	2	8
10	1	9

表 8-4 外接滤波器电容器的容量与截止频率的关系

截止频率	C_{EXT}
1 MHz	不连接
100 kHz	47 pF
10 kHz	525 pF

在本设计中，采用 LPC214x 的 I/O 引脚端控制 PGA202 引脚端 A_0 和 A_1 的输入电平，从而改变放大器输入级的增益。采用 LPC214x 控制芯片内部 DAC 的输出，改变引脚端 4 的 V_{REF}，调节输出级的增益。

(2) 电压偏置电路

LPC214x 的内部 ADC 只能够对正的电压值进行采样，所以需要一个电压偏置电路，用来改变信号电压的偏置。为了保证偏置电压的精度，使用基准芯片 MAX6029 提供 2.5 V 基准电压，将基准电压与 PGA202 放大后的信号进行求和运算(求和运算电路可以采用运算放大器芯片 μA741，如图 8-29 所示)，从而改变信号电压(PGA202 的输出)的偏置，使信号电压能够以正值输入到 LPC214x 的内部 ADC，进行 A/D 转换。

MAX6029 是微功耗、低压差带隙电压基准，具有超低工作电流和低漂移特性，提供 6 种输出电压：2.048 V、2.5 V、3 V、3.3 V、4.096 V 和 5 V。MAX6029 采用 5 引脚的 SOT23 或 8 引脚的 SO 封装，温度范围为 −40～+85 ℃。MAX6029 的典型应用电路如图 8-30 所示。

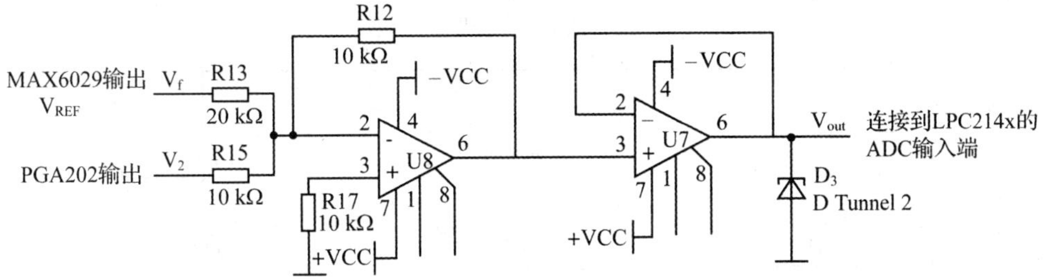

图 8-29　采用 μA741 组成的求和运算电路

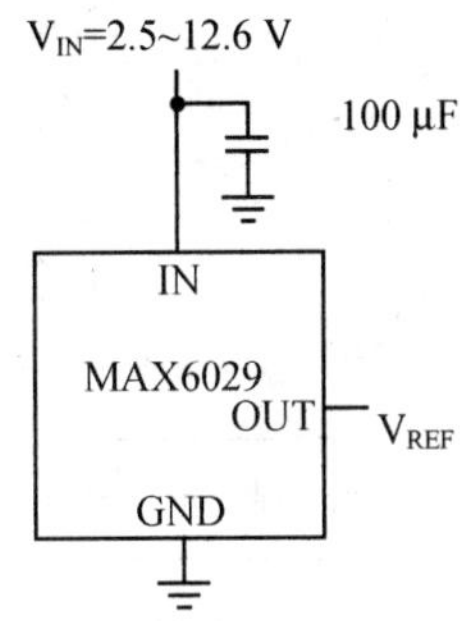

图 8-30　MAX6029 的典型应用电路

4. 系统程序设计

系统程序设计的核心是 FFT 算法。哈尔滨工业大学的杨场、周进和羊绛军采用 LPC2138 制作的作品(2007 年全国一等奖作品),使用定点运算及 FFT 算法,每秒可完成 115 次 1 024 点 32 位精度的 FFT 运算,可满足信号带宽20 Hz～10 kHz、频率分辨率 20 Hz 的要求。

该设计采用时间抽取基 4 定点 FFT 算法。使用基 4 算法,在进行碟形单元运算时可以充分利用 ARM7 内核的寄存器和乘法累加器。由于 ARM7 内核不含浮点运算单元,软件模拟浮点运算效率较低,而 ARM7 是 32 位内核,使用定点数计算,既可以减少运算量又能保证计算精度。

设计中 FFT 的核心算法(碟形单元、位翻转等)采用汇编语言编写,以获得最高的效率。由于 FFT 算法是对复数进行的,在处理输入数据时,将 1 024 点纯实数采样数据转换成 512 点复数再进行运算,可使运算量减少近一半。

通过以上优化,经测试,在 60 MHz 的 ARM7TDMI 内核上进行 1 000 次 1 024 点 FFT 运算只需要 8.7 s,比优化前的浮点 FFT 算法快了 126 倍。

测量过程软件流程图如图 8-31 所示。

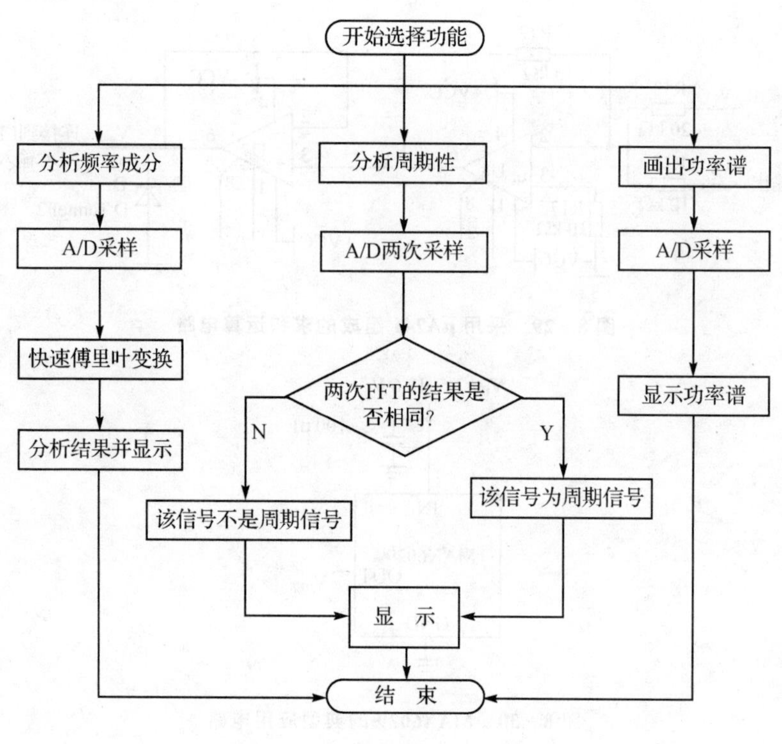

图 8-31 测量过程软件流程图

8.3 正弦波信号发生器

8.3.1 AD9850/51 DDS 模块简介

采用 AD 公司的高性能 DDS 芯片 AD9850/51 制成的 DDS 模块，可产生 0～50 MHz/70 MHz、调节分辨率为 0.04 Hz 的正弦信号和 CMOS 电平波信号，其中正弦信号幅度的峰-峰值为 1.5～350 mV，输出阻抗 200 Ω，可用作高精度、高稳定度的可编程信号源，产生时序电路的标准时钟，通信电路中的本振信号及 PLL 锁相环的参考时钟。模块使用 3.3 V 或 5 V 供电，功耗为 650 mW，当使用 5 V 供电时，在输出频率为 50/70 MHz 时的工作电流为 113 mA，模块时钟频率高达 125/180 MHz。

AD9850/51 DDS 模块电路如图 8-32 所示，支持串行或并行总线模式，AD9850/51 芯片内部包含有 DAC 及高速比较器，该 DDS 模块输出采用 2 阶 LC 椭圆低通滤波器，能有效抑制 DDS 的输出杂散，使输出波形非常稳定，同时还使用了钽电解电容对电源进行退耦，有效消除了模块电源回路高频成分对其他外设的影响。

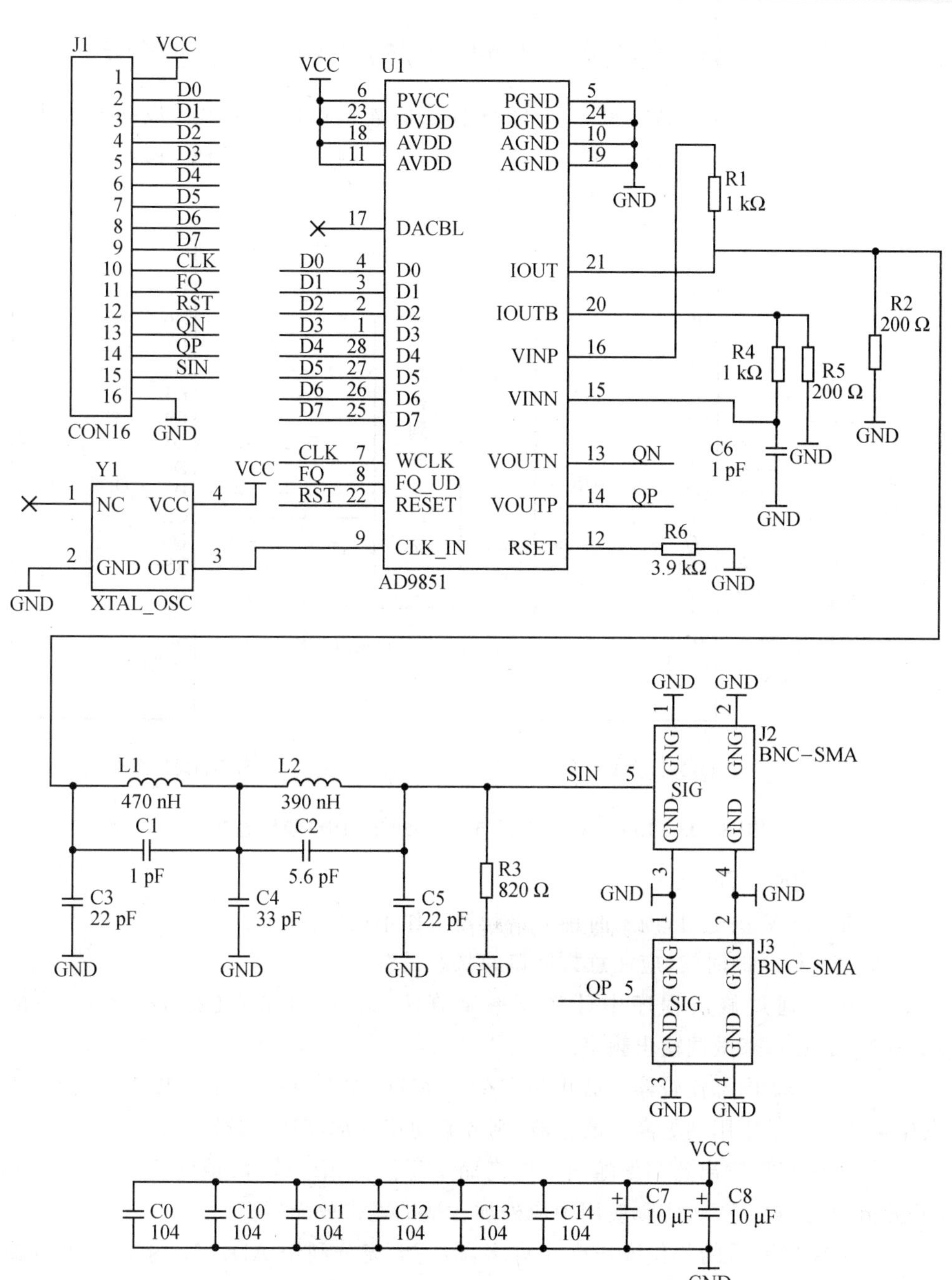

图 8-32 AD9850/51 DDS 模块电路

8.3.2 LPC214x 开发板与 AD9850/51 DDS 模块的连接

LPC214x 开发板与 AD9850/51 DDS 模块的连接可以采用串行或并行总线模式，如图 8-33 所示。本设计中，为节约开发板控制器引脚，LPC214x 开发板与 AD9850/51 DDS 模块的连接采用串行模式。

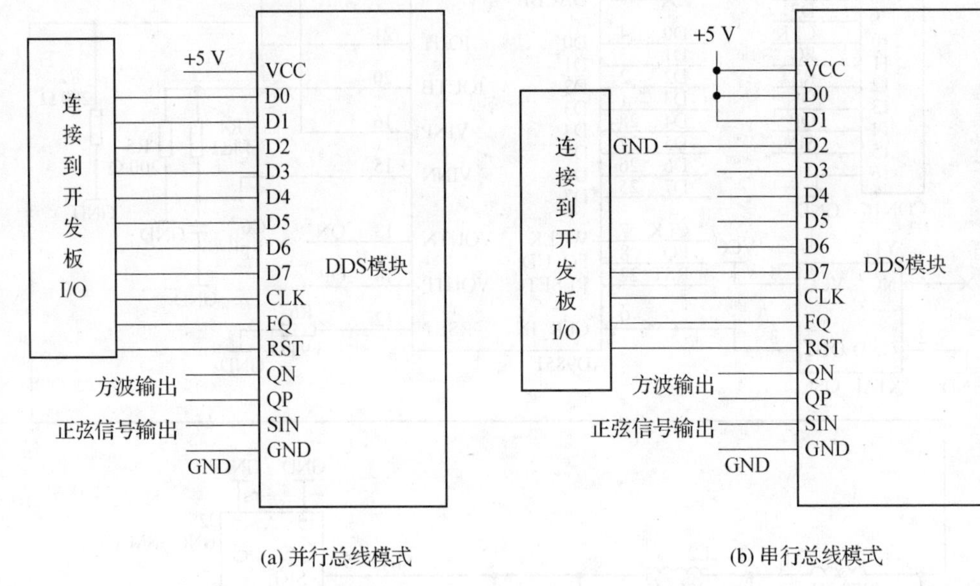

图 8-33 LPC214x 开发板与 AD9850/51 DDS 模块的连接

模块使用时应注意：

① 在 3.3 V 供电时，内部时钟 6 倍频器工作不稳定。

② 串行模式的时候，应注意数据口的状态设置。

③ 可以通过修改程序中计算频率字算法 dds = 23.860 929 422 × f_{req} 中的 23.860 929 422，来微调输出频率。

④ AD9851 内部比较器的截止频率在 30 MHz 左右，在 10 MHz 以上，输出波形发生畸变。如果要用到更高频的方波，则务必使用外部高速比较器。

⑤ DDS 正弦输出的幅度随频率增高而下降。低频端峰-峰值约为 1 V，高频端峰-峰值约为 200 mV。实际应用中应外加合适的宽带放大器。

⑥ 关掉倍频后的相位噪声会小许多，但不能充分利用 AD9851 的 180 MHz 高时钟频率特性，请用户自行斟酌。

8.3.3 AD9850/51 DDS 模块的编程示例

本示例使用 LPC2148 以串行通信方式控制 DDS 模块输出，设定默认输出频率 10 MHz，且可通过按键设定频率。启动 ADS 1.2，使用 ARM Executable Image for lpc2148 工程模板创建一个工程。开发板使用中，用到了 ZLG7290 控制数码管和检测按键，故需要将 I^2C 软件包文件 I2CINT.c、I2CINT.h 包含进工程，并且在 config.h 中将 I2CINT.h 包括进去，且还需要 Startup.s 修改系统模式，堆栈设置为 0x5f，开启中断。I^2C 软件包可以在周立功公司网站(http://www.zlgmcu.com)上下载。软件流程图如图 8-34 所示，具体程序如程序清单 8.1 所示。

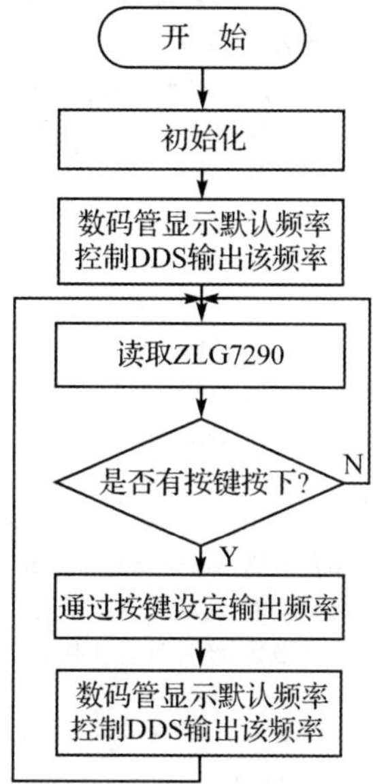

图 8-34　示例程序流程图

程序清单 8.1　DDS 模块串行模式输出(频率可调)

```
#include "config.h"

#define   clk0     (IO0CLR = clk)
#define   clk1     (IO0SET = clk)

#define   load0    (IO0CLR = load)
#define   load1    (IO0SET = load)

#define   dat0     (IO0CLR = Dat)
#define   dat1     (IO0SET = Dat)

/*********************************************************************
* 名称:mDelaymS()
* 功能:延时 ms 毫秒
* 入口参数:ms   延时参数,值越大,延时越久
* 出口参数:无
*********************************************************************/
void mDelaymS( uint32 ms )
{
    uint32     i;
    while ( ms -- ) for ( i = 25000; i != 0; i -- );
}
/*********************************************************************
* 函数名称:init_dds_Serial()
* 函数功能:初始化 DDS 模块
* 入口参数:无
* 出口参数:无
*********************************************************************/
void init_dds_Serial(void)
{
clk0;
load0;
clk1;
clk0;
load1;
load0;
}
```

```
/*************************************************************************
 * 函数名称:write_freq_Serial()
 * 函数功能:DDS 模块写入频率
 * 入口参数:需要输出的频率
 * 出口参数:无
*************************************************************************/
void write_freq_Serial(unsigned long freq)
{
unsigned long dds;
dds = 34.35943 * freq;                                    // 频率字计算,可进行微调
write_dds_Serial(dds);
}
/*************************************************************************
 * 函数名称:write_dds_Serial()
 * 函数功能:初始化 DDS 模块
 * 入口参数:经处理后的频率字
 * 出口参数:无
*************************************************************************/
void write_dds_Serial(unsigned long dds)
{
    uchar i;
    load0;
    clk0;
    for(i = 0;i<40;i++)
    {
        clk0;
        if(dds & 0x00000001)
        dat1;
        else dat0;
        clk1;
        dds = dds>>1;
    }
    load1;
    clk0;
    load0;
}
/*************************************************************************
 * 函数名称:main()
 * 函数功能:控制 DDS 模块输出设定频率
 * 入口参数:无
 * 出口参数:无
*************************************************************************/
int main(void)
{
    uint8   key;
    uint32 AD9850_F = 10000000;                           // 设定默认频率为 1 MHz
    PINSEL0 = 0;
```

```
PINSEL1 = 0;
PINSEL2 &= ~(0x00000006);
IO0DIR = 0;
AD9850_Serial_init();                    // 串行模式初始化
I2C_Init(40000);
mDelaymS(200);
Display(AD9850_F);                       // 显示默认频率
init_dds_Serial();                       // 串行模式初始化 DDS
write_freq_Serial(AD9850_F);             // 输出频率
while(1)
{
    key = 0;
    key = Read_Key();                    // 读取按键
    if(key! = 0)
    {
        switch(key)
        {
            case K1: AD9850_F += 100000;     // 若按下第 1 个按键,频率加 100 000
                     break;
            case K2: AD9850_F += 10000;      // 若按下第 2 个按键,频率加 10 000
                     break;
            case K3: AD9850_F += 1000;       // 若按下第 3 个按键,频率加 1 000
                     break;
            case K4: AD9850_F += 100;        // 若按下第 4 个按键,频率加 100
                     break;
            case K5: AD9850_F += 10;         // 若按下第 5 个按键,频率加 10
                     break;
            case K6: AD9850_F += 1;          // 若按下第 6 个按键,频率加 1
                     break;
            case K7: AD9850_F -= 100000;     // 若按下第 7 个按键,频率减 100 000
                     break;
            case K8: AD9850_F -= 10000;      // 若按下第 8 个按键,频率减 10 000
                     break;
            case K9: AD9850_F -= 1000;       // 若按下第 9 个按键,频率减 1 000
                     break;
            case K10:AD9850_F -= 100;        // 若按下第 10 个按键,频率减 100
                     break;
            case K11:AD9850_F -= 10;         // 若按下第 11 个按键,频率减 10
                     break;
            case K12:AD9850_F -= 1;          // 若按下第 12 个按键,频率减 1
                     break;
        }
        Display(AD9850_F);                   // 显示设定频率
        init_dds_Serial();                   // 串行模式初始化 DDS
        write_freq_Serial(AD9850_F);         // 输出设定频率
        mDelaymS(20);
```

```
        }
    }
    return 0;
}
```

8.4 基于ARM微控制器的声音导引系统

8.4.1 设计要求

设计并制作一声音导引系统,示意图如图8-35所示。

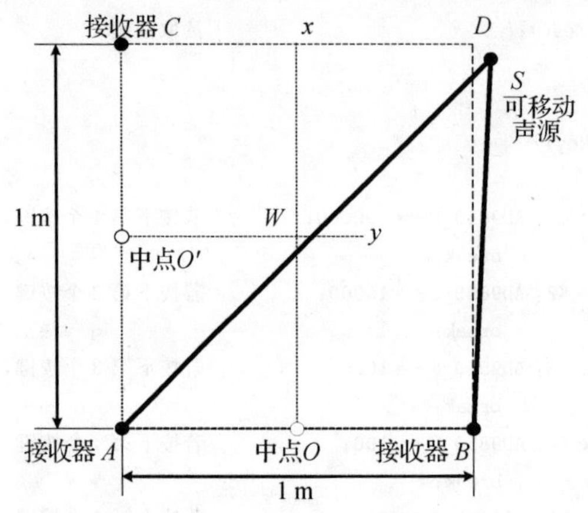

图8-35 系统示意图

图8-35中,AB与AC垂直,Ox是AB的中垂线,$O'y$是AC的中垂线,W是Ox和$O'y$的交点。

声音导引系统有一个可移动声源S,三个声音接收器A、B和C,声音接收器之间可以有线连接。声音接收器能利用可移动声源和接收器之间的不同距离,产生一个可移动声源离Ox线(或$O'y$线)的误差信号,并用无线方式将此误差信号传输至可移动声源,引导其运动。

可移动声源运动的起始点必须在Ox线右侧,位置可以任意指定。

1. 基本要求

① 制作可移动的声源。可移动声源产生的信号为周期性音频脉冲信号,如图8-36所示,声音信号频率不限,脉冲周期不限。

② 可移动声源发出声音后开始运动,到达Ox线并停止,这段运动时间为响应时间,测量响应时间,用下列公式计算出响应的平均速度,要求平均速度大于5 cm/s。

$$平均速度=\frac{可移动声源的起始位置到Ox线的垂直距}{响应时间}$$

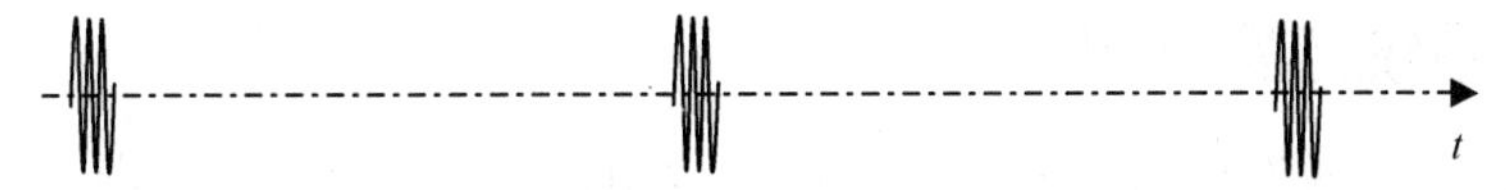

图 8 - 36　信号波形示意图

③ 可移动声源停止后的位置与 Ox 线之间的距离为定位误差，定位误差小于 3 cm。

④ 可移动声源在运动过程中任意时刻超过 Ox 线左侧的距离小于 5 cm。

⑤ 可移动声源到达 Ox 线后，必须有明显的光和声指示。

⑥ 功耗低，性价比高。

2. 发挥部分

① 将可移动声源转向 180°(可手动调整发声器件方向)，能够重复基本要求。

② 平均速度大于 10 cm/s。

③ 定位误差小于 1 cm。

④ 可移动声源在运动过程中任意时刻超过 Ox 线左侧距离小于 2 cm。

⑤ 在完成基本要求部分移动到 Ox 线上后，可移动声源在原地停止 5～10 s，然后利用接收器 A 和 C，使可移动声源运动到 W 点，到达 W 点以后，必须有明显的光和声指示并停止，此时声源距离 W 的直线距离小于 1 cm。整个运动过程的平均速度大于 10 cm/s。

$$\text{平均速度}=\frac{\text{可移动声源在 } Ox \text{ 线上重新启动位置到移动停止点的直线距离}}{\text{两次运动时间}}$$

⑥ 其他。

3. 说　明

① 本题必须采用组委会提供的电机控制 ASSP 芯片(型号 MMC - 1)实现可移动声源的运动。

② 在可移动声源两侧必须有明显的定位标志线，标志线宽度 0.3 cm 且垂直于地面。

③ 误差信号传输采用的无线方式、频率不限。

④ 可移动声源的平台形式不限。

⑤ 可移动声源开始运行的方向应和 Ox 线保持垂直。

⑥ 不得依靠其他非声音导航方式。

⑦ 移动过程中不得人为对系统施加影响。

⑧ 接收器和声源之间不得使用有线连接。

4. 评分标准

评分标准如表 8 - 5 所列。

8.4.2　系统方案设计

经分析,将系统分为两部分,一部分是主控制台,另一部分是以小车为载体的可移动声源。两部分所包含的模块如图 8-37 所示。可移动声源的音频信号发生模块用于产生一定频率的音频信号,声光提示模块用于定位后的提示;主控制台的 3 个接收器模块用于接收音频信号,人机界面模块用于输入指示信号和显示系统数据,语音模块用于播报提示信息,无线收发模块用于两部分之间数据通信。

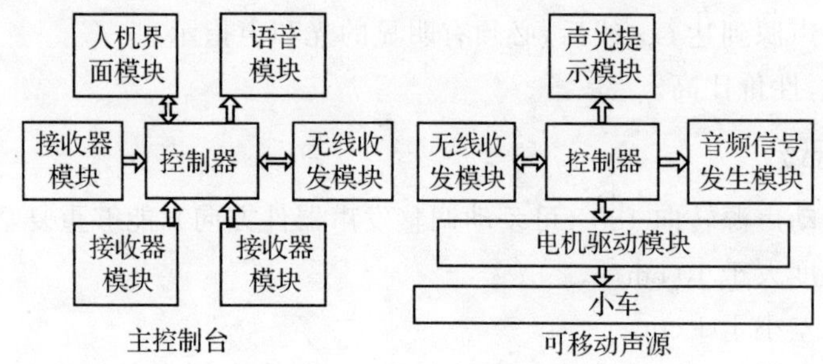

图 8-37　系统结构示意图

表 8-5　评分标准

	项　目	主要内容	分　数
设计报告	系统方案	整体方案比较 控制方案	7
	设计与论证	设计、计算 误差信号产生 控制理论简单计算	12
	电路设计	系统组成 各种电路图	3
	测试结果	测试数据完整性 测试结果分析	3
	设计报告	摘要 正文结构完整性 图表的规范性	5
	总分		30

续表 8－5

基本要求	基本要求实际完成情况	50
发挥部分	完成第①项	5
	完成第②项	10
	完成第③项	10
	完成第④项	10
	完成第⑤项	10
	完成第⑥项	5
	总分	50

8.4.3　系统主要单元的选择与论证

1. 控制器方案的选择与论证

方案一：采用 51 系列单片机。如 STC89C52，共 40 个引脚，其中 32 个 I/O 口，功能相对单一，时钟频率为 11.059 2 MHz，芯片典型工作电压为 5 V，典型工作电流为 4～7 mA。

方案二：采用 ARM7TDMI－S 微控制器 LPC2148。32 位的 LPC2148 工作电压为3.3 V，典型工作电流为 53 mA，工作频率可高达 60 MHz，具有 45 个可承受 5 V 电压的 I/O 口，内置宽范围的串行通信接口，采用 3 级流水线工作模式，具有掉电和空闲两种低功耗工作模式。

方案选择：LPC2148 处理速度比 51 系列单片机快，功能更加强大，故系统主控制台和可移动声源两部分的控制器都选择方案二。

2. 音频信号发生模块方案选择与论证

方案一：控制器输出 20 Hz～20 kHz 的 PWM 信号，经功率放大后送扬声器输出。

方案二：直接使用大功率蜂鸣器。

方案选择：方案二直接简单可行，只需控制蜂鸣器的导通与截止便可产生周期性的音频信号，故选择方案二，使用 DC 24 V 供电的电子蜂鸣器。

3. 音频信号接收模块方案选择与论证

方案一：直接购买声音接收器模块。该模块只在接收到声音信号后输出高电平。

方案二：使用驻极体电容传声器，即咪头，接收音频信号，使用 LM358 放大，LM567 音调解码器对 3 kHz 的接收信号进行处理，只在接收到所需音频信号时输出低电平。

方案选择：方案二电路简单，且可通过调节锁相环的频率来匹配系统所产生的音频信号，故选择方案二。

4. 车体方案选择与论证

方案一：购买小车车轮及配件，装上电机和驱动，搭建小车模型。

方案二：购买小车车体成品。只需装上电机驱动，便可运行。

方案选择：考虑到搭建模型耗时太长，且搭建不好会严重影响精度，故选择方案二。最终购买的车体是北京亿学通电子生产的大功率履带式直流电机驱动车体，其具有动力性能强，底盘稳定性高，转弯灵活的特点，最快速度可达 30 cm/s，满足题目要求。

5. 直流电机驱动模块方案选择与论证

方案一：使用 N 沟道增强型场效应管构建 H 桥式电路驱动直流电机。电路简单，但要注意对场效应管死区时间的控制，不然很容易烧坏场效应管。

方案二：选用 L298N 驱动直流电机。L298N 属于 H 桥集成电路，输出电流为 2 A，最高工作电压 50 V，驱动能力强。

方案选择：方案二电路简单、稳定度高，故选择方案二。配合 ASSP 芯片（型号 MMC－1）控制电机的示意图如图 8－38 所示。

图 8－38　直流电机驱动控制电路示意图

6. 无线收发模块方案选择与论证

方案一：使用 nRF24L01 无线收发模块。其工作频段为 2.4 GHz，最高工作速率为 2 Mbit/s，最远传输距离为 100 m，低功耗 1.9～3.6 V 工作，待机模式下为 22 μA，可通过 SPI 接口方便地与控制器相连。

方案二：使用 nRF905 无线收发模块。其工作频段为 433 MHz，最高工作速率为50 kbit/s，最远传输距离可达 1 000 m，稳定性非常好，具有低功耗模式，工作电压为 1.9～3.6 V，电流消耗很低，待机模式下仅为 2.5 μA，通过 SPI 接口可方便地编程配置其工作模式。

方案选择：nRF24L01 和 nRF905 都是完全集成的单片无线收发器芯片，但后者传输数据时抗干扰能力更强，通信更加稳定，且传输距离更远，故选择方案二。

7. 人机界面模块的方案选择与论证

方案一：使用数码管进行显示，按键用于信息输入。可采用周立功公司生产的 ZLG7290 芯片来配合控制器对数码管和按键进行控制，该芯片具有 I^2C 串行接口，只需占用控制器的 3 个引脚，便可方便地控制数码管显示和检测按键。

方案二：使用迪文触摸屏人机交互模组(HMI)。型号为 DMT32240S035_01WT,工作电压为 DC 5～28 V,功耗为 1 W,使用异步、全双工串口与控制器通信,不需额外电路。

方案选择：数码管不能显示汉字,而迪文人机交互模组(HMI)带触摸功能,可省去按键,使画面更生动有趣。故人机交互界面模块选择方案二。

8.4.4　系统组成

经分析,决定系统最终方案如下,方框图如图 8－39 所示。

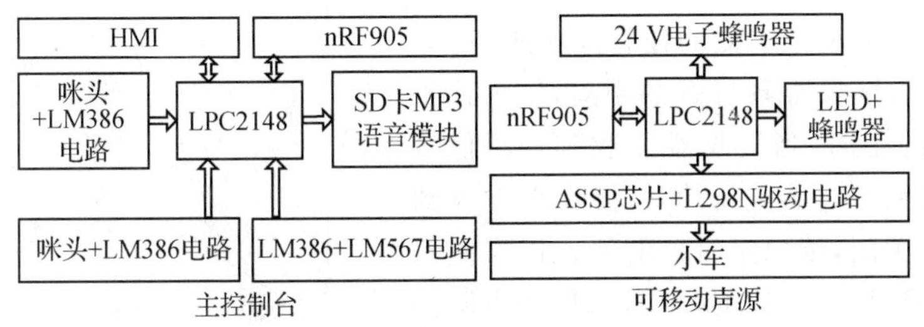

图 8－39　系统组成

① 主控制台和可移动声源控制器选择：LPC2148。

② 音频信号产生模块：24 V 大功率电子蜂鸣器。

③ 音频信号接收器模块：咪头＋LM386＋LM567 电路。

④ 直流电机驱动模块：L298N。

⑤ 人机界面模块：迪文触摸屏人机交互模组(HMI),型号为 DMT32240S035_01WT。

⑥ 无线收发模块：nRF905 无线收发模块。

此外,声光提示模块使用高亮度的 LED 和蜂鸣器,语音模块使用 SD 卡 MP3 模块。

8.4.5　理论分析及计算

1. 误差信号产生方法分析

可移动声源 S 通过处理误差信号来控制电机的运作,故误差信号的产生在一定程度上决定了系统性能,经认真分析,误差信号的产生有以下两种方法：

方法一：主控制台根据接收器信息计算出 S 当前时刻距定位点的距离作为误差信号。

方法二：主控制台将音频信号到达不同接收器的时间差值作为误差信号,需定位 OX 线上时,以接收器 A 和接收器 B 接收到音频信号的到达时间差值作为误差信号;需定位 W 点时,以接收器 A 和接收器 C 接收到音频信号的到达时间差值作为误差信号。

对于本系统,一个脉冲可使小车前进 1.153 cm,若使用直接测距的方法,则不能达到系统要求;而第二种方法只需用到控制器的一个定时器开启和关闭,便可获取误差信号,因控制器时钟频率为 12 MHz,定位精度可达到 0.287 5 cm,故选择方法二,以到达时间差作为误差信号来进行定位。

2. 系统定位的分析与计算

(1) OX 线定位的分析与计算

系统定位示意图如图 8-40 所示,将可移动声源放置在 OX 线右侧的任意位置,因 $L_2 > L_1$,其中 L_1 为接收器 B 与 S 的距离,L_2 为接收器 A 与 S 的距离,当接收器 B 收到音频信号时控制器开启定时器重新计时,接收器 A 收到音频信号时关闭,读取定时器的值 t,即误差信号。不考虑环境对声速的影响,可移动声源的控制器根据误差信号可算出 ΔL(单位 m):

$$\Delta L = L_2 - L_1 = t \times 345 \quad (8.4.1)$$

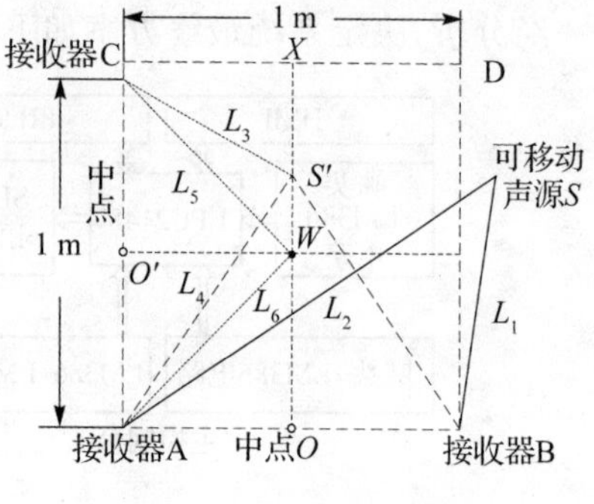

图 8-40 系统定位示意图

且依据此误差信号 t 来控制可移动声源的运行速度,当 t 大于某个值时,控制小车快速行驶,随着 t 值的减小,对应减小可移动声源的速度,直到最后停止,这个门限值需要经过调试来决定。而判断可移动声源是否停止的 t 值范围也需要通过调试决定,理想状态下,可移动声源在收到的 t 为 0 时停止,此时刚好处在 OX 线上。在此过程中,若接收器 A 比接收器 B 先检测到音频信号,说明小车已驶过 OX 线,则定时器在接收器 A 收到信号时开启,在接收器 B 收到音频信号关闭,再读取误差信号 t_1,根据 t_1 值控制小车后退。

(2) W 点定位的分析与计算

与 OX 线的定位分析方法相同,接收器 C 或 A 中任一个收到音频信号后开启定时器,另一个接收器收到音频信号后关闭,读取误差信号 t_2,控制小车运行,当从 S' 运动到 W 点时,$L_5 = L_6$,小车停止,$t_2 = 0$,在实际情况下要通过调试给出让车停止时的 t_2 范围。

8.4.6 系统电路设计

1. 主控制台的主控板电路设计

主控制台的主控制器电路、无线收发模块接口、MP3 语音模块全部设计在主控板电路中,电路原理图请参考 8.1 节基于 ARM 微控制器的随动控制系统的有关部分。使用时,将各模块插入主控板对应的接口便可,避免了使用杜邦线连接电路时可能导致的通信不稳定等问题。迪文触摸屏控制器与无线收发模块使用模拟 SPI 通信;与语音模块使用串口通信,可滤除杂波,提高语音音质,语音模块的输出端接 TEA2025B 双声道功放电路,此功放电路功耗低、声道分离度高。

2. 音频信号发生电路设计

音频信号发生模块电路原理图如图 8-41 所示,CN 接控制器 I/O 口,当需要发出信号时,控制该 I/O 口输出高电平,使三极管 Q_1 导通便可,输出低电平则关闭蜂鸣器。

3. 音频信号接收电路设计

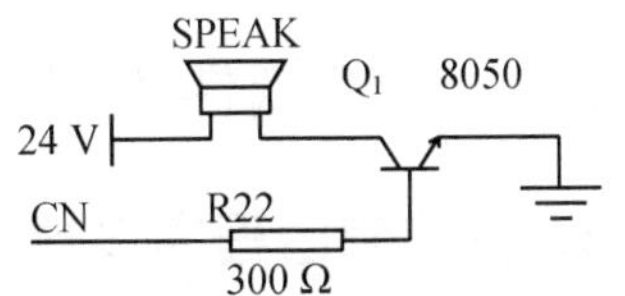

图 8-41　音频信号发生模块电路

音频信号接收器电路主要由采样部分、放大部分和解码部分组成。驻极体电容话筒采样声音,经 LM358 放大后,3 kHz 的音频信号传给 LM567 音调解码,OUT 脚输出的方波信号接控制器中断脚,电路原理图如图 8-42 所示。

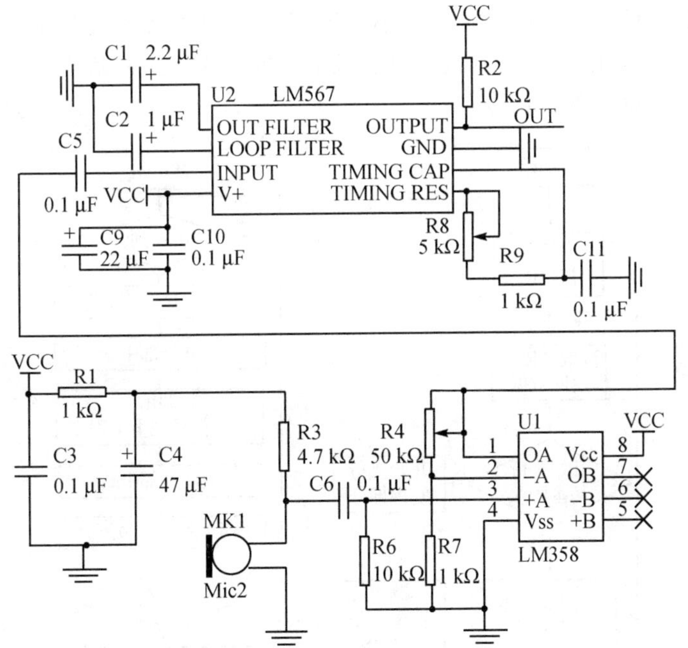

图 8-42　音频信号接收电路

4. 直流电机驱动模块电路设计

使用 L289N 作为电机驱动芯片,电路请参考第 4 章的电机控制。A、B 接 ASSP 芯片通道 1 方向和 PWM 输出引脚,C、D 接通道 2 方向和 PWM 输出引脚,对 ASSP 写入数据可控制电机正反转、停止、调速,故控制器把电机控制命令写入 ASSP,ASSP输出控制信号给 L298N 增加驱动能力后便可驱动电机按设定模式运转。为保护 ASSP 芯片，在其与 L298N 中间接光耦电路，为保护 L298N 芯片，使用 8 个快速恢复二极管泄放绕组电流。

除了完成系统基本要求和发挥部分外，还添加了以下 3 个功能，使系统更加完善。

① 使用迪文触摸屏进行测试项选择，显示结果和运行时间，使系统更加直观。

② 使用语音模块，实现语音播报，使系统更加人性化。

③ 使用舵机自动完成可移动声源的发声器件扬声器转向 180°。

8.4.7　系统软件设计

1. 主控制台程序设计

主控制台的控制器通过读取触摸屏返回的数据来判断被选择的命令，当选择执行基本要求时，调用基本要求子程序，选择执行发挥部分时，调用发挥部分子程序，而这两个子程序又是通过调用误差信号产生子程序来完成的，控制总流程图和这几个子程序流程图如图 8－43 所示。为减小功耗，在进入基本要求或发挥部分子程序后，控制器被设置为空闲模式，直到接收到音频信号的接收器给控制器一个中断信号，由此来唤醒控制器继续执行指令。

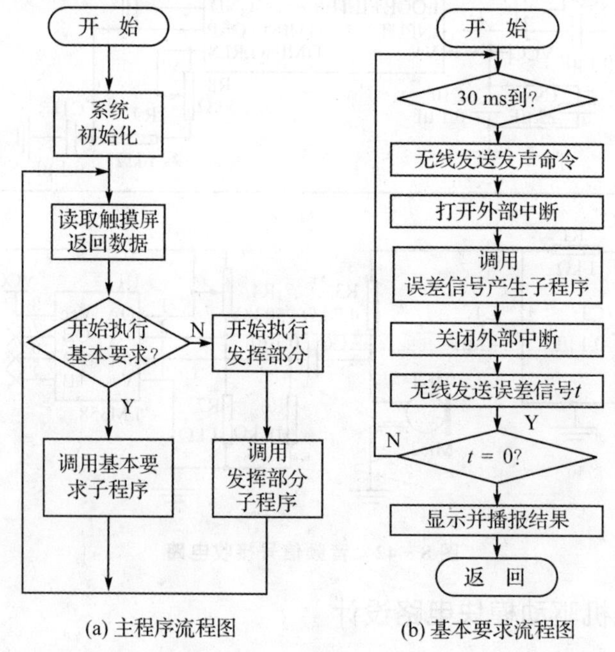

图 8－43　主控制台主要程序流程图

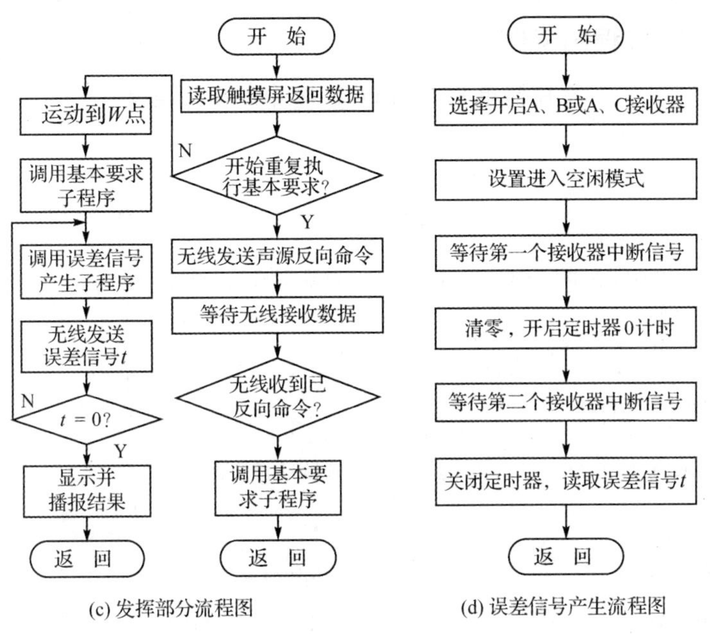

图 8-43　主控制台主要程序流程图(续)

2. 可移动声源程序设计

可移动声源的控制器需产生频率为 3 kHz 的音频信号，然后便开始等待无线接收主控制台数据，若收到的为误差信号 t，则通过对误差信号 t 的处理，控制电机运行；若收到的为声源反向命令，则需控制舵机旋转 180°，且给控制台返回已反向的确认数据，可移动声源主程序流程图如图 8-44 所示。

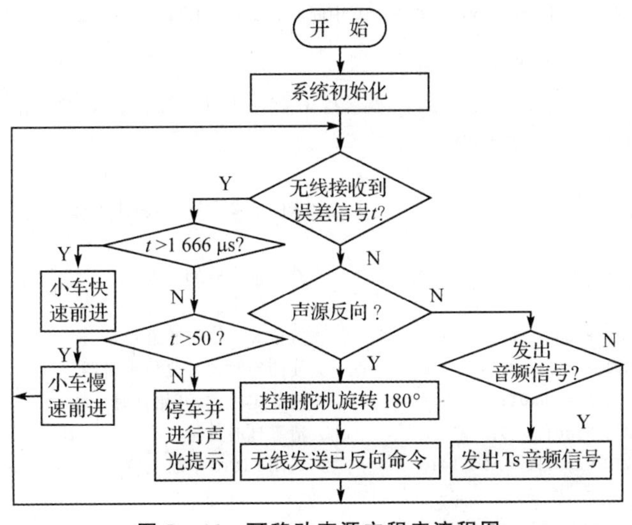

图 8-44　可移动声源主程序流程图

3. 声音定位程序实例

程序清单 8.2 LPC2148 中断相关程序

```
#include "config.h"
volatile uint32 A_num = 0,B_num = 0;
/*************************************************************************
 * 名称:MIC_Done_AB()
 * 功能:开启中断引脚,检测声音信号,同时开启定时器,计时声音时间差值
 * 入口参数:无
 * 出口参数:无
*************************************************************************/
void MIC_Done_AB(void)
{
    AB_OR_AC = 1;                          // 选择 AB 两个通道
    MIC_Flag = 0;
    MIC_Time1 = 0;
    T0TCR = 0x00;                          // 关闭定时器
    EXTINT = 0x01;                         // 清零 EINT0 中断标志
    EXTINT = 0x04;                         // 清零 EINT2 中断标志
    EXTINT = 0x08;                         // 清零 EINT3 中断标志
    VICIntEnable = 1<<14;                  // 使能外部中断 EINT0,开启 A 声音传感器
    VICIntEnable = 1<<16;                  // 使能外部中断 EINT2,开启 B 声音传感器
    VICIntEnClr  = 1<<17;                  // 关闭外部中断 EINT3,关闭 C 声音传感器
    while(MIC_Time1 == 0);                 // 等待 AB 两路中断信号到达
}
/*************************************************************************
 * 名称:AB_MIC_Channel_A_IRQ()
 * 功能:MIC 接收器 1 中断执行函数,被 EINT0 中断服务函数调用
 * 入口参数:无
 * 出口参数:无
*************************************************************************/
void AB_MIC_Channel_A_IRQ(void)
{
    if(A_num>2)                            // 去除前 2 个脉冲,防止干扰信号
    {
        if(MIC_Flag == 0)                  // 若之前未收到脉冲
        {
            MIC_Flag = 1;                  // 置位 MIC 接收通道号
            T0TCR = 0x03;                  // 使能定时器并复位定时器
            T0TCR = 0x01;                  // 使能定时器,开始计时
            MIC_Time1 = 0;                 // 清零计时变量
            VICIntEnClr = 1<<14;           // 关闭 EINT0 中断
        }
        else if((MIC_Flag&0x00f0) == 0)    // 若第二次收到声音脉冲
        {
```

```
            MIC_Flag =
            (MIC_Flag & 0xff0f) | 1<<4;     // 置位 MIC 接收通道号
            T0TCR = 0x00;                   // 关闭定时器
            MIC_Time1 = T0TC;               // 读取时间差值
        }
        VICIntEnClr = 1<<14;                // 关闭 EINT0 中断
        A_num = 0;
    }
    else A_num++;
}
/********************************************************************
* 名称:AB_MIC_Channel_B_IRQ()
* 功能:MIC 接收器 1 中断执行函数,被 EINT0 中断服务函数调用
* 入口参数:无
* 出口参数:无
********************************************************************/
void AB_MIC_Channel_B_IRQ(void)
{
    if(B_num>2)                             // 去除前 2 个脉冲,防止干扰信号
    {
        if(MIC_Flag == 0)                   // 若之前未收到脉冲
        {
            MIC_Flag = 2;                   // 置位 MIC 接收通道号
            T0TCR = 0x03;                   // 使能定时器并复位定时器
            T0TCR = 0x01;                   // 使能定时器,开始计时
            MIC_Time1 = 0;                  // 清零计时变量
        }
        else if((MIC_Flag&0x00f0) == 0)     // 若第二次收到声音脉冲
        {
            MIC_Flag =
            (MIC_Flag & 0xff0f) | 2<<4;     // 置位 MIC 接收通道号
            T0TCR = 0x00;                   // 关闭定时器
            MIC_Time1 = T0TC;               // 读取时间差值
        }
        VICIntEnClr = 1<<16;                // 关闭 EINT2 中断
        B_num = 0;
    }
}
uint8 Eint0_flag = 0;
uint8 Eint1_flag = 0;
uint8 Eint2_flag = 0;
uint8 Eint3_flag = 0;
/********************************************************************
* 名称:Eint0_init()
* 功能:初始化外部中断 0(EINT1)为向量中断,并设置为下降沿触发模式
* 入口参数:无
```

```
* 出口参数:无
* 说明:在 STARTUP.S 文件中使能 IRQ 中断(清零 CPSR 中的 I 位)
*************************************************************************/
void Eint0_init(void)
{
    PINSEL1 = (PINSEL1 &
    0xfffffffc) | 0x00000001;                    // 设置 P0.16 为外部中断 0
    EXTMODE | = (0x01);                           // 设置外部中断为边沿触发模式
    EXTPOLAR &= ~0x01;                            // 设置为下降沿触发
    /*采用向量 IRQ 开启 EXINT1*/
    VICIntSelect    = 0x00000000;                 // 设置所有中断为 IRQ 中断
    VICVectCntl0 = 0x2e;
    VICVectAddr0 = (int)IRQ_Eint0;                // 设置中断服务程序地址
    EXTINT = 0x01;                                // 清零 EINT0 中断标志
    VICIntEnable = 1<<14;                         // 1<<14 使能 EINT1 中断
}
/*************************************************************************
* 名称:Eint2_init()
* 功能:初始化外部中断 2(EINT1)为向量中断,并设置为下降沿触发模式
* 入口参数:无
* 出口参数:无
* 说明:在 STARTUP.S 文件中使能 IRQ 中断
*************************************************************************/
void Eint2_init(void)
{
    PINSEL0 = (PINSEL0 & 0xffff3fff) | 0x0000c000;   // 设置 P0.7 为外部中断 2 引脚
    EXTMODE | = 1 << 2;                           //设置外部中断为边沿触发模式
     EXTPOLAR &= ~(1 << 2);                       // 设置为下降沿触发
    /*采用向量 IRQ 开启 EXINT1*/
    VICIntSelect = 0x00000000;                    // 设置所有中断为 IRQ 中断
    VICVectCntl2 = 0x30;
    VICVectAddr2 = (int)IRQ_Eint2;                // 设置中断服务程序地址
    EXTINT = 0x04;                                // 清零 EINT2 中断标志
    VICIntEnable = 1<<16;                         // 1<<16 使能 EINT1 中断
}
/*************************************************************************
* 名称:IRQ_Eint0()
* 功能:外部中断 EINT0 服务函数
* 入口参数:无
* 出口参数:无
*************************************************************************/
void __irq IRQ_Eint0(void)
{
    Eint0_flag = 1;
    AB_MIC_Channel_A_IRQ();
```

```
    EXTINT = 0x01;                        // 清零 EINT0 中断标志
    VICVectAddr = 0;                      // 结束中断
}
/*******************************************************************
* 名称:IRQ_Eint2()
* 功能:外部中断 EINT2 服务函数
* 入口参数:无
* 出口参数:无
*******************************************************************/
void __irq IRQ_Eint2(void)
{
    Eint2_flag = 1;
    AB_MIC_Channel_B_IRQ();
    EXTINT = 0x04;                        // 清零 EINT2 中断标志
    VICVectAddr = 0;                      // 结束中断
}
```

程序清单 8.3 声音定位主程序的示例程序

```
/*******************************************************************
* 功能:声音定位系统主控台主函数
*       测量 A、B 两个声音传感器的数据,计算传感器时间差值,再控制小车完成声音定位,使小车运
*       动到 AB 的中线上
* 编译环境:ADS 1.2
*******************************************************************/
#include   "config.h"
#define   BEEP     1<<28                        // 蜂鸣器引脚
#define   MIN_Difference       50
#define   Middle_Difference    600
#define   MAX_Difference       2000

#define   Fowrd    1
#define   Back     2
#define   T_L      3
#define   T_R      4
#define   STOP     5
#define   Sound    6

uint8   TxBuf[28] = {0};
extern   uint8   RxBuf[28];

extern volatile uint8   Time0_Flag;              // 定时器 0 中断标志
extern volatile uint16 MIC_Flag;                 // 3 个 MIC 声音检测中断状态标志
extern volatile uint32 MIC_Time1;                // 第一次脉冲到第二次脉冲的时间间距
extern volatile uint32 MIC_Time2;                // 第三次脉冲到第二次脉冲的时间间距

extern   uint8   Eint0_flag;
extern   uint8   Eint1_flag;
```

```
extern   uint8   Eint2_flag;
extern   uint8   Eint3_flag;
uint8 TASK;
/*************************************************************************************
 * 名称:AB_Sound_Check()
 * 功能:通过测量 A、B 两点的声音传感器收到小车声音的时间差值,控制小车到达 A、B 的中线,
 *      即 OX 线上刚好停止
 * 入口参数:无
 * 出口参数:无
 * 说明:使用 LPC2148 的中断引脚检测声音传感器信号,用 Timer0 计时,测得的时间差值
 *      存放在 MIC_Time1 全局变量里
**************************************************************************************/
void AB_Sound_Check(void)
{
    mDelaymS(10);                              // 防止前一次小车蜂鸣器的声音造成干扰
    Car_nRF905_Send(Sound,8,0);                // 控制小车蜂鸣器响 6 ms
    MIC_Done_AB();                             // MIC 传感器测量 AB 两点声音时间差差值
    mDelaymS(10);
    if((MIC_Flag&0x0f) == 2)                   // 若 B 点先收到声音
    {
        if(MIC_Time1>MAX_Difference)           // 若测得 A、B 距离差较大
        {
            Car_nRF905_Send(Fowrd,2,0);        // 小车以最快速度前进
        }
        if(MIC_Time1>MIN_Difference)           // 若测得 A、B 距离差适中
        {
            Car_nRF905_Send(Fowrd,1,0);        // 小车以慢速前进
        }
        else                                   // 若小于最小距离
        {
            Car_nRF905_Send(STOP,2,0);         // 控制小车停止,并声光指示
            TASK = 0;                          // 清除标志位,结束小车运动
        }
    }
    else                                       // 若 A 点先收到声音,则刚好过 OX 线
    {
        Car_nRF905_Send(STOP,2,0);             // 控制小车停止,并声光指示
        TASK = 0;                              // 清除标志位,结束小车运动
    }
}
/*************************************************************************************
 * 函数名称:main()
```

```
* 函数功能:声音定位系统小车主控台主函数
*****************************************************************/
int  main(void)
{
    uint8 key;
    PINSEL0 = 0;
    PINSEL1 = 0;
    PINSEL2 = 0;
    IO1DIR = 0;
    IO0DIR = 0;
    IO1DIR | = BEEP;                        // 蜂鸣器引脚设置为输出
    IO1SET = BEEP;                          // 关闭蜂鸣器
    Time0Init();                            // Timer0 初始化,用于声音识别计时
    RTCIni();                               // RTC 时钟初始化
    UART0_Init(115200);                     // 初始化 UART0
    UART1_Init(9600);                       // 初始化 UART1
    I2C_Init(400000);                       // I²C 初始化,用于读取按键数据
    Eint0_init();                           // 设置 P0.16 为 A 通道声音传感器中断引脚
    Eint2_init();                           // 设置 P0.7 为 B 通道声音传感器中断引脚
    Eint3_init();                           // 设置 P0.20 为 C 通道声音传感器中断引脚
    VICIntEnClr = 1≪14;                     // 关闭 EINT0 中断
    VICIntEnClr = 1≪16;                     // 关闭 EINT2 中断
    VICIntEnClr = 1≪17;                     // 关闭 EINT3 中断
    nRF905_IO_Init();                       // nRF905 引脚初始化
    nRF905Init();                           // NRF905 初始化
    Config905();                            // 配置 nRF905 寄存器
    SetTxMode();                            // 设置 nRF905 为发送模式
    TxPacket(TxBuf);                        // 发送数据缓存区
    SetRxMode();                            // 设置 nRF905 为接收模式
    mDelaymS(200);
    UART1_MP3display_Folder_song(2,0);
    while(1)
    {
        if(TASK == 1)
        {
            AB_Sound_Check();               // 控制小车到达 A、B 两个传感器中线上
            if(TASK == 0)                   // 判断是否定位完成
            {
                Disp_Picture(80);           // 显示基本要求结果
                DelayNS(100);
```

```
                UART1_MP3display_Folder_song(1,1);    //语音播报到达目标
            }
        }
        /* 读取并处理按键数据 */
        key = 0;
        key = Read_Key();
        if(key! = 0)
        {
            switch(key)
            {
                case K1: TASK = 1;                    //按下 K1 键,开始执行声音定位任务
                        break;

                case K2:
                        break;
            }
            mDelaymS(150);
        }
    }
    return 0;
}
```

第 9 章

开发环境及 ISP 下载

本章介绍了在 MDK 开发环境里如何建立、编译连接工程，以及对工程 HEX 文件进行下载的基本方法。

9.1 MDK 集成开发环境

9.1.1 MDK 集成开发环境简介

RealView MDK 开发套件源自德国 Keil 公司，是 ARM 公司目前最新推出的针对各种嵌入式处理器的软件开发工具。RealView MDK 集成了业内最领先的技术，包括 μVision3 集成开发环境与 RealView 编译器。支持 ARM7、ARM9 和最新的 Cortex - M3 核处理器，自动配置启动代码，集成 Flash 烧写模块，强大的 Simulation 设备模拟，以及性能分析等功能，与 ARM 之前的工具包 ADS 等相比，RealView 编译器的最新版本可将性能改善超过 20%。RealView MDK 的组成如表 9 - 1 所列，其主窗口如图 9 - 1 所示。

有关 MDK 使用的更多内容请登录 http://www.keil.com/arm 查阅有关技术文档，或者参考以下文献：

[1] 李宁. ARM MCU 开发工具 MDK 使用入门[M]. 北京：北京航空航天大学出版社，2012.

表 9 - 1 RealView MDK 的组成

RealView MDK 的组成部分名称	描 述
启动代码配置向导	自动生成完善的启动代码，并提供图形化的窗口
μVision3 设备模拟器	● 高效指令集仿真； ● 中断仿真； ● 片内外围设备仿真； ● ADC，DAC，EBI，Timers； ● UART，CAN，I^2C…； ● 外部信号和 I/O 仿真

续表 9-1

RealView MDK 的组成部分名称	描　述
性能分析器	性能分析器可给所有的 MCU 实现如程序运行时间统计、被调用次数统计、代码覆盖率统计等高端功能
RealView 编译器(RVCT)	RealView MDK 集成的 RealView 编译器
MicroLib	为进一步改进基于 ARM 处理器的应用代码密度，RealView MDK 采用了新型 microlib C 库
ULINK2 仿真器	ULINK2 是 ARM 公司最新推出的配套 RealView MDK 使用的仿真器，是 ULink 仿真器的升级版本

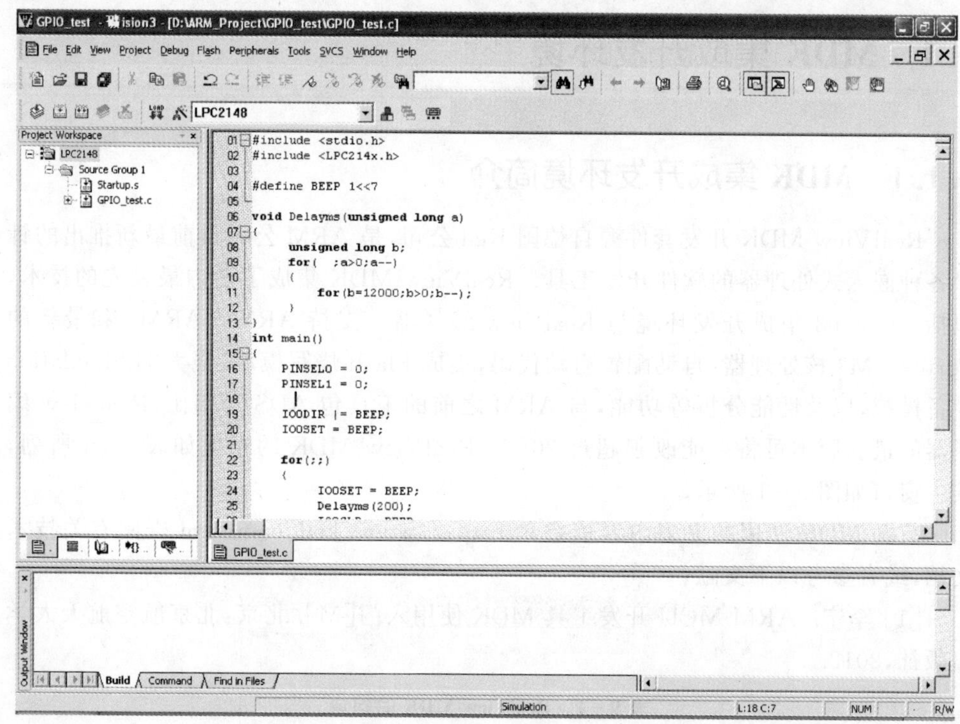

图 9-1　Keil μVision3 开发环境

[2] 任哲，等. 嵌入式系统基础：ARM 与 RealView MDK(Keil for ARM)[M]. 北京：北京航空航天大学出版社，2012.

9.1.2　工程的编辑

1. 建立工程

单击 Windows 操作系统的“开始”→“程序”→Keil μVision3 启动 Keil μVision3；或双击 Keil μVision3 桌面快捷方式，启动 Keil μVision3，如图 9-2 所示。

图 9－2　启动 Keil μVision3

然后，单击 Project 菜单，选择 New μVision Project 命令即弹出 Create New Project 对话框，如图 9－3 所示。

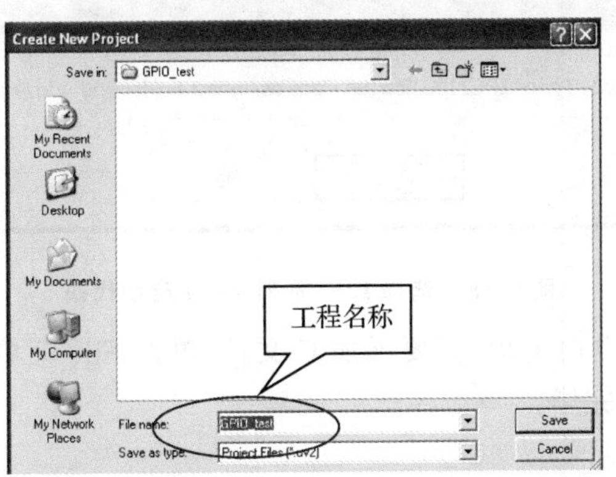

图 9－3　Create New Project 对话框

接着，开始新建工程，在工程名称处输入所建工程的名称(名称不能使用中文)，最后选择工程的存放途径(路径不能包含中文),确定后即建立了一个新的工程,出现如图9-4所示的选择芯片型号窗口。选择NXP-LPC2148,然后单击OK按钮。

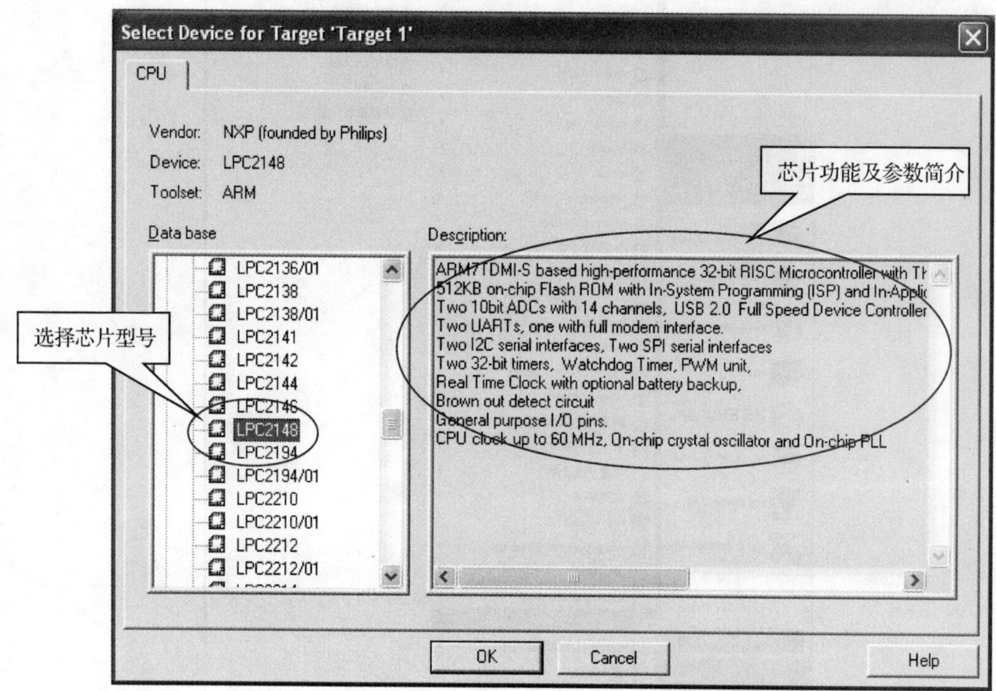

图9-4 芯片型号选择窗口

接下来,会出现图9-5所示的对话框,显示是否添加LPC系列ARM启动代码,单击Yes按钮。

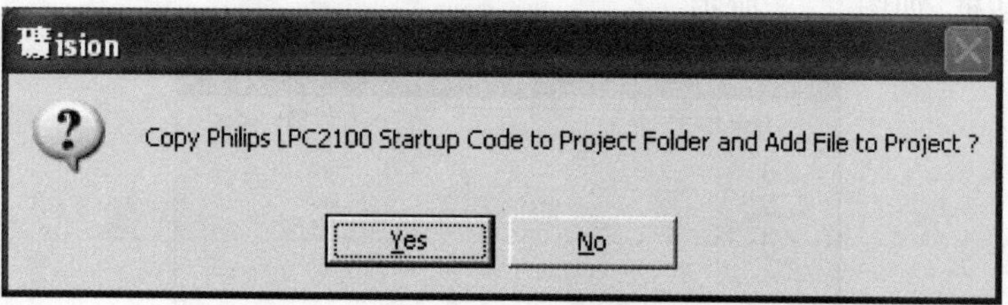

图9-5 选择LPC系列ARM启动代码

最后,出现一个空白工程,需要添加C文件,单击File菜单,选择New,保存为GPIO.c,工程即新建完毕。

2. 程序编译

在GPIO.c中输入C程序代码,保存,然后单击Project菜单,选择Options for Target在Output菜单中的Create HEX File选项上打勾(见图9-6),然后在

Utilities菜单中选择 Use Target Driver for Flash Programming 项（见图 9－7）。单击 OK 按钮，即可完成编译设置。

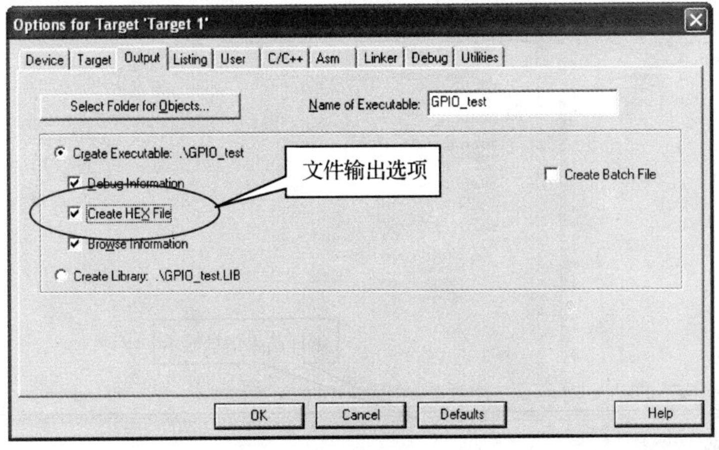

图 9－6　文件输出选择窗口

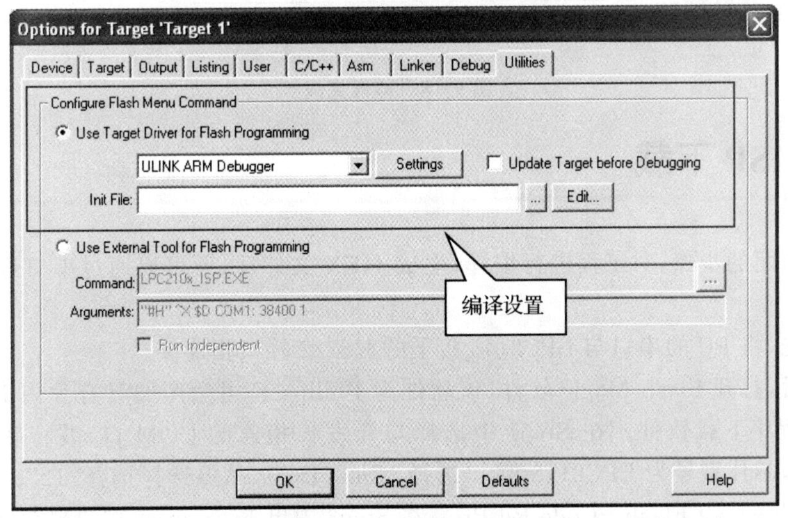

图 9－7　编译设置窗口

完成设置后，单击 Project 菜单，选择 Build Target，即可编译整个工程。编译完成后在 Output Window 中显示工程编译结果的信息（见图 9－8），如有无错误、警告及代码所占空间大小等。在工程所在目录下会生成 Hex 文件。

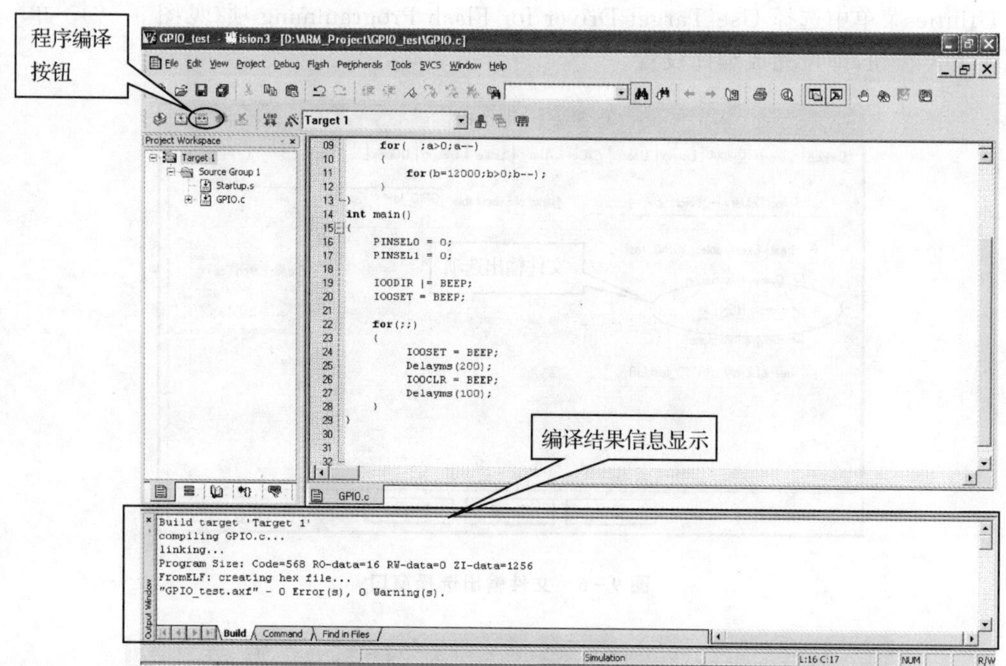

图 9-8 编译完成

9.2 ISP 下载

按前面的步骤,对工程进行编译,生成HEX文件后,便可以通过串口进行程序的下载。

首先,将PC的串口与LPC2148板子的RS232接口相连接。

然后,打开Flash Magic软件,该软件为Philips公司生产的具有片内Flash的ARM7程序下载软件。在Step1中选择与开发板相连的COM口,波特率设置为“38400”,芯片型号为LPC2148,接口选择None(ISP),晶振选择选择“12”;在Step2中选择Erase all Flash+Code Rd Prot;在Step3中单击Browser按钮浏览目录并选中已经生成的HEX文件,如图9-9所示。

最后,将LPC2148的P0.14引脚接GND,按下开发板的复位键,单击Flash Magic软件Step5的Start按钮,即可开始下载,正在执行程序下载的界面如图9-10所示。

当程序下载完成之后,Flash Magic的下载提示处将显示finished,此时,断开P0.14引脚,再次复位LPC2148,程序便开始运行。

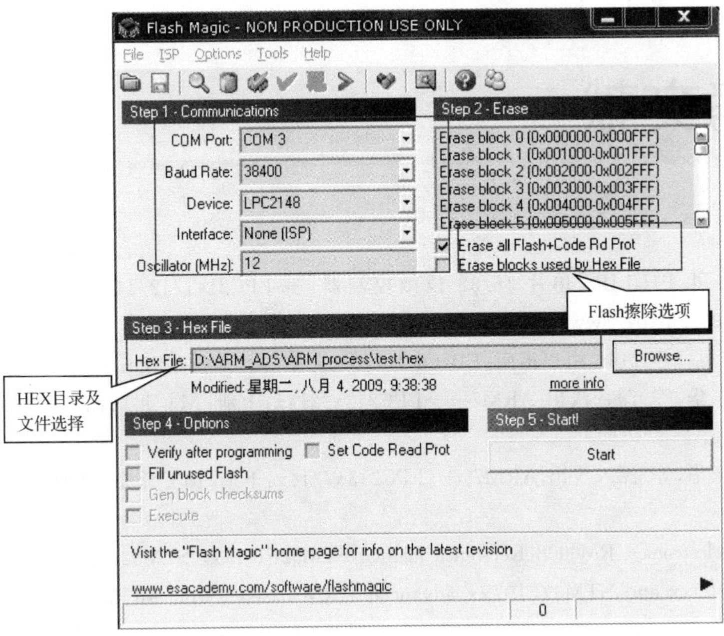

图 9－9　Flash Magic 软件设置

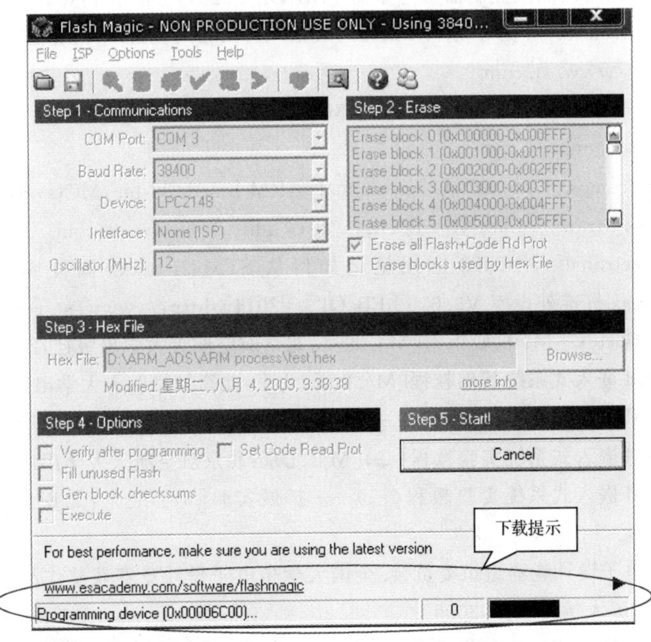

图 9－10　Flash Magic 正在下载界面

参考文献

[1] 周立功公司. PHILIPS 单片 16/32 位微控制器——LPC2141/42/44/46/48 数据手册[EB/OL]. (2009). http://www.zlgmcu.com.

[2] 周立功公司. LPC214x 用户指南[EB/OL]. (2009). http://www.zlgmcu.com.

[3] 周立功,张华,等. 深入浅出 ARM7——LPC213x/214x:上册[M]. 北京:北京航空航天大学出版社,2005.

[4] 周立功,张华,等. 深入浅出 ARM7——LPC213x/214x:下册[M]. 北京:北京航空航天大学出版社,2006.

[5] ST Microelectronics. RM0008 Reference manual STM32F101xx, STM32F102xx, STM32F103xx, STM32F105xx and STM32F107xx advanced ARM-based 32-bit MCUs [EB/OL]. [2014]. http://www.st.com.

[6] ST Microelectronics. STM32F101xx、STM32F102xx、STM32F103xx、STM32F105xx 和 STM32F107xx, ARM 内核 32 位高性能微控制器参考手册 [EB/OL]. [2014]. http://www.st.com.

[7] ST Microelectronics. 数据手册 STM32F103xC STM32F103xD STM32F103xE [EB/OL]. [2014]. http://www.st.com.

[8] ST Microelectronics. AN2586 应用笔记 STM32F10xxx 硬件开发使用入门 [EB/OL]. [2014]. http://www.st.com.

[9] ST Microelectronics. UM0427 User manual ARM-based 32-bit MCU STM32F101xx and STM32F103xx firmware library [EB/OL]. [2014]. http://www.st.com.

[10] ST Microelectronics. AN2953 应用笔记如何从 STM32F10xxx 固件库 V2.0.3 升级为 STM32F10xxx 标准外设库 V3.0.0 [EB/OL]. [2014]. http://www.st.com.

[11] Joseph Yiu. Cortex-M3 权威指南[M]. 北京:北京航空航天大学出版社,2009.

[12] 周立功. ARM 嵌入式系统基础教程[M]. 2 版. 北京:北京航空航天大学出版社,2008.

[13] 周立功. ARM 嵌入式系统实验教程(一)[M]. 北京:北京航空航天大学出版社,2004.

[14] 周立功. ARM 嵌入式系统实验教程(二)[M]. 北京:北京航空航天大学出版社,2005.

[15] 周立功. ARM 嵌入式系统实验教程(三)——扩展实验[M]. 北京:北京航空航天大学出版社,2006.

[16] 全国大学生电子设计竞赛组织委员会. 全国大学生电子设计竞赛获奖作品选编(2007)[M]. 北京:北京理工大学出版社,2009.

[17] 全国大学生电子设计竞赛组织委员会. 全国大学生电子设计竞赛章程[EB/OL]. (2009-5). http://www.nuedc.com.cn/news.asp? bid=1.

[18] 全国大学生电子设计竞赛组织委员会. 历届题目[EB/OL]. (2009-5). http://www.nuedc.com.cn/news.asp? bid=1.

[19] 吴华波,钱春来. 基于 AT89C2051 的多路舵机控制器设计[J]. 单片机与嵌入式系统应用,2006(8):55-59.
[20] 黄智伟. LED 驱动电路设计[M]. 北京:电子工业出版社,2014.
[21] 黄智伟. 电源电路设计[M]. 北京:电子工业出版社,2014.
[22] 黄智伟. 嵌入式系统中的模拟电路设计[M]. 2 版. 北京:电子工业出版社,2014.
[23] 黄智伟. 全国大学生电子设计竞赛基于 TI 器件的模拟电路设计[M]. 北京:北京航空航天大学出版社,2014.
[24] 黄智伟. 印制电路板(PCB)设计技术与实践[M]. 2 版. 北京:电子工业出版社,2013.
[25] 黄智伟,等. ARM9 嵌入式系统基础教程[M]. 2 版. 北京:北京航空航天大学出版社,2013.
[26] 黄智伟. 高速数字电路设计入门[M]. 北京:电子工业出版社,2012.
[27] 黄智伟,王兵,朱卫华. STM32F 32 位微控制器应用设计与实践[M]. 北京:北京航空航天大学出版社,2012.
[28] 黄智伟. 低功耗系统设计——原理、器件与电路[M]. 北京:电子工业出版社,2011.
[29] 黄智伟. 超低功耗单片无线系统应用入门[M]. 北京:北京航空航天大学出版社,2011.
[30] 黄智伟,等. 32 位 ARM 微控制器系统设计与实践[M]. 北京:北京航空航天大学出版社,2010.
[31] 黄智伟. 基于 NI mulitisim 的电子电路计算机仿真设计与分析[M]. 修订版. 北京:电子工业出版社,2011.
[32] 黄智伟. 全国大学生电子设计竞赛系统设计[M]. 2 版. 北京:北京航空航天大学出版社,2011.
[33] 黄智伟. 全国大学生电子设计竞赛电路设计[M]. 2 版. 北京:北京航空航天大学出版社,2011.
[34] 黄智伟. 全国大学生电子设计竞赛技能训练[M]. 2 版. 北京:北京航空航天大学出版社,2011.
[35] 黄智伟. 全国大学生电子设计竞赛制作实训[M]. 2 版. 北京:北京航空航天大学出版社,2011.
[36] 黄智伟. 全国大学生电子设计竞赛常用电路模块制作[M]. 北京:北京航空航天大学出版社,2011.
[37] 黄智伟,等. 全国大学生电子设计竞赛 ARM 嵌入式系统应用设计与实践[M]. 北京:北京航空航天大学出版社,2011.
[38] 黄智伟. 全国大学生电子设计竞赛培训教程[M]. 修订版. 北京:电子工业出版社,2010.
[39] 黄智伟. 射频小信号放大器电路设计[M]. 西安:西安电子科技大学出版社,2008.
[40] 黄智伟. 锁相环与频率合成器电路设计[M]. 西安:西安电子科技大学出版社,2008.
[41] 黄智伟. 混频器电路设计[M]. 西安:西安电子科技大学出版社,2009.
[42] 黄智伟. 射频功率放大器电路设计[M]. 西安:西安电子科技大学出版社,2009.
[43] 黄智伟. 调制器与解调器电路设计[M]. 西安:西安电子科技大学出版社,2009.
[44] 黄智伟. 单片无线发射与接收电路设计[M]. 西安:西安电子科技大学出版社,2009.
[45] 黄智伟. 无线发射与接收电路设计[M]. 2 版. 北京:北京航空航天大学出版社,2007.
[46] 黄智伟. GPS 接收机电路设计[M]. 北京:国防工业出版社,2005.

[47] 黄智伟.单片无线收发集成电路原理与应用[M].北京:人民邮电出版社,2005.
[48] 黄智伟.无线通信集成电路[M].北京:北京航空航天大学出版社,2005.
[49] 黄智伟.蓝牙硬件电路[M].北京:北京航空航天大学出版社,2005.
[50] 黄智伟.射频电路设计[M].北京:电子工业出版社,2006.
[51] 黄智伟.通信电子电路[M].北京:机械工业出版社,2007.
[52] 黄智伟.FPGA系统设计与实践[M].北京:电子工业出版社,2005.
[53] 黄智伟.凌阳单片机课程设计[M].北京:北京航空航天大学出版社,2007.
[54] 黄智伟.单片无线数据通信IC原理应用[M].北京:北京航空航天大学出版社,2004.
[55] 黄智伟.射频集成电路原理与应用设计[M].北京:电子工业出版社,2004.
[56] 黄智伟.无线数字收发电路设计[M].北京:电子工业出版社,2004.